Self-Doped Conducting Polymers

Self-Doped Conducting Polymers

Michael S. Freund and Bhavana A. Deore
Department of Chemistry,
University of Manitoba,
Canada

John Wiley & Sons, Ltd

Copyright © 2007   John Wiley & Sons Ltd, The Atrium, Southern Gate, Chichester,
West Sussex PO19 8SQ, England

Telephone (+44) 1243 779777

Email (for orders and customer service enquiries): cs-books@wiley.co.uk
Visit our Home Page on www.wileyeurope.com or www.wiley.com

All Rights Reserved. No part of this publication may be reproduced, stored in a retrieval system or transmitted in any form or by any means, electronic, mechanical, photocopying, recording, scanning or otherwise, except under the terms of the Copyright, Designs and Patents Act 1988 or under the terms of a licence issued by the Copyright Licensing Agency Ltd, 90 Tottenham Court Road, London W1T 4LP, UK, without the permission in writing of the Publisher. Requests to the Publisher should be addressed to the Permissions Department, John Wiley & Sons Ltd, The Atrium, Southern Gate, Chichester, West Sussex PO19 8SQ, England, or emailed to permreq@wiley.co.uk, or faxed to (+44) 1243 770620.

Designations used by companies to distinguish their products are often claimed as trademarks. All brand names and product names used in this book are trade names, service marks, trademarks or registered trademarks of their respective owners. The Publisher is not associated with any product or vendor mentioned in this book.

This publication is designed to provide accurate and authoritative information in regard to the subject matter covered. It is sold on the understanding that the Publisher is not engaged in rendering professional services. If professional advice or other expert assistance is required, the services of a competent professional should be sought.

The Publisher and the Author make no representations or warranties with respect to the accuracy or completeness of the contents of this work and specifically disclaim all warranties, including without limitation any implied warranties of fitness for a particular purpose. This work is sold with the understanding that the Publisher is not engaged in rendering professional services. The advice and strategies contained herein may not be suitable for every situation. In view of ongoing research, equipment modifications, changes in governmental regulations, and the constant flow of information relating to the use of experimental reagents, equipment, and devices, the reader is urged to review and evaluate the information provided in the package insert or instructions for each chemical, piece of equipment, reagent, or device for, among other things, any changes in the instructions or indication of usage and for added warnings and precautions. The fact that an organization or Website is referred to in this work as a citation and/or a potential source of further information does not mean that the author or the publisher endorses the information the organization or Website may provide or recommendations it may make. Further, readers should be aware that Internet Websites listed in this work may have changed or disappeared between when this work was written and when it is read. No warranty may be created or extended by any promotional statements for this work. Neither the Publisher nor the Author shall be liable for any damages arising herefrom.

Other Wiley Editorial Offices

John Wiley & Sons Inc., 111 River Street, Hoboken, NJ 07030, USA

Jossey-Bass, 989 Market Street, San Francisco, CA 94103-1741, USA

Wiley-VCH Verlag GmbH, Boschstr. 12, D-69469 Weinheim, Germany

John Wiley & Sons Australia Ltd, 42 McDougall Street, Milton, Queensland 4064, Australia

John Wiley & Sons (Asia) Pte Ltd, 2 Clementi Loop #02-01, Jin Xing Distripark, Singapore 129809

John Wiley & Sons Canada Ltd, 6045 Freemont Blvd, Mississauga, Ontario, L5R 4J3, Canada

Wiley also publishes its books in a variety of electronic formats. Some content that appears in print may not be available in electronic books.

Library of Congress Cataloging-in-Publication Data:

Freund, Michael S.
  Self-doped conducting polymers / Michael S. Freund and Bhavana Deore.
    p. cm.
  Includes bibliographical references.
  ISBN-13: 978-0-470-02969-5
  ISBN-10: 0-470-02969-2
  1. Conducting polymers. 2. Doped semiconductors. 3. Polymerization. 4. Electric apparatus and appliances – Materials. I. Deore, Bhavana. II. Title.
  QD382.C66F74 2006
  547'.70457 – dc22
                                                                                            2006032502

British Library Cataloguing in Publication Data

A catalogue record for this book is available from the British Library

ISBN: 978-0-470-02969-5 (HB)

Typeset in 10.5/13pt Sabon by Laserwords Private Limited, Chennai, India.
Printed and bound in Great Britain by TJ International, Padstow, Cornwall
This book is printed on acid-free paper responsibly manufactured from sustainable forestry
in which at least two trees are planted for each one used for paper production.

Contents

About the Authors ix

Preface xi

1 Introduction 1
  1.1 Conducting Polymers 1
    1.1.1 History of Conjugated Conducting Polymers 1
    1.1.2 Concept of Doping in Intrinsically Conducting Polymers 9
    1.1.3 Conduction Mechanism 10
    1.1.4 Synthesis 20
    1.1.5 Processability 22
  1.2 Self-Doped Conducting Polymers 25
  1.3 Types of Self-Doped Polymers 29
  1.4 Doping Mechanism in Self-Doped Polymers 33
    1.4.1 p-Type Doping 33
    1.4.2 n-Type Doping 36
    1.4.3 Auto Doping 38
  1.5 Effect of Substituents on Properties of Polymers 42
    1.5.1 Solubility 43
    1.5.2 DC Conductivity 44
    1.5.3 Molecular Weight 45
    1.5.4 Redox Properties 46
    1.5.5 Electronic and Spectroscopic Properties 46
    1.5.6 Mechanical and Thermal Properties 47
  1.6 Applications of Self-Doped Polymers 48
    1.6.1 Molecular Level Processing 48
    1.6.2 Transistors 50
    1.6.3 Biosensors 52
    1.6.4 e-Beam Lithography 53
    1.6.5 Electrochromic Devices 54
    1.6.6 Ion Exchangers 55

1.6.7 Rechargeable Batteries	55
1.6.8 Dip-Pen Nanolithography	56
References	58

2 Self-Doped Derivatives of Polyaniline — 75

2.1 Introduction	75
2.2 Chemical Synthesis of Sulfonic Acid Derivatives	77
2.2.1 Post-Polymerization Modification	77
2.2.2 Polymerization of Monomers	86
2.3 Electrochemical Synthesis of Sulfonic Acid Derivatives	92
2.3.1 Aqueous Media	92
2.3.2 Non-Aqueous Media	97
2.4 Enzymatic Synthesis of Sulfonic Acid Derivatives	98
2.5 Properties of Sulfonic Acid Derivatives	100
2.5.1 Solubility	101
2.5.2 Conductivity	102
2.5.3 pH Dependent Redox Behavior	107
2.5.4 Electronic and Spectroscopic Properties	109
2.5.5 Molecular Weight	118
2.5.6 Thermal Stability	119
2.5.7 Morphology	121
2.6 Synthesis and Characterization of Carboxylic Acid Derivatives	123
2.6.1 Chemical Synthesis	123
2.6.2 Electrochemical Synthesis	128
2.7 Synthesis and Characterization of Phosphonic Acid Derivatives	130
2.8 Self-Doped Polyaniline Nanostructures	132
References	140

3 Boronic Acid Substituted Self-Doped Polyaniline — 156

3.1 Introduction	156
3.2 Synthesis	158
3.2.1 Electrochemical Synthesis	158
3.2.2 Chemical Synthesis	165
3.3 Properties of Self-Doped PABA	166
3.3.1 pH Dependent Redox Behavior	166
3.3.2 Spectroscopy	172
3.3.3 Molecular Weight	175
3.4 Self-Crosslinked Self-Doped Polyaniline	177
3.4.1 Introduction	177

	3.4.2 Synthesis and Characterization	179
	3.4.3 Mechanical Properties	181
	3.4.4 ^{11}B NMR	182
	3.4.5 Thermal Properties	184
	3.4.6 Temperature Dependent Conductivity	185
3.5	Applications	187
	3.5.1 Saccharide Sensor	187
	3.5.2 Nucleotide Sensors	189
	3.5.3 Amine Sensors	196
	3.5.4 Molecular Level Processing for Controlled Release of RNA	199
	References	206
4	**Self-Doped Polythiophenes**	**219**
4.1	Sulfonic Acid Derivatives	220
	4.1.1 Electrochemical Polymerization	220
	4.1.2 Chemical Polymerization	234
	4.1.3 Post Polymerization Modification	249
4.2	Carboxylic Acid Derivatives	249
4.3	Phosphonic Acid Derivatives	255
	References	258
5	**Miscellaneous Self-Doped Polymers**	**262**
5.1	Self-Doped Polypyrrole	262
	5.1.1 Electrochemical Polymerization	263
	5.1.2 Chemical Polymerization	277
	5.1.3 Polycondensation	282
5.2	Carboxylic Acid Derivatives	284
5.3	Self-Doped Poly(3,6-(carbaz-9-yl)propanesulfonate)	290
5.4	Self-Doped Poly(*p*-phenylene)s	292
5.5	Self-Doped Poly(*p*-phenylenevinylene)s	296
5.6	Self-Doped Poly(indole-5-carboxylic acid)	299
5.7	Self-Doped Ionically Conducting Polymers	303
	References	304
Index		**315**

About the Authors

Bhavana Deore is a Research Associate in the Chemistry Department at the University of Manitoba. She has been at the University of Manitoba since 2002. She received her PhD in 1998 from Pune University, India. Subsequently, she became a Postdoctoral Fellow in the Yamaguchi University, Faculty of Engineering, and Department of Applied Chemistry, Japan under the Venture Business Laboratory and Japan Society for Promotion of Science Fellowship Program. Her research interests include synthesis, characterization and applications of conducting polymers, including chemical and biological sensors. She has authored over 34 publications in leading peer reviewed journals.

Michael Stephen Freund is an Associate Professor of Chemistry and Adjunct Professor of Electrical Engineering at the University of Manitoba, Canada. He holds a Canada Research Chair in Conducting polymers and Electronics Materials, and has authored over 45 publications in leading journals and has 20 issued patents. He serves on the editorial board of the *Proceedings of the Royal Society A: Mathematical, Physical*

and Engineering Sciences and has been invited to participate in special journal editions devoted to emerging investigators and future leaders in analytical chemistry, including the 1999 issue of *Analytica Chimica Acta* 'Looking to the Future of Analytical Chemistry' and an issue of the *Analyst* in 2003 devoted to 'Global Emerging Investigators.'

Professor Freund received a BS Degree in Chemistry from Florida Atlantic University in 1987 and his PhD in Analytical Chemistry from the University of Florida in 1992. During his graduate studies, he was awarded the Shell Fellowship in Chemistry (1992) and the Electrochemical Society's Joseph W. Richards Fellowship (1991). Subsequently, he became a Postdoctoral Fellow in the Department of Chemistry at the California Institute of Technology where his research contributions aided in the establishment of a multi-investigator, interdisciplinary research program on the development olfactory-inspired sensor arrays. Upon completion of his postdoctoral fellow, he established himself in both analytical chemistry and material science as an Assistant Professor of Chemistry at Lehigh University and as the Director of the Molecular Materials Research Center in the Beckman Institute at Caltech, respectively. He has been at the University of Manitoba since 2002.

Preface

The importance of organic conducting polymers is reflected in the growing number of scientists and engineers who are entering the field. These researchers focus on diverse aspects ranging from developing quantum mechanics based models in order to understand the physics of these unique materials, to developing new synthetic approaches to enhance properties, and expand the range of mechanisms through which the electronic properties can be coupled to chemical systems. The excitement surrounding the area has resulted from the tremendous possibilities presented by the merging of the vast knowledge base of organic chemistry and polymer science with the critically important areas of electronic materials and solid-state physics. This rapidly growing field presents opportunities for revolutionizing materials science and electronics in ways we are just beginning to imagine.

One particularly important area of development has been the functionalization of organic conducting polymers. This approach is the most direct way of creating an electronic material that can interact and respond to its environment (i.e., a smart material) and as a way to tune the electronic properties exhibited by these materials. The creation of these kinds of modified polymers has sparked many new technological applications. For example, the development of organic light emitting diode (OLED) displays has emerged from this area and promises to revolutionize the electronic display industry. Functionalized conducting polymers have also been used in bioanalytical applications such as sensing, biomolecule preconcentration and controlled drug release. The impact of this approach is expected to grow as these materials make their way into other technically important areas.

Due to the role of charge carriers (for conductivity and optical properties) as well as counterions (to maintain charge neutrality), interactions of functional groups that can impact the creation and/or elimination of charge play a critical role in controlling the properties of these polymers. Organic conducting polymers that contain covalently bound, charged, functional groups that in turn impact the properties of the polymer are referred to as 'self-doped' conducting polymers. In particular, the presence of these groups can impact on the stability of the doped structure as well as influencing the nature of charge compensation and ion movement during redox switching. Since the discovery of this class of conducting polymer there has been a growing recognition of its importance and potential impact on a wide range of technologies ranging from electrochromic devices to batteries. As a result, there is a growing number of researchers and companies focused on the creation and implementation of this type of conducting polymer.

The purpose of this book is to cover the rapidly developing area of self-doped conducting polymers, with the goal of describing the wide range of approaches that have been developed to synthesize, characterize and utilize them. Our intent is to provide an up-to-date, detailed overview of developments in the field and, in turn, provide researchers and students with a useful reference book. We have made every effort to provide a comprehensive overview of the field; however, the field is rapidly growing and involves researchers from many disciplines publishing in a wide range of journals, proceedings and books, making some unintended omissions likely.

We wish to express our sincere appreciation to the teachers, mentors and supporters who have made our work in this field possible. We thank the undergraduate, graduate and postdoctoral researchers who have contributed to our research in this area. Finally, we would like to thank the following people for making significant contributions to this book: Dr Eiichi Shoji, Ms Insun Yu, Ms Sarah Hachey, Ms Carmen Recksiedler, and Mr Joseph English.

Bhavana Deore
Michael Freund
University of Manitoba, Canada

1
Introduction

1.1 CONDUCTING POLYMERS

1.1.1 History of Conjugated Conducting Polymers

Polymers have emerged as one of the most important materials in the twentieth century. The twenty-first century will undoubtedly see the use of polymers move from primarily passive materials such as coatings and containers to active materials with useful optical, electronic, energy storage and mechanical properties. Indeed, this development has already begun with the discovery and study of conducting polymers. Electronically conducting polymers possess a variety of properties related to their electrochemical behavior and are therefore active materials whose properties can be altered as a function of their electrochemical potential. The importance and potential impact of this new class of material was recognized by the world scientific community when Hideki Shirakawa, Alan J. Heeger and Alan G. MacDiarmid were awarded the Nobel Prize in Chemistry in 2000 for their research in this field [1–5]. Although these materials are known as new materials in terms of their properties, the first work describing the synthesis of a conducting polymer was published in the nineteenth century [6]. At that time 'aniline black' was obtained as the product of oxidation of aniline, however, its electronic properties were not established.

It has been known for more than 40 years that the electronic conductivity of conjugated organic polymer chains, is orders of magnitude higher than that of other polymeric materials [7–9] although they are not metallic, however, the possibility of producing polymers

---

Self-Doped Conducting Polymers   M.S. Freund and B.A. Deore
© 2007 John Wiley & Sons, Ltd

with conductivities approaching those of metals was not recognized. A key discovery that changed the outlook for producing highly conducting polymers was the finding in 1973 that the inorganic polymer polysulfur nitride $(SN)_x$ is highly conducting [10]. The room-temperature conductivity of $(SN)_x$ is of the order of 10^3 S/cm, approaching the conductivity of copper, $\sim 10^5$ S/cm. Below a critical temperature of about 0.3 K, $(SN)_x$ becomes a superconductor [11]. These discoveries were of particular importance because they proved the possibility of generating highly conducting polymers, and stimulated the enormous amount of focus and activity necessary for the discovery of other polymeric conductors. In the period of 1976 and 1977, it was observed that the room-temperature conductivity of $(SN)_x$ could be enhanced by an order of magnitude following exposure to bromine or other similar oxidizing agents [12], suggesting that it was possible to increase the number of charge carriers in the material *via* doping.

In 1958, polyacetylene was first synthesized by Natta *et al.* as a black powder and found to be a semiconductor with conductivity in the range of 10^{-11} to 10^{-3} S/cm, depending upon how the polymer was processed and manipulated [13]. This polymer remained a scientific curiosity until 1967, when a coworker of Hideki Shirakawa at the Tokyo Institute of Technology was attempting to synthesize polyacetylene, and a silvery thin film was produced as a result of a mistake. It was found that the amount of Ziegler–Natta catalyst, $Ti(O-n-But)_4-Et_3Al$, was three orders of magnitude higher than required. When this film was investigated it was found to possess a higher conductivity than previously observed, approaching that of the best carbon black (graphite) powders. In the years between 1971 and 1975, Shirakawa and coworkers prepared crystalline polyacetylene films using refinements of the technique in the presence of Ziegler catalyst; however, the nature of conductivity was not pursued [14–16]. The real breakthrough in the development of conjugated organic conducting polymers was only reached after the discovery of metallic conductivity in crystalline polyacetylene films with p-type dopants such as halogens during collaborative research involving Shirakawa, MacDiarmid and Heeger in 1977 [1, 2]. A year later, it was discovered that analogous effects could be induced by electron donors (n-type dopant) [17]. Following this work there has been an explosion of activity around the characterization, synthesis and use of conducting polymers in a wide range of fields from electronics to medicine.

An organic polymer that possesses the electrical and optical properties of a metal while retaining its mechanical properties and processability, is termed an 'intrinsically conducting polymer' (ICP). These properties are intrinsic to the 'doped' form of the polymer. The conductivity of ICPs lies above that of insulators and extends well into the region of common metals; therefore, they are often referred to as 'synthetic metals.' The common feature of most ICPs is the presence of alternating single and double bonds along the polymer chain, which enable the delocalization or mobility of charge along the polymer backbone. The conductivity is thus assigned to the delocalization of π-bonded electrons over the polymeric backbone, exhibiting unusual electronic properties, such as low energy optical transitions, low ionization potentials and high electron affinities [18].

1.1.1.1 Conducting Polymer Composites

ICPs are 'doped conjugated polymers' and are fundamentally different from 'conducting polymer composites' [19], 'redox polymers' [20] and 'ionically conducting polymers' such as polymer/salt electrolytes [21]. Conducting polymer composites are typically a physical mixture of a nonconductive polymer and a conducting material such as a metal or carbon powder distributed throughout the material. Conductive carbon blacks, short graphite fibers, and metal coated glass fibers, as well as metal particles or flakes, were used in early experiments for the preparation of such composites. Their conductivity is governed by percolation theory, which describes the movement of electrons between metallic phases and exhibits a sudden drop in conductivity (percolation threshold) at the point where the dispersed conductive phase no longer provides a continuous path for the transport of electrons through the material. The conductivity above the percolation threshold of these materials can be as high as 10^{-1} S/cm, at 10–40 wt% fractions of the conductive filler [22, 23]. There are a number of drawbacks associated with such composite materials (fillers), namely: (i) their conductivity is highly dependent on processing conditions, (ii) there is often an insulating surface layer formed on the conductor and (iii) the composite may become mechanically unstable due to heavy loading of the conducting particles. More recently, composites as well as blends and grafting of ICP materials have been utilized in order to impart processability and improve the mechanical properties of these composites [24].

1.1.1.2 Redox Polymers

In 'conjugated conducting polymers', the redox sites are delocalized over a conjugated π system; however, 'redox polymers' have localized redox sites. The redox polymers are well known to transport electrons by hopping or self-exchange between donor and acceptor sites. The redox conductivity is comparatively lower than that of conjugated conducting polymers, likely due to slow electron transport to/from the redox centre. Apart from conjugated organic polymers such as polypyrrole, polythiophene and polyaniline, the first generation of redox polymers included the following main group materials:

(i) Saturated organic polymers with pendent transition metal complexes, such as polyvinylferrocene **1**[†] and metal complexes of polyvinylpyridine **2**[†][25].

(ii) Electrochemically polymerized transition metal complexes with multiple polymerizable ligands, such as poly[ruthenium(4-vinyl-4'-methyl-2,2'-bipyridine)$_3$$^{2+}$] **3**[†] [26], poly[iron(4-(2-pyrrol-l-ylethyl)-4'-methyl-2,2'-bipyridine)$_3$$^{2+}$] **4**[†] [27], and poly[tetra(4-pyrrole-l-ylphenyl)porphyrin] (**poly-5**[†]) [28]. These differ from the polymers in group (i) in that they do not contain extended

---

[†] *Journal of Materials Chemistry*, 1999, 9, 1641, P.G. Pickup. Reproduced by permission of the Royal Society of Chemistry.

organic chains. The complexes are linked in a 3-D network primarily by intermolecular dimerization of the ligands.
(iii) Saturated organic polymers with pendent electroactive organic moieties [29], such as poly(4-nitrostyrene) **6**.[#]
(iv) Ion exchange polymers containing electrostatically bound electroactive ions, such as Nafion containing Ru(bpy)$_3^{2+}$ (bpy = 2,2′-bipyridine) and quaternized polyvinylpyridine containing Fe(CN$_6$)$^{3-/4-}$.

More recently, there has been growing interest in a new type of redox polymer that is a hybrid of materials from conjugated organic polymers and group (ii), referred to as conjugated metallopolymers [30, 31]. Examples include metal complexes of poly(2,2′-bipyridine) **7**[#] and the polyferrocenes **8**[#]. The key feature of this class of material is that the metal is coordinated directly to the conjugated backbone of the polymer, or forms a link in the backbone, such that there is an electronic interaction between the electroactive metal centers and the electroactive polymer backbone. This can enhance electron transport in the polymer, enhance its electrocatalytic activity, and lead to novel electronic and electrochemical properties.

[#] *Journal of Materials Chemistry*, 1999, **9**, 1641, P.G. Pickup. Reproduced by permission of the Royal Society of Chemistry.

1.1.1.3 Ionically Conducting Polymers

Ionically conducting polymers (polymer/salt electrolytes) are of great interest because they exhibit ionic conductivity in a flexible but solid membrane. Ionic conductivity is different than the electronic conductivity of metals and conjugated conducting polymers, since current is carried through the movements of ions. They have been critical to the development of devices such as all-solid-state lithium batteries. The research and development of solid polymer electrolytes began when Wright *et al.* found ion conductivity in a PEO-alkaline metal ion complex in 1973 [32]. The ionic conductivity at that time was 10^{-7} S/cm at room temperature. Since then many salts, including those containing di- and trivalent cations, have been combined with a variety of polymers in order to form polymer electrolytes [33–35]. The considerable potential of these materials as solid ionic conductors was first recognized by Armand in 1979 [36]. Since that time there has been intense interest in the synthesis and characterization of this class of material, as well as considerable focus on their potential use as solid electrolytes in electrochemical devices such as rechargeable lithium batteries, electrochromic displays and smart windows [37]. Polymer electrolytes also represent a fascinating class of coordination compounds such as the oxo-crown ethers [38]. Amorphous polymer electrolytes have been studied intensively for 30 years, and although the conductivities have increased substantially over that period, they remain too low ($<10^{-4}$ S/cm) for many applications. The recently discovered crystalline polymer electrolytes represent a new class of solid ionic conductors and offer a different approach to ionic conductivity in the solid state. It is only in recent years that substantial progress in understanding the structure of these materials has become possible through the methods of crystal structure determination from powders [39, 40]. In general, polymer electrolytes have lower conductivity than liquid electrolytes and cannot deliver high power at room temperature and, in particular, low temperature batteries. Still, polymer electrolytes have many excellent properties such as ease of battery fabrication in various shapes, and better safety than conventional organic liquid electrolytes.

1.1.1.4 Intrinsically Conducting Polymers

Intrinsically conducting polymers offer a unique combination of ion exchange characteristics and optical properties that make them distinctive. They are readily oxidized and reduced at relatively low potentials, and the redox process is reversible and accompanied by large changes in the composition, conductivity and color of the material. These polymers are made conducting, or 'doped', by the reaction of conjugated semiconducting polymer with an oxidizing agent, a reducing agent or a protonic acid, resulting in highly delocalized polycations or polyanions [41]. The conductivity of these materials can be tuned by chemical manipulation of the polymer backbone, by the nature of the dopant, by the degree of doping and by blending with other polymers. In addition, polymeric materials are lightweight, processable and flexible.

Since Shirakawa, Heeger and MacDiarmid [1, 2] discovered that polyacetylene can reach extremely high electronic conductivities, the field of conducting polymers has attracted the interest of thousands of scientists and engineers. Much of the combined research efforts of industrial, academic and government researchers have been directed toward developing materials that are stable (mechanically and electronically) for use in applications, easily processable and can be produced simply and at a low cost. These goals have been surprisingly difficult to reach and only recently have companies been able to bring products to market. Examples of commercial providers of bulk ICP include Zipperling Kessler & Co., who provide polyaniline powder suspensions sold under the name 'ORMECON', the Bayer Corporation who provide highly transparent and stable poly(3,4-ethylendioxythiophene) (PEDOT) water dispersion with variable conductivity under the name 'Baytron,' and Panipol who provide polyaniline powder dispersible in polyolefines.

Recent advances in the field of ICPs have led to a variety of materials with great potential for commercial applications such as rechargeable batteries, light emitting diodes (LEDs), photovoltaics, membranes, electronic noses and sensors, etc. The field of conducting polymers is broad, involving synthesis, characterization and implementation. Over the past three decades there has been explosive growth in conducting polymer research. Since 1977, thousands of articles on conducting polymers have been published in scientific journals every year. The most cited journals in this field include *Synthetic Metals, Applied Physics Letters, Journal of Applied Physics, Advanced Materials, Macromolecules* and *Chemistry of Materials*. There have been excellent reviews on conducting polymers that cover synthesis [42–46], applications [4, 5, 47–49] and the current understanding of electronic structure [50–52]. Also, approximately 58 books and 8 handbooks to date cover the field in detail.

Of the many interesting conducting polymers that have been developed over the past 30 years, those based on polyanilines, polypyrroles,

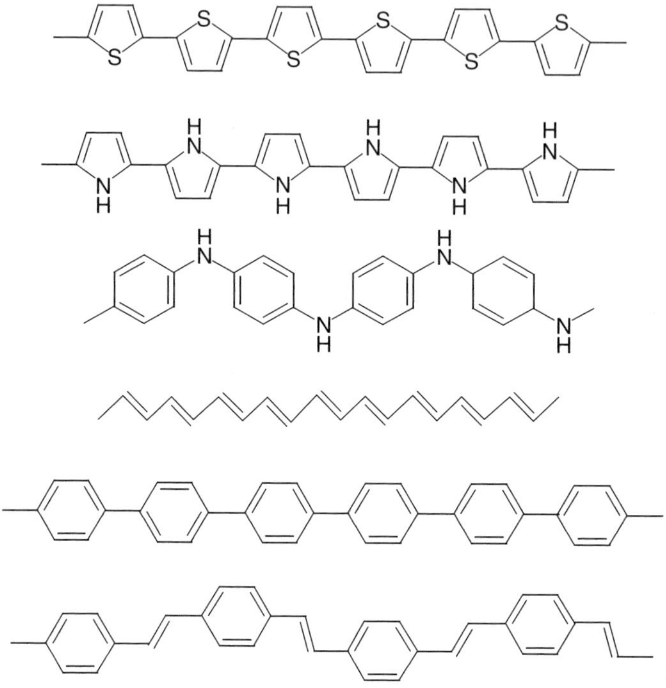

Figure 1.1 Chemical structure of some conjugated polymers.

polythiophenes, polyphenylenes and poly(*p*-phenylene vinylene)s have attracted the most attention. Figure 1.1 shows the structure of some conjugated polymers in their neutral insulating form. In order to make them electronically conductive, it is necessary to introduce mobile carriers into the conjugated system; this is achieved by oxidation or reduction reactions and the insertion of counterions (called 'doping'). Dedoped conjugated polymers are semiconductors with band gaps ranging from 1 to several eV, therefore their room temperature conductivities are very low, typically 10^{-8} S/cm or lower. However, by doping, conductivity can increase by many orders of magnitude. The concept of doping is unique and distinguishes conducting polymers from all other types of polymer [1, 53]. During the doping process, an organic polymer, either an insulator or semiconductor having small conductivity, typically in the range of 10^{-10} to 10^{-5} S/cm, is converted to a polymer which is in a 'metallic' conducting regime (1 to 10^4 S/cm). The highest value reported to date has been obtained in iodine-doped polyacetylene ($>10^5$ S/cm)

CONDUCTING POLYMERS

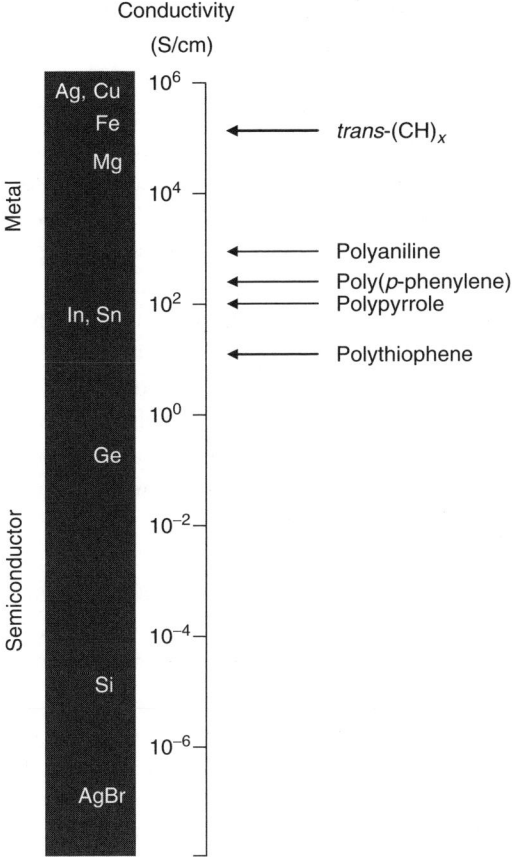

Figure 1.2 Conductivity of some metals and doped conjugated polymers.

and the predicted theoretical limit is about 2×10^7, more than an order of magnitude higher than that of copper [1]. Conductivity of other conjugated polymers reaches up to 10^3 S/cm [2, 3, 54, 55] as shown in Figure 1.2.

1.1.2 Concept of Doping in Intrinsically Conducting Polymers

Doping in a conventional semiconductor such as silicon is very different from that in a conjugated organic polymer. In conventional semiconductors, the dopant is a small amount of a donor or acceptor that is introduced into the atomic lattice resulting in a change in the occupancy of the electronic states in the solid as a result of thermal ionization of the dopant. The band structure and the density of states are essentially unchanged by the introduction of the dopant [56].

Doping of a conducting polymer, on the other hand, involves the introduction of a large amount of a donor or acceptor (in the range of a few up to 30 wt%). The presence of such a large amount of dopant and structural changes in the polymer result in a material that is significantly different from the nondoped material. The dopant perturbs the polymer extensively not only because of its significant physical size and the fact that it does not incorporate into the molecular structure, but also because of the extensive charge transfer that takes place between the polymer chain and the dopant, causing both to become ionic and leading to changes in the geometry of the chain. The doping level can also be reversibly controlled to obtain conductivities anywhere between the insulating nondoped form to the fully doped, highly conducting form of the polymer.

Doping involves either oxidation or reduction of the polymer backbone. Oxidation removes electrons and produces a positively charged polymer and is described as 'p-doping.' Similarly, reduction produces a negatively charged backbone and is known as 'n-doping.' The oxidation and reduction reactions can be induced either by chemical species (e.g., iodine, sodium amalgam or sodium naphthaline) or electrochemically by attaching the polymer to an electrode. The electrochemical doping process proceeds in much the same way as with chemical doping, with the exception that the driving force for the oxidation and reduction is provided by an external voltage source (i.e., by the electrochemical potential of the working electrode). Electrochemical p- and n-doping can be accomplished under anodic and cathodic conditions by immersing polymer film in contact with an electrode in an electrolyte solution. In these p- and n-doping processes, the positive and negative charges on polymers remain delocalized and are balanced by the incorporation of counterions (anions or cations) which are referred to as dopants. The chemical doping process of polyacetylene is shown in Figure 1.3. Upon p-doping, an ionic complex consisting of positively charged polymer chains and counteranions (I_3^-) is formed. In the case of n-type doping, an ionic complex consisting of negatively charged polymer chains and countercations (Li^+) is formed. The electronic conductivity can be controlled by the amount of dopant present.

1.1.3 Conduction Mechanism

The electronic properties of any material are determined by its electronic structure. The theory that most reasonably explains electronic structure

CONDUCTING POLYMERS

$$(CH)_n + 3/2\ ny(I_2) \longrightarrow [(CH)^{y+}(I_3^-)_y]_n$$
p-doping

$$(CH)_n + y(Li) \longrightarrow [(CH)^{y-}(Li^+)_y]_n$$
n-doping

Figure 1.3 Chemical and electrochemical doping of polyacetylene.

of materials is band theory. Quantum mechanics stipulates that the electrons of an atom can only have specific or quantized energy levels. However, in the lattice of a crystal, the electronic energy of individual atoms is altered. When the atoms are closely spaced, the energy levels are form bands. The highest occupied electronic levels constitute the valence band and the lowest unoccupied levels, the conduction band (Figure 1.4). The electrical properties of conventional materials depend on how the bands are filled. When bands are completely filled or empty no conduction is observed. If the band gap is narrow, at room temperature, thermal excitation of electrons from the valence band to the conduction band gives rise to conductivity. This is what happens in the case of classical semiconductors. When the band gap is wide, thermal energy at room temperature is insufficient to excite electrons across the gap and the solid is an insulator. In conductors, there is no band gap since the valence band overlaps the conduction band and hence their high conductivity.

Conducting polymers are unusual in that they do not conduct electrons *via* the same mechanisms used to describe classical semiconductors and hence their electronic properties cannot be explained well by standard band theory. The electronic conductivity of conducting polymers results from mobile charge carriers introduced into the conjugated π-system

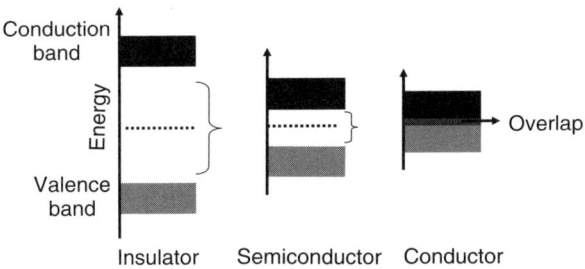

Figure 1.4 Energy band in solid.

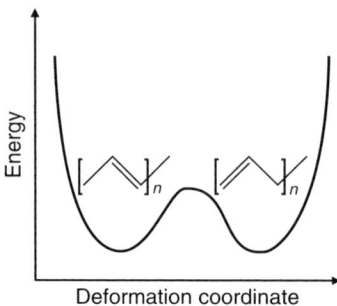

Figure 1.5 Energetically equivalent forms of degenerate polyacetylene. (Reproduced from *Polymer International*, 2004, 53, 1397, A. Moliton, R. C. Hiorns. Copyright Society of Chemical Industry. Permission is granted by John Wiley & Sons, Ltd on behalf of the SCI.)

through doping. To explain the electronic phenomena in these organic conducting polymers, new concepts including solitons, polarons and bipolarons [57–61] have been proposed by solid-state physicists. The electronic structures of π-conjugated polymers with degenerate and nondegenerate ground states are different. In π-conjugated polymers with degenerate ground states, solitons are the important and dominant charge storage species. Polyacetylene, $(CH)_x$, is the only known polymer with a degenerate ground state due to its access to two possible configurations as shown in Figure 1.5. The two structures differ from each other by the exchange of the carbon–carbon single and double bonds. While polyacetylene can exist in two isomeric forms: *cis*- and *trans*-polyacetylene, the *trans*-acetylene form is thermodynamically more stable and the *cis–trans* isomerization is irreversible [52].

Oxidative (p-type) doping of polyacetylene involves the chemical or anodic oxidation of the polymer to produce carbonium cations and radicals with simultaneous insertion of an appropriate number of anions between the polymer chains that neutralize the charge as shown in Figure 1.6 [62]. Two radicals can then recombine to give a spinless dication referred to as a positive soliton, which can act as the charge carrier [57]. Each soliton constitutes a boundary which separates domains that differ in the phase of their π-bonds. The ground state structure of polyacetylene is twofold degenerate and, therefore, the charged cations are not bound to each other by a higher energy bonding configuration and can freely separate along the chain. The effect of this is that the charged defects are independent of one another and can form domain walls that separate two phases of opposite orientation

CONDUCTING POLYMERS 13

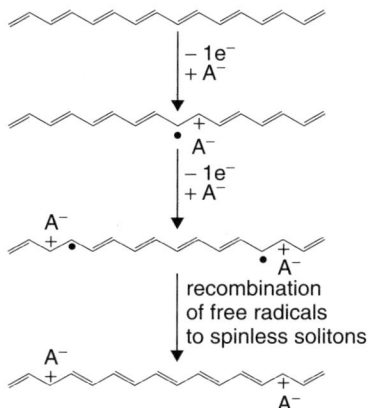

Figure 1.6 p-Type doping in polyacetylene. (Reprinted from *Progress in Polymer Science*, 27, A. Pron, P. Rannou, 135. Copyright (2002), with permission from Elsevier.)

and identical energy. In solid-state physics a charge associated with a boundary or domain wall is called a soliton, because it has the properties of a solitary wave that can move without deformation and dissipation [63]. A soliton can also be viewed as an excitation of the system that leads from one potential well to another well of the same energy (see Figure 1.5 degenerate polyacetylene).

A neutral soliton occurs in pristine *trans*-polyacetylene when a chain contains an odd number of conjugated carbons, in which case there remains an unpaired π-electron, a radical, which corresponds to a soliton (Figure 1.7). In a long chain, the spin density in a neutral soliton (or charge density in a charged soliton) is not localized on one carbon but spread over several carbons [57, 64, 65], which gives the soliton a width. Starting from one side of the soliton, the double bonds become gradually longer and the single bonds shorter, so that arriving at the other side, the alternation has completely reversed. This implies that the bond lengths do equalize in the middle of a soliton. The presence of a soliton leads to the appearance of a localized electronic level at mid-gap, which is half occupied in the case of a neutral soliton and empty (doubly occupied) in the case of a positively (negatively) charged soliton (Figure 1.7). Similarly, in n-type doping, neutral chains are either chemically or electrochemically reduced to polycarbonium anions and simultaneously charge-compensating cations are inserted into the polymer matrix. In this case, negatively charged, spinless solitons are charge carriers.

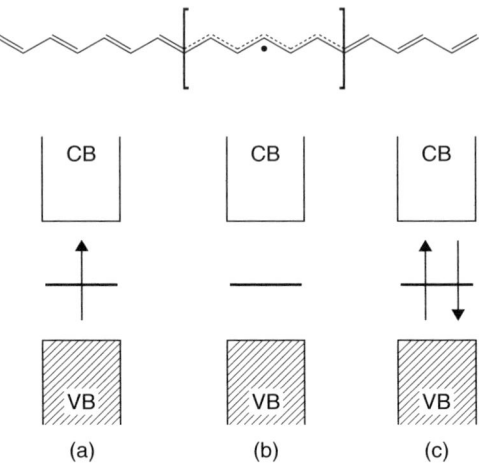

Figure 1.7 Top: schematic illustration of the geometric structure of a neutral soliton on a *trans*-polyacetylene chain. Bottom: band structure for a *trans*-polyacetylene chain containing (a) a neutral soliton, (b) a positively charged soliton and (c) a negatively charged soliton. (Reprinted with permission from *Accounts of Chemical Research*, **18**, 309. Copyright (1985) American Chemical Society.)

The π-conjugated systems based on aromatic rings, such as polythiophene, polypyrrole, polyaniline, polyparaphenylene and their derivatives have nondegenerate ground states. In these polymers, the ground-state degeneracy is weakly lifted (Figure 1.8) so that polarons and bipolarons (confined soliton pairs) are the important and dominant charge storage configurations. For example, the oxidative doping of polypyrrole is shown in Figure 1.9. The removal of one electron from the π-conjugated system of polypyrrole results in the formation of a radical cation. In solid-state physics, a radical cation that is partially delocalized over a segment of the polymer is called a polaron. It is stabilized through the polarization of the surrounding medium, hence the name. Since it is really a radical cation, a polaron has spin 1/2. The radical and cation are coupled to each other via local resonance of the charge and the radical. The presence of a polaron induces the creation of a domain of quinone-type bond sequence within the polypyrrole chain exhibiting an aromatic bond sequence. The lattice distortion produced by this is of higher energy than the remaining portion of the chain. The creation and separation of these defects cost energy, which limits the number of quinoid-like rings that can link these two species, i.e., radical and cation, together. In the case of polypyrrole it is believed that the distortion extends over four pyrrole rings.

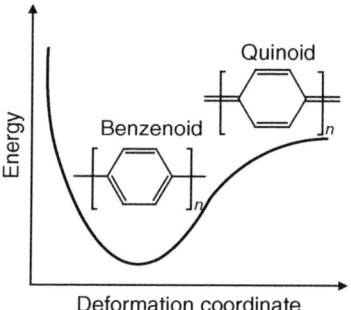

Figure 1.8 Non-degenerate ground state polyparaphenylene. Reproduced from *Polymer International*, 2004, 53, 1397, A. Moliton, R. C. Hiorns. Copyright Society of Chemical Industry. (Reproduced with permission. Permission is granted by John Wiley & Sons Ltd on behalf of the SCI.)

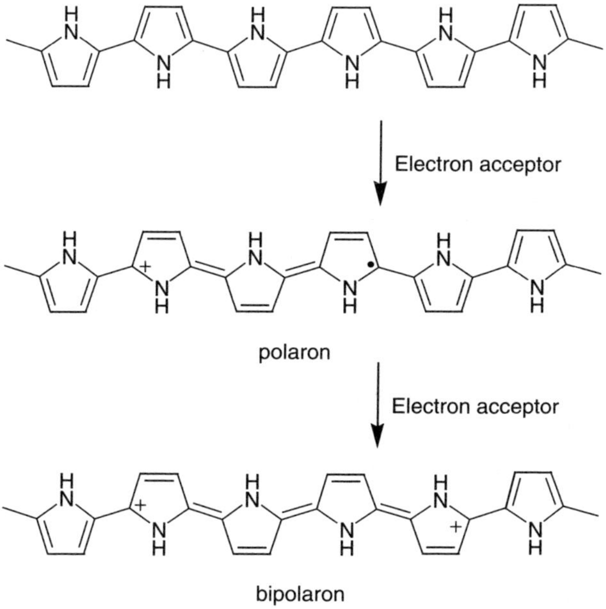

Figure 1.9 Polaron and bipolaron formation on π-conjugated backbone of polypyrrole.

Upon further oxidation, the subsequent loss of another electron can result in two possibilities: the electron can come from either a different segment of the polymer chain thus creating another independent polaron, or from a polaron level (removal of an unpaired electron) to create a dication separating the domain of quinone bonds from the sequence of aromatic-type bonds in the polymer chain, referred to as a bipolaron. This is of lower energy than the creation of two distinct polarons; therefore, at higher doping levels it becomes possible for two polarons to combine to form a bipolaron, thereby replacing polarons with bipolarons [52, 66]. Bipolarons also extend over four pyrrole rings.

Experimental and theoretical investigations of the evolution of the electronic and transport properties as a function of doping level have been conducted on polyacetylene [67–70], polypyrrole [71–74], polythiophene [74, 75] and polyparaphenylene [61, 74, 76]. Theoretical studies of the evolution of the polypyrrole electronic band structure as a function of doping level have been performed using methods ranging from highly sophisticated *ab initio* techniques to simple Hückel theory with σ compressibility. They all converge on the same picture [73, 74]. The polypyrrole band structure upon doping is shown in Figure 1.10 [52]. In the nondoped state, the band gap of polypyrrole is 3.2 eV. The presence of a polaron creates a new localized electronic state in the gap, with the lower energy states being occupied by a single unpaired electron. The polaron levels are approximately 0.5 eV away from the band edges. The polaron binding energy is 0.12 eV, constituting the difference between the 0.49 eV decrease in ionization energy and the 0.37 eV $\pi + \sigma$ energy needed for the change in geometry. The geometry relaxation in the bipolaron is stronger than in the polaron case (i.e., the geometry within the bipolaron is more quinoid-like than within the polaron), so that the empty bipolaron electronic levels in the gap are $\sim$0.75 eV away from the band edges. The bipolaron binding energy is 0.69 eV, meaning that a bipolaron is favoured over two polarons by 0.45 (0.69 $-$ 2 $\times$ 0.12) eV. This evolution is supported by electron spin resonance measurements on oxygen-doped polypyrrole [77]. At low doping, the electron spin resonance signal grows, in accordance with the fact that polarons with spin 1/2 are formed. At intermediate doping, the electron spin resonance signal saturates and then decreases, consistent with polarons recombining to form spinless bipolarons. At high doping, in electrochemically cycled samples, no electron spin resonance signal is observed although the system is highly conducting, indicating that the charge carriers in that regime are spinless. Analysis of the Pauli contribution to the susceptibility indicates that the density of states at the Fermi level is extremely small, <0.03 states eV^{-1} per monomer.

The band structure for a doping level of 33 mol % (based on polymer repeat unit), which is usually achieved in the electrochemically grown

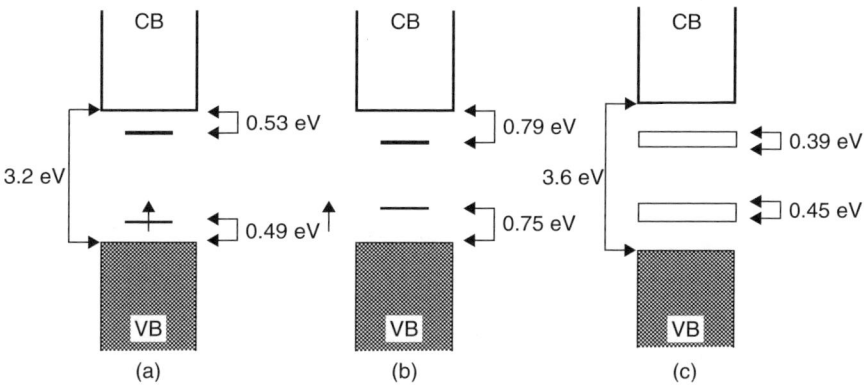

Figure 1.10 Evolution of the polypyrrole band structure upon doping: (a) low doping level, polaron formation; (b) moderate doping level, bipolaron formation; (c) high doping level (33 mol%), formation of bipolaron bands. (Reprinted with permission from *Accounts of Chemical Research*, **18**, 309. Copyright (1985) American Chemical Society.)

polypyrrole films is shown in Figure 1.10. With continued doping, the overlap between the bipolaron states forms two ~0.4 eV continuous bipolaron bands in the gap. The band gap increases from 3.2 eV in the neutral state to 3.6 eV in the highly doped state. This is due to the fact that the bipolaron states forming in the gap are at the expense of states in the valence and conduction band edges. For a very heavily doped polymer, it is conceivable that the upper and the lower bipolaron bands will merge with the conduction and valence bands respectively to produce partially filled bands and metal-like conductivity.

Among conjugated polymers, polyaniline represents a special case where doping of the neutral, unoxidized form of the polymer involves electron and proton transfer. The individual steps are shown in Figure 1.11. Polyaniline exists in three well defined oxidation states: leucoemeraldine, emeraldine and pernigraniline. Leucoemeraldine and pernigraniline are the fully reduced (all amine nitrogens) and the fully oxidized (all imine nitrogens) forms, respectively, and the emeraldine form has an amine/imine ratio of ~0.5 [78, 79]. Starting from the electronically insulating leucoemeraldine, electronically conducting emeraldine can be obtained by standard chemical or electrochemical oxidation similar to other conjugated polymers. Upon further oxidation a fully oxidized pernigraniline can be obtained.

In addition, these various oxidation states are pH sensitive and (except leucoemeraldine) can also be readily switched between doped salt and

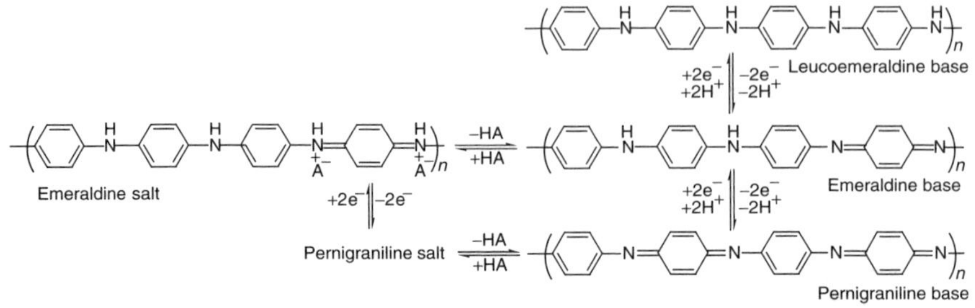

Figure 1.11 Oxidative and protonic acid (nonoxidative) doping of polyaniline.

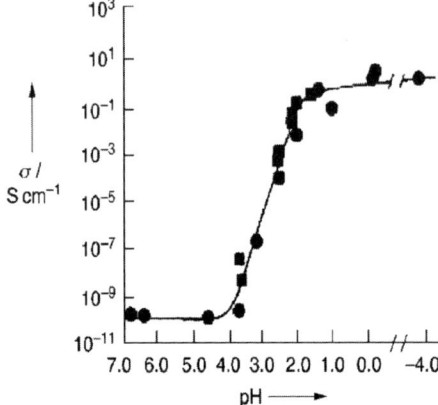

Figure 1.12 Conductivity of emeraldine base as a function of the pH of the HCl dopant solution as it undergoes protonic acid doping (● and □ represent two independent series of experiments). (Reproduced from *Angewandte Chemie – International Edition*, 2001, 40, 2581; A. MacDiarmid, with permission from Wiley-VCH.)

nondoped base forms. Out of five forms of polyaniline, only the emeraldine salt form is conducting. Protonic acid doping occurs by proton addition to the polymer chain rather than by partial oxidation or reduction of the polymer π-system. The number of electrons associated with the polymer backbone does not change during the protonation step. Protonic acid doping of emeraldine base units with, for example, 1 M aqueous HCl results in complete protonation of the imine nitrogen atoms to give the fully protonated emeraldine hydrochloride salt. As shown in Figure 1.12, protonation is accompanied by a 9–10 orders of magnitude increase in conductivity, reaching saturation in *ca* 1 M

HCl [5, 78–80]. The process is reversed [78, 80, 81] when the protonated polymer is treated with aqueous alkali. Further, this behavior has been observed in other systems, such as poly(heteroaromatic vinylenes) [82].

In their doped, electronically conducting forms, ICPs typically possess positive charges (polarons or bipolarons) along their polymer chains, whose charge is balanced by the incorporation of anions. These anions may consist of a wide variety of chemical groups ranging from simple anions such as Cl^-, HSO_4^-, ClO_4^-, NO_3^- [41], to bulkier species such as p-toluenesulfonate (pTS^-) or camphor-10-sulfonate (CSA^-) [55, 83, 84], to large polyelectrolytes such as poly(styrenesulfonate) [85–89], as well as amino acids and biopolymers, including proteins and DNA [90–93]. The incorporated anions provide a powerful additional element for tuning the properties of the resulting ICPs, giving rise to a wide range of properties and applications. The nature and extent of doping not only increase electronic conductivity of the polymer but also induce desired properties such as improved processability, environmental stability and special catalytic, optical or spectroscopic, and redox properties useful for optical pH sensors, heterogeneous catalysis, gas separation membrane, drug delivery and biosensors [62, 94, 95]. The nature of the dopant anion also strongly influences the morphology of the polymer [62]. Recently, it has been shown that the surfactant function of the dopant plays an important role in the formation of nanostructures [96, 97]. Wan *et al.* have reported that the morphology of polyaniline nanostructures prepared in the presence of inorganic acids is dependent on dopant structures [98].

Apart from oxidative and protonic acid doping processes, there are some doping processes where no counter dopant ion is involved, such as photodoping [99] and charge injection [100–102]. In photodoping, when the polymer is exposed to radiation of greater energy than its band gap, electrons are promoted across the gap. For example, in *trans*-polyacetylene, solitons are observed under appropriate experimental conditions [99]. After irradiation is discontinued, charge carriers disappear rapidly because of the recombination of electrons and holes. If a potential is applied during irradiation, then the electrons and holes separate and photoconductivity is observed. Charge injection is most conveniently carried out using a metal/insulator/semiconductor configuration involving a metal and a conducting polymer separated by a thin layer of a high dielectric-strength insulator. In the case of charge injection at a metal–semiconductor interface, the polymer is oxidized or reduced (electrons are added to π^*-band or removed from the π-band). However, the polymer is not doped in the sense of chemical and electrochemical doping, for there are no counterions. In the case of charge injection at a metal–semiconductor interface, electrons reside in the π^*-band and/or holes reside in the π-band only as long as a biasing voltage is applied. Because of the self-localization associated with the formation of solitons, polarons and bipolarons, charge injection leads to the formation of localized structural distortions and electronic states in the energy gap. This approach has resulted in the observance of superconductivity in a polythiophene derivative [100–102].

1.1.4 Synthesis

Intrinsically conducting polymers are generally synthesized *via* chemical or electrochemical oxidation of a monomer where the polymerization reaction is stoichiometric in electrons. However, several other synthetic approaches exist, such as photochemical-initiated or biocatalytic oxidative polymerizations using naturally occurring enzymes.

1.1.4.1 Chemical Synthesis

Chemical synthesis has the advantage of being a simple process capable of producing bulk quantities of ICPs on a batch basis. To date it has been the major commercial method of producing such materials. Chemical polymerization is typically carried out with relatively strong chemical oxidants like ammonium peroxydisulfate, ferric ions, permanganate or bichromate anions, or hydrogen peroxide. These oxidants are able to oxidize the monomers in solution, leading to the formation of cation radicals. These cation radicals further react with other monomers or *n*-mers, yielding oligomers or insoluble polymer. There are two main limitations of the chemical oxidation technique, both related to the limited range of chemical oxidants available. The counterion of the oxidants ultimately ends up as a dopant or codopant in the polymer. Hence it is difficult to prepare ICPs with different dopants. The limited range of oxidants also makes it difficult to control the oxidizing power in the reaction mixture and in turn the degree of overoxidation during synthesis. Both the type of dopant and the level of doping are known to impact upon final properties of the polymer such as molecular weight, crosslinking and, ultimately, conductivity.

1.1.4.2 Electrochemical Synthesis

The electrochemical synthesis of conducting polymers, first demonstrated with polypyrrole [103], has proven important in the development of the field. Using this approach, semiconducting polymers have been obtained from a wide variety of monomers including thiophene, furan, carbazole, aniline, indole, azulene and polyaromatic monomers such as pyrene and fluoranthene. In general, chemical oxidation provides ICPs as powders, while electrochemical synthesis leads to films deposited on a working electrode. A wide range of anodes may be employed, including platinum, gold, carbon and indium-doped tin oxide (ITO)-coated glass. The ITO-coated glass electrodes, being transparent in the visible/near-infrared region, are very useful for *in situ* spectral investigations involving absorption or circular dichroism of the deposited polymer films.

Table 1.1 shows a list of dopant ions and their source electrolytes which are currently being used in the electrochemical synthesis of conducting polymers. All of these dopant ions with the exception of the last two (marked by *) are anions and are associated with electrochemical oxidation of the polymers at the anode. Aizawa *et al.* [104] reported the first example of the reductive doping of an electrochemically synthesized polythioenylene film with cations like tetraethyl ammonium (Et_4N^+)

Table 1.1 List of dopant ions and their source of electrolyte[a] (Reprinted from *Materials Chemistry and Physics*, 61, K. Gurunathan, A. V. Murugan, R. Marimuthu, U. P. Mulik, D. P. Amalnerkar, 173. Copyright (1999) with permission from Elsevier.)

Dopant ion	Supporting electrolyte
BF_4^-	$R_4N^+BF_4^-$, MBF_4
PF_6^-	R_4NPF_6, MPF_6
ClO_4^-	R_4NClO_4, $MClO_4$
Cl^-	R_4NCl, HCl, MCl
Br^-	R_4NBr, MBr
I^-	R_4NI, MI
AsF_6^-	$MAsF_6$
HSO_4^-	$MHSO_4/R_4NHSO_4$
$CF_3SO_3^-$	$MCF_3SO_3/R_4NCF_3SO_3$
$CH_3C_6H_4SO_3^-$	$MCH_3C_6H_4SO_3$
SO_4^{2-}	Na_2SO_4, H_2SO_4
(Et_4N^+)*	Et_4NPF_6
(Bu_4N^+)*	Bu_4NPF

[a] R = Alkyl; Et = Ethyl; But = Butyl; M = Metal (Li^+, Na^+, Ag^+, K^+).

and tetrabutyl ammonium (Bu_4N^+). They observed a red to green electrochromic transition as a result of cation doping.

The advantage of electrochemical polymerization is that by selecting an appropriate electrolyte, a much wider choice of cations and anions for use as 'dopant ions' is possible. Also, in electrochemical polymerization, doping and processing take place simultaneously with polymerization. However, in conventional methods, first polymer synthesis and doping are carried out, followed by processing. The electrochemical oxidation route also has greater control over the electrochemical potential of the system (thereby limiting overoxidation) and in turn the rate of polymerization as well as the film thickness.

1.1.4.3 Photochemical and Biocatalyzed Synthesis

Alternative routes to ICPs, including photochemical, are reportedly advantageous in some applications over the conventional chemical and electrochemical approaches. For example, pyrrole has been successfully polymerized to polypyrrole by irradiation with visible light using either $[Ru(bipy)_3]^{2+}$ (bipy = 2,2'-bipyridine) or $[Cu(dpp)_2]^+$ (dpp = 2,9-diphenyl-1,10-phenanthroline) as the photosensitizer and an appropriate electron acceptor (sacrificial oxidant) [105, 106]. Recently, the enzyme horseradish peroxidase has been used for the polymerization of anilines in the presence of hydrogen peroxide through oxidative free radical coupling reactions [91, 107]. The advantage of this synthetic approach is that the polymerization of aniline can be carried out at environmentally mild conditions compared to the chemical and electrochemical methods.

1.1.5 Processability

Over the last three decades, ICPs such as polythiophene, polypyrrole and polyaniline have attracted a great deal of research interest because their electrical, electrochemical and optical properties show great promise for commercial applications in rechargeable batteries, light-emitting diodes, electrochromic display devices, sensors, etc. In these applications, polymers are normally used as films or coatings, with different thicknesses, depending on the requirements of the application. Key hurdles to the use of these polymers in commercial products have been the lack of facile synthetic methods for producing useful quantities as well as simple approaches for processing bulk polymer into useful forms. In addition, the stability, including mechanical, thermal, conductivity (with time and at high temperature) and overoxidation, has been a concern.

The limitations in postsynthesis processability are due to the chain stiffness and interchain interactions that render these materials insoluble in common solvents. For example, polymers can become crosslinked, highly branched, or electrostatically crosslinked due to polaron/bipolaron charge interactions. The chemical or ionic crosslinking renders the polymer intractable. However, several approaches have been adopted to facilitate solution processability including:

(i) *Reduction of polymer to nonconducting state to remove interchain charge interactions.* A unique property of polyaniline is that in the deprotonated form, emeraldine base (EB), it is readily soluble in some solvents such as N-methylpyrrolidinone, dimethyl formamide and dimethyl sulfoxide [108]. Also, polyaniline has been processed without changing molecular structure in certain amines like pyrrolidine and tripropylamine [109], fluorinated solvents [110], acetic acid or in concentrated sulfuric acid, and other strong acids [111, 112].

(ii) *Substitution of alkyl chains (polypyrrole, polythiophene and polyaniline).* A major breakthrough in the field of ICPs was the synthesis of soluble and conducting poly(3-alkylthiophene) [113, 114]. These polymers have straight chain alkyl groups, such as hexyl, octyl, dodecyl and octadecyl, and a degree of polymerization ranging from 100 to 200, depending on the alkyl group. The neutral polythiophenes substituted by flexible and long alkyl side chains (with more than three carbon atoms) are rendered soluble in common organic solvents (chloroform, toluene, etc.) through reduced interchain interactions and favorable substituent–solvent interactions. However, steric interactions along the polymer backbone result in reduced orbital overlap and a corresponding lowered conductivity. Many attempts have been made to improve the processability of polyaniline by the introduction of an alkyl group on the N-substituent [115, 116] or alkoxy substituent [117, 118].

(iii) *Counterion induced processability.* Organic solvent solubility can be imparted to conducting polyaniline and polypyrrole salts by the incorporation of large functionalized protonic acids such as camphor sulfonic or dodecylbenzenesulfonic acid as a counterion during synthesis of these polymers [55, 83]. The long alkyl chains of the dodecylbenzene functional group lead to solubility in common solvents such as toluene, xylenes, decalin, chloroform, etc.

(iv) *Enzymatic synthesis.* Water soluble polyaniline [107, 119, 120] and polypyrrole [121] have been prepared using a template-guided enzymatic approach. Strong acid polyelectrolytes, such as sulfonated polystyrene, and strong acid surfactant molecules, such as sodium dodecylbenzenesulfonate, are used as templates because they provide a lower local pH environment for the formation of the conducting polymer.

(v) *Colloidal dispersions.* The most commercially successful method of producing processable forms of ICPs has been the aqueous colloidal dispersion route. In this approach, aqueous colloidal dispersions of conducting polymers have been prepared in the presence of a water-soluble polyelectrolyte such as polystyrene sulfonic acid [122] and polymeric stabilizers such as poly(vinyl alcohol-co-acetate), methyl cellulose, poly(vinyl pyrrolidone), poly(vinyl methyl ether), etc. and surfactant stabilizers [123, 124]. The polyelectrolyte and anionic surfactants act as the charge compensating dopant for the polymer and also render the resulting complex a colloidal dispersion as these polyelectrolyte and surfactants are themselves water soluble.

(vi) *In situ polymerization of metastable monomer–oxidant mixtures.* Selecting a chemical oxidizing agent with a formal potential near to, but below, the oxidation potential of the monomer, results in a metastable reaction mixture that polymerizes upon solvent evaporation. This behavior allows facile preparation of well defined conducting polymer films on any substrate [125, 126].

(vii) *Self-doping.* Recently, environmental concerns have placed restrictions on the commercial use of many organic solvents. These concerns in turn have encouraged the use of polymers that can be processed in aqueous media. 'Self-doping' is the most successful approach for increasing the solubility of conductive polymers in aqueous solution. When ionizable functional groups that form negatively charged sites are attached to the polymer chain to make the polymer conducting, it is referred to as 'self-doping' or when the group is an acid, 'self-acid-doping'. The distinctive properties of self-doped conducting polymers are their water solubility, electroactivity and conductivity over a wider pH range (in the case of polyaniline), and thermal stability. The ionizable groups on the backbone give the polymer certain polyelectrolyte properties, i.e., these groups dissociate into aqueous solvent. The solubility of self-doped conducting polymers in aqueous solutions can be attributed to the hydrophilic interactions between the covalently attached ionized group on the polymer backbone and polar molecules of water. In water, the steric and ionic repulsive interactions overcome the interchain interactions and allow for the rapid solvation of polymer backbone.

1.2 SELF-DOPED CONDUCTING POLYMERS

Charge transfer doping and the associated changes in electronic properties of conjugated organic polymers [41] such as polyacetylene, polythiophene and polypyrrole, are well known. As discussed above, in such systems, the injection of charge into the delocalized π-electron system (either chemically or electrochemically) requires that dopants or counterions diffuse into the polymer during the charge injection process in order to maintain charge neutrality. In the case of polyaniline, doping can be achieved without electron transfer by protonation of the polymer (Figure 1.11). In this case as well, however, the doping process requires (and is limited by) diffusion of counterions into the structure to preserve charge neutrality. This reversible exchange of large anions between the active polymer mass and electrolyte limits many important characteristics such as electrochromic switching, charging rates, etc.

The driving force behind the extensive research on self-doped polymers has been to improve the processability in aqueous media, to increase the speed of electrochromic switching and to enhance the charge storage performance of polymer-based batteries. In 1987, Wudl *et al.* first reported the novel concept of 'self-doping' in conducting polymers [127, 128]. In this work, a self-doped conducting polymer was described as a conjugated polymer where a significant fraction of monomer units contained a covalently attached ionizable, negatively charged, functional group acting as a stable/immobile dopant anion. This principle of self-doping in a conjugated polymer is shown in Figure 1.13. As shown in the figure, by immobilizing the larger anion, upon oxidation, the smaller mobile proton or other monovalent cation is ejected from the polymer into the electrolyte to maintain charge neutrality. Due to the higher mobility of the smaller cations, the rate of the charging (redox) process is significantly increased. Sodium salts and acid forms of poly(3-thiophene ethanesulfonate) and poly(3-thiophene

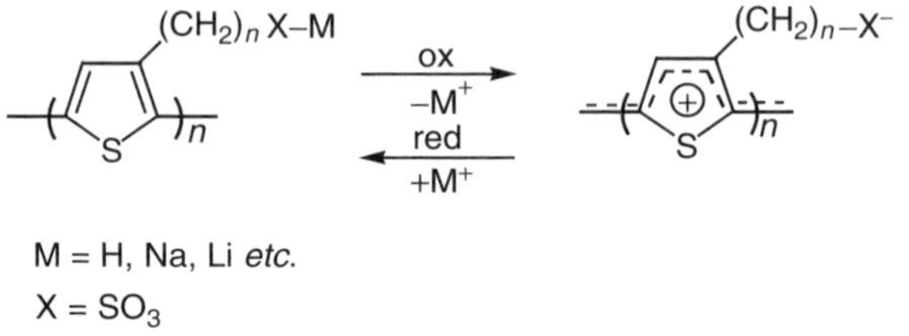

Figure 1.13 Oxidation–reduction reactions of polythiophene derivatives showing self-doping during the oxidation reactions. (Reprinted from *Synthetic Metals*, 20, A. O. Patil, Y. Ikenoue, N. Basescu, N. Colaneri, F. Wudl, A. J. Heeger, 151. Copyright (1987), with permission from Elsevier.)

butanesulfonate) were the first self-doped conducting polymer materials to show water solubility in neutral (insulating) and doped (conducting) states [127]. The films cast from water solution exhibit electronic conductivity in the range of $10^{-7}-10^{-2}$ S/cm depending on the relative humidity. Since these polymers are prepared by the electropolymerization of 3-substituted thiophene monomers in the presence of 'foreign' electrolyte, various experiments and arguments were put forward to prove self-doping, including cyclic voltammetry experiments combined with pH measurements in organic solvents [129].

Following these initial reports, Havinga *et al.* electropolymerized a potassium salt of 3′-propylsulfonate 3′-propylsulfonate 2,2′:5′,5″-terthienyl in the absence of foreign electrolyte [130, 131]. In this case, the monomer itself acts as the electrolyte during electrochemical polymerization, resulting in solid evidence for self-doping. The self-doped poly(3′-propylsulfonate 2,2′:5′,5″-terthienyl) shown in Figure 1.14 is soluble in water. However, the polymer solutions are unstable. Ikenoue *et al.* reported the first chemical synthesis of self-doped polymer by chemically oxidizing sodium 3-(3′-thienyl)propanesulfonate using ferric chloride in an aqueous medium. This polymer shows self-doped behavior

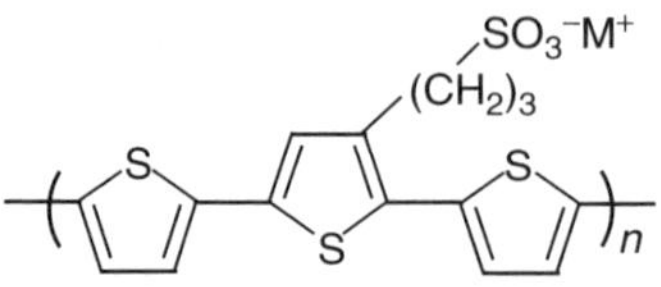

Figure 1.14 Chemical structure of self-doped poly(3′-propylsulfonate 2,2′:5′,5″-terthienyl). (Reprinted with permission from *Chemistry of Materials*, **1**, 650. Copyright (1989) American Chemical Society.)

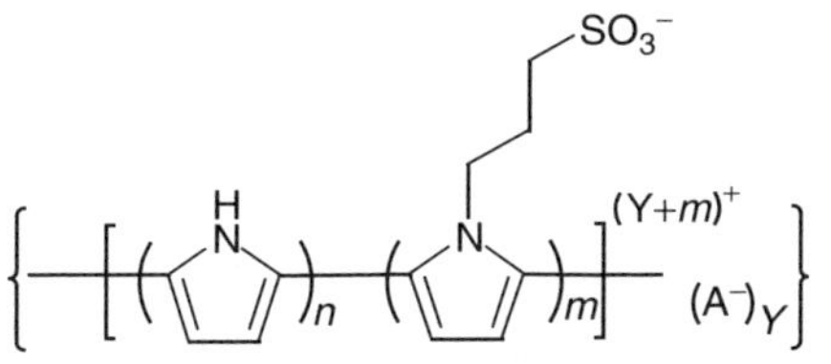

Figure 1.15 Chemical structure of self-doped poly(pyrrole-co(3-(pyrrol-lyl)propanesulfonate). (*Chemical Communications*, 1987, **621**, N. S. Sundarsan, S. Basak, M. Pomerantz, J. R. Reynolds. Reproduced by permission of the Royal Society of Chemistry.)

both in water and as a film [132]. The spin-coated films of poly(3-(3'-thienyl)propanesulfonate) exhibit faster optical switching response and higher stability.

This approach of introducing ionic substituents has been extended to other ICPs like polypyrrole [130, 133–135], polyaniline [136–138], polyphenylenes [139–142] and polyphenylenevinylenes [143]. For the polypyrrole system, the self-doping concept has been extended by Reynolds *et al.* [133, 134] and Havinga *et al.* [130, 135]. Reynolds *et al.* chemically polymerized the N-substituted pyrrole and potassium-3-(pyrrol-l-yl)propanesulfonate in aqueous solution with ferric chloride as oxidizing agent and electropolymerized it or copolymerized with pyrrole in n-(Bu)$_4$NBF$_4$/CH$_3$CN or LiClO$_4$/CH$_3$CN solution, electrochemically forming a poly{pyrrole-co(3-(pyrrol-lyl)propanesulfonate} (Figure 1.15). Havinga *et al.* polymerized the ring substituted pyrrole, sodium (3-pyrrolyl)alkanesulfonate (Figure 1.16), chemically in aqueous solution with ferric chloride as the oxidizing agent and electrochemically in acetonitrile without any electrolyte. These water soluble self-doped forms of polypyrrole have electronic conductivities in the range of 10^{-3}–0.5 S/cm and good stability under ambient conditions.

The first self-doped polyaniline was produced by Hany and Genies in 1989 [136] by heating the emeraldine salt of polyaniline in dimethyl sulfoxide with either 1,3-propane sulfone or 1,4-butane sulfone to produce the corresponding poly(aniline-N-propylsulfonic acid) or poly(aniline-N-butylsulfonic acid). However, these polymers had poor solubility and significantly lower conductivity (10^{-9} S/cm). In 1990, Yue and Epstein [137] and Bergeron *et al.* [138] reported the first water-soluble, self-doped, conducting polyaniline derivatives. The ring-sulfonated polyaniline (Figure 1.17 A) was prepared by the sulfonation of the emeraldine base of polyaniline with fuming sulfuric acid for 2 h at room temperature followed by precipitation

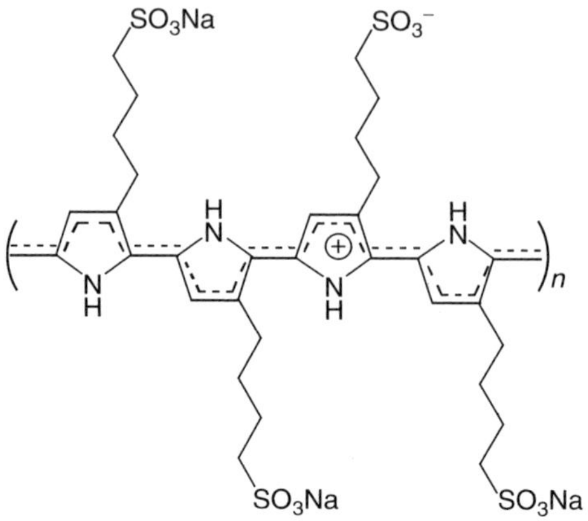

Figure 1.16 Chemical structure of self-doped poly(sodium (3-pyrrolyl)alkanesulfonate). (Reprinted with permission from *Chemistry of Materials*, **1**, 650. Copyright (1989) American Chemical Society.)

of the reaction product with methanol and acetone. In the resulting sulfonated polyaniline, only approximately half of aromatic rings contained sulfonate groups. This degree of modification was sufficient to compensate all positive charges of the protonated nitrogen sites in the emeraldine salt form of this polymer. The polymer obtained exhibited a significantly lower conductivity (approximately 0.1 S/cm) compared with unmodified polyaniline (typically 1–10 S/cm), however, the conductivity was independent of pH up to 7, unlike polyaniline. Electrostatic interactions prevent the proton from diffusing away from

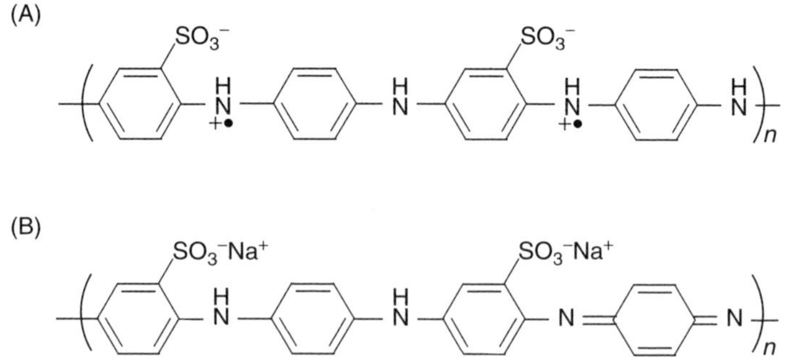

Figure 1.17 Chemical structures of (A) self-doped ring-sulfonated polyaniline and (B) dedoped (insulating) salt form of ring-sulfonated polyaniline. (Reprinted with permission from *Journal of the American Chemical Society*, **112**, 2800. Copyright (1990) American Chemical Society.)

the ring-bound, negatively charged, polymer chain. This high local concentration of protons in the vicinity of the polymer backbone has been claimed to be responsible for the retention of doping at neutral pH [137]. In contrast to the parent polyaniline, self-doped sulfonated polyaniline was found to be soluble in diluted aqueous base, forming the dedoped (insulating) salt form (Figure 1.17 B). Since these initial reports, many papers have been published, dealing with synthesis, properties and applications of self-doped conducting polymers.

1.3 TYPES OF SELF-DOPED POLYMERS

Self-doped conducting polymers have been generated by functionalizing the monomer (prepolymerization modification) and polymer (postpolymerization modification) with ionizable, negatively charged, moieties. These self-doped conducting polymer derivatives have been prepared with various functional groups such as sulfonic acid, alkyl and alkoxysulfonic acid, carboxylic acid, phosphonic acid, acetic acid and boronic acid. In general, two kinds of self-doped conducting polymer derivatives have been prepared. One of them contains ionizable functional groups (usually sulfonate), attached directly to the aromatic rings of the polymer backbone. The other kind contains ionizable functional groups, bound through a spacer consisting of alkane or aromatic groups to heteroatoms in the conjugated backbone. Some examples of polyaniline, polypyrrole and polythiophene derivatives are shown in Figure 1.18 ((1) [144], (2) [136, 138, 145] (3) [146, 147], (4) [148], (5) [149], (6) [150, 151], (7) [144, 152–154], (8) [147], (9) [148], (10) [155], (11) [147], Figure 1.19 ((1) [130, 135], (2) [156], (3) [157], (4) [158, 159], (5) [159], (6) [134] and Figure 1.20 ((1) [127–129], (2) [160], (3) [161, 162], (4) [163], (5) [164], (6) [165], (7) [166].

The common feature of all self-doped polymers is that the dopant is attached covalently to the conjugated backbone. Conducting polymers prepared with large dopants containing functional protonic acids, for example, camphorsulfonic acid, dodecylbenzene sulfonic acid, and polyelectrolytes or polymeric acids, are not technically self-doped polymers, although in many cases they can exhibit similar properties including solubility, pH stability, fast switching, etc. The lack of mobility associated with these large anions, as well as the fact that polyelectrolytes are entangled in the conducting polymer structure, make their expulsion difficult during the dedoping process. Consequently cations are incorporated to maintain electroneutrality. In 1992 Cao and Heeger *et al.* [55, 167] reported the concept of counterion-induced processibility

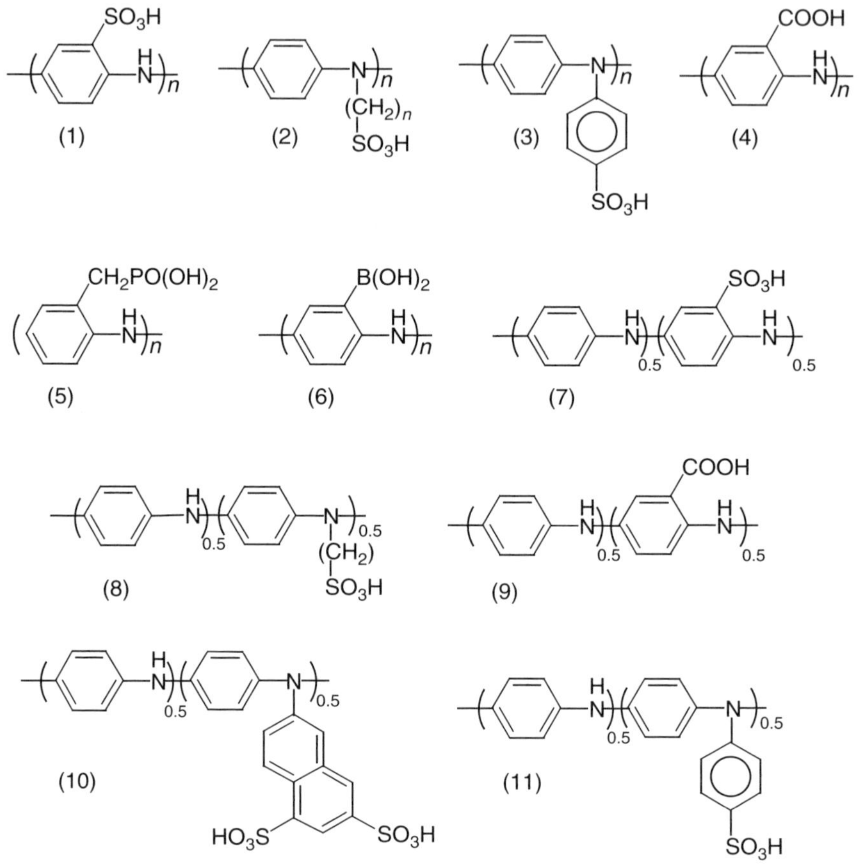

Figure 1.18 Various self-doped polyaniline derivatives.

of conducting polyaniline. Doping polyaniline with functionalized protonic acids renders the polymer conducting and soluble in common organic solvents such as xylene, decalin, chloroform or m-cresol. The polyelectrolytes or polymeric acids that have been used as dopants to prepare water-soluble conducting polymers include poly(ethanesulfonic acid) [85], poly(acrylic acid) [85, 86], poly(styrenesulfonic acid) [85, 86, 88, 89], poly(2(acrylamido)-2-methyl-1-propanesulfonic acid) [168] and poly(amic acid) [169]. These dopants can be incorporated into the polymer during either chemical or electrochemical polymerization or by doping in their aqueous solution. Recently, an enzymatic approach has been developed for the polymerization of aniline in the presence of a polyelectrolyte template such as poly(4-styrenesulfonate)

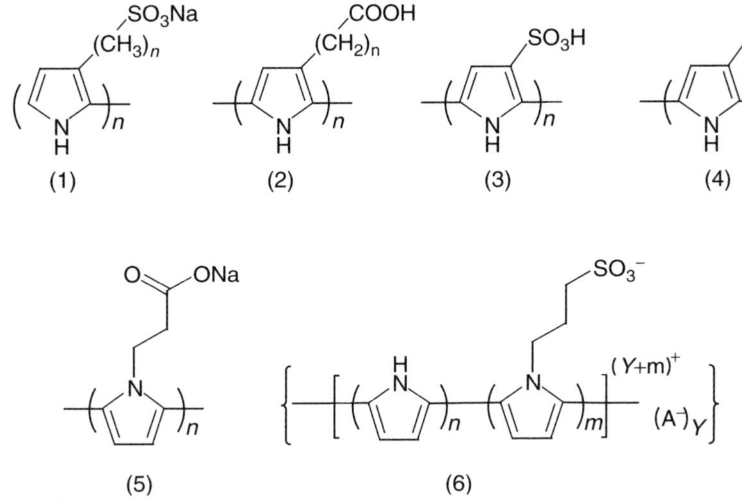

TYPES OF SELF-DOPED POLYMERS

Figure 1.19 Various self-doped polypyrrole derivatives.

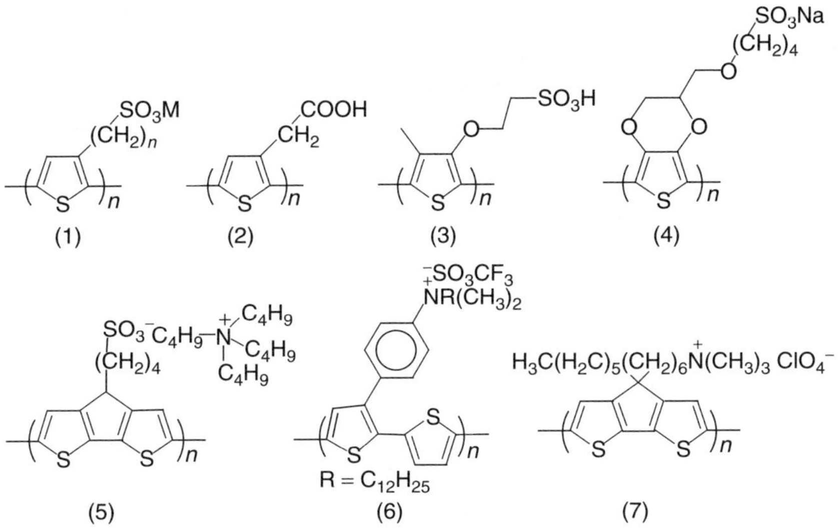

Figure 1.20 Various self-doped polythiophene derivatives.

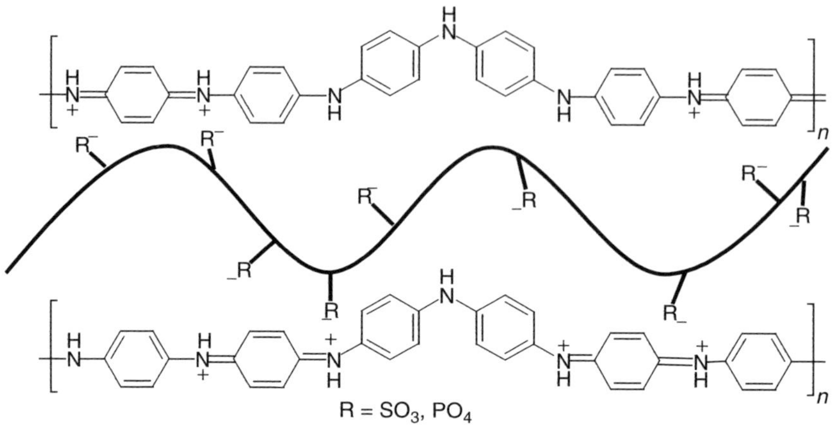

Figure 1.21 Complex formation of macromolecular polyelectrolyte template along with polyaniline chain. (Reprinted with permission from *Macromolecules*, 37, 4130. Copyright (2004) American Chemical Society.)

or poly(vinylphosphonic acid) under mild aqueous pH 4.3 buffer conditions [107, 119, 120]. The polyelectrolytes act as a template upon which the aniline monomers and oligomers preferentially align themselves through electrostatic and hydrophobic interactions and form complex, as well as large, molecular counterions, which are integrated and essentially locked to the conducting polymer chains (Figure 1.21) [170], thus making doped (as synthesized) polyaniline water soluble. Similarly water-soluble polypyrrole [121] and water-dispersible poly(thieno[3,4-b])thiophene have been prepared in the presence of poly(styrenesulfonic acid) [122].

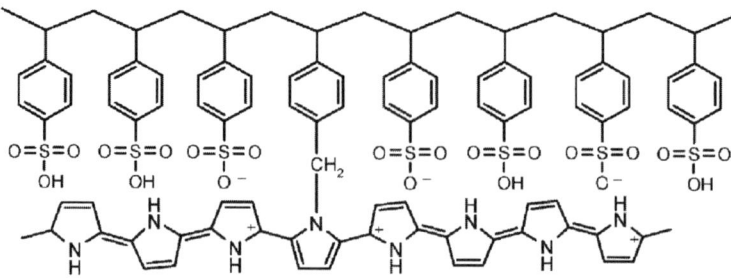

Figure 1.22 The structure of self-doped polypyrrole graft copolymer (Reprinted with permission from *Macromolecules*, 38, 1044. Copyright (2005) American Chemical Society.)

Recently, Bae and Jo et al. prepared water-soluble self-doped polyaniline [171] and polypyrrole [172] graft copolymers. In their study, the polymeric dopant poly(styrenesulfonic acid) is covalently bonded to the conjugated polymer backbone as shown in Figure 1.22. Water-soluble moieties (sulfonic acid groups) that do not participate in self-doping, make polyaniline and polypyrrole graft copolymers soluble in water and polar organic solvents. The conductivities of these polymers are reportedly 4.8×10^{-1} and 1.2×10^{-1} S/cm, for polypyrrole and polyaniline copolymers, respectively. These conductivity values are higher than those of water-soluble conducting polymers prepared using polyelectrolytes as a template [170], and other benzene ring or nitrogen atom substituted water-soluble polyanilines [173–175].

1.4 DOPING MECHANISM IN SELF-DOPED POLYMERS

1.4.1 p-Type Doping

Self-p-doping is by far the most common type of doping found in the literature and there are many examples based on conjugated polymers derived from polyaniline [137, 152, 153], polypyrrole [130, 135], polythiophene [127–129] and polyphenylenes [139–142]. In general, when unsubstituted conjugated polymers are oxidized (p-doping), anions are incorporated to preserve charge neutrality in the materials as the positive charges develop along the polymer backbone. Upon reduction, or dedoping, the anions are ejected. In contrast, when a self-doped polymer is reduced, a cation must move into the polymer to charge compensate the immobilized anion. Upon oxidation that cation is ejected from the polymer. For example, in the case of self-doped polythiophenes with an alkanesulfonic acid substituent, upon electrochemical or chemical oxidation, charge transfer resulting in the generation of charge carriers occurs concomitant with the ejection of protons from the bonded dopant to maintain charge neutrality as shown in Figure 1.23 [176, 177].

The self-protonation of sulfonic acid ring substituted polyaniline (in emeraldine oxidation state) differs from the 'self-doping' mechanism of poly(3-(4-alkanesulfonate)thiophene shown in Figure 1.23 [127]. Figure 1.24 shows the oxidation of sulfonated polyaniline by ejection of electrons from nitrogen and dedoping of protons. The removal of protons produces negative charges on SO_3^-, which compensates positive charges of nitrogen. According to Epstein et al. [144, 153, 173], in

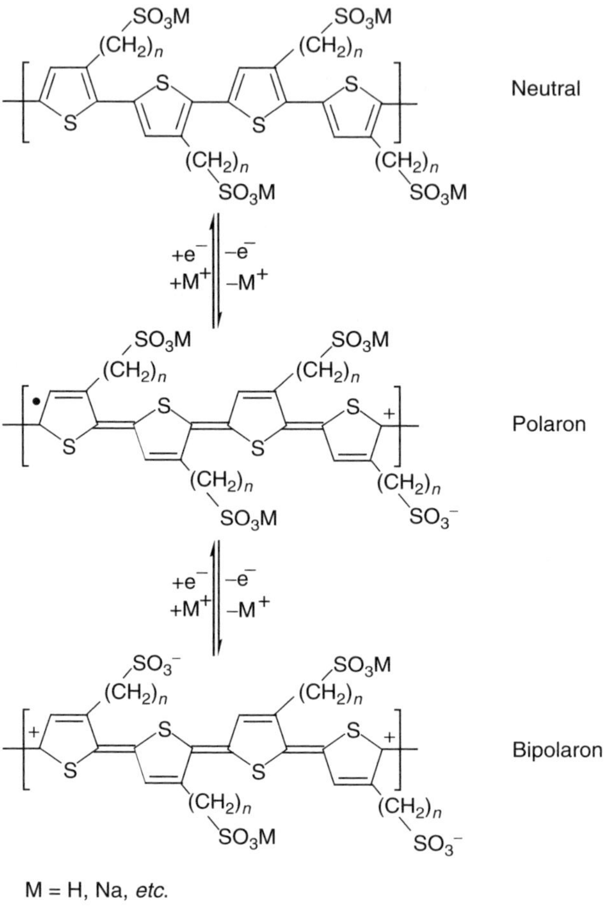

M = H, Na, *etc.*

Figure 1.23 Chemical or electrochemical oxidation of polythiophene derivatives. (Reproduced from *Advanced Materials*, 1997, 9, 1087, M. Leclerc, K. Faid, with permission from Wiley-VCH.)

the oxidized state of sulfonated polyaniline the positive charge carriers are more localized on nitrogen atoms than are those of the parent polyaniline emeraldine salt. This is due to strong electrostatic interaction between SO_3^- groups and cation radical nitrogen atoms or amine hydrogens to form five- or six-member rings, which are in an energetically favorable configuration as shown in Figure 1.25 A. This configuration can effectively localize the positive charge around the nitrogen atoms. The interaction between SO_3^- groups and cation radical nitrogen atoms or amine hydrogens is also likely take place between two adjacent chains (Figure 1.25 B) in addition to intrachain interaction shown in

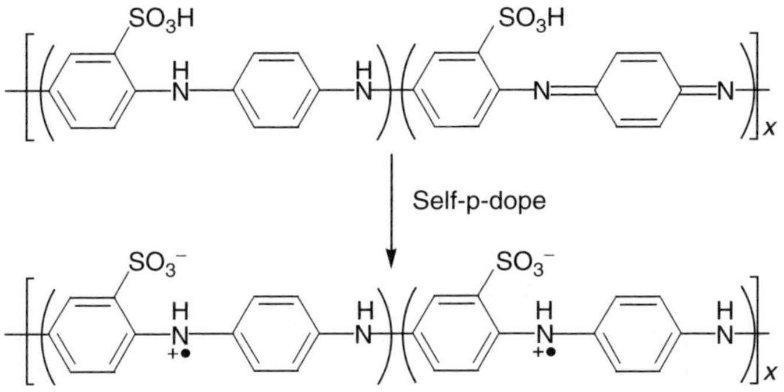

DOPING MECHANISM IN SELF-DOPED POLYMERS 35

Figure 1.24 Structural changes of sulfonated polyaniline by oxidation. (Reprinted with permission from *Journal of the American Chemical Society*, **112**, 2800. Copyright (1990) American Chemical Society.)

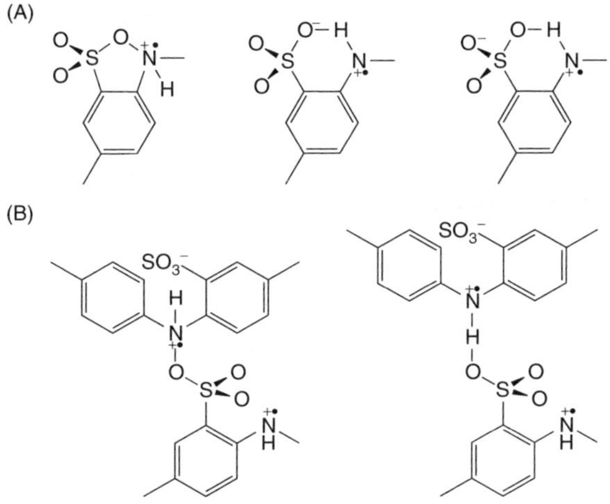

Figure 1.25 Intrachain (A) and interchain (B) interactions in sulfonated polyaniline. (Reprinted with permission from *Journal of the American Chemical Society*, **113**, 2665. Copyright (1991) American Chemical Society.)

Figure 1.25 A. These interchain interactions may play an important role in increasing the three-dimensionality of the charge and spin motion [153].

In contrast with its parent form, introduction of the $-SO_3H$ group on the polyaniline backbone brings new acid–base chemistry within the polymer. The acid–base equilibrium can be changed externally either chemically or electrochemically. The conductivity of sulfonated polyaniline is independent of external protonation over a broad pH range. This clearly indicates that the internal acid–base chemistry in sulfonated polyaniline is not affected by the external medium within this pH range. In the case of highly sulfonated polyaniline, conductivity is pH independent in the range 0 to 12 [173]. This is likely due to the enhancement of the doping strength of protons from sulfonic acid groups on the imine nitrogen atoms by the formation of six-member-ring complexes. As a result, doped imines are more difficult to dedope. Even after exchange of protons with cations, the six-member ring conformation may still exist and thus imines may still be doped by weaker metal cation Lewis acids. Therefore, the polymer treated with alkaline aqueous solutions is still highly conducting.

1.4.2 n-Type Doping

All self-p-doped polymers contain covalently attached negatively charged counterions. However, in self-n-doped polymers, cationic sites act as dopant and are incorporated into the polymer. The concept of self-n-doping in conjugated polymers was introduced by Wudl et al. [178]. As shown in Figure 1.26A, the polymer poly(dipropargylhexylamine) was prepared from the cyclopolymerization of N-hexyldipropargylamine. Treatment of the polymer with methyl trifluoromethanesulfonate resulted in the corresponding poly(dipropargyl-N-hexyl-N-methyl ammonium triflate) as shown in Figure 1.26B. The UV-vis results of the methyl trifluoromethanesulfonated polymer clearly support n-doping. Upon reduction with sodium sulfide in tetrahydrofuran, the conductivity of methyl trifluoromethanesulfonated polymer ($<10^{-6}$ S/cm) was increased to approximately 0.1 S/cm. This polymer was reportedly environmentally stable. However, the polymer is electrochemically unstable, decomposing after a few cathodic cycles.

Ferraris et al. [165] chemically synthesized a processable, self-n-doped, poly(2-(2'-thienyl)3-(4-dimethyl dodecyl ammonium phenyl)thiophene triflate) using the monomer shown in Figure 1.27A. This polymer was reportedly soluble in N-methylpyrrolidone to a concentration of 53 mg/mL. Similarly, the polymer was prepared electrochemically in tetramethyl ammonium trifluromethane sulfonate. The polymer prepared chemically and electrochemically exhibited reversible redox behaviour in the p- and n-doping regions. The presence of bound cations to the polymer backbone provides the ability to p- and n-

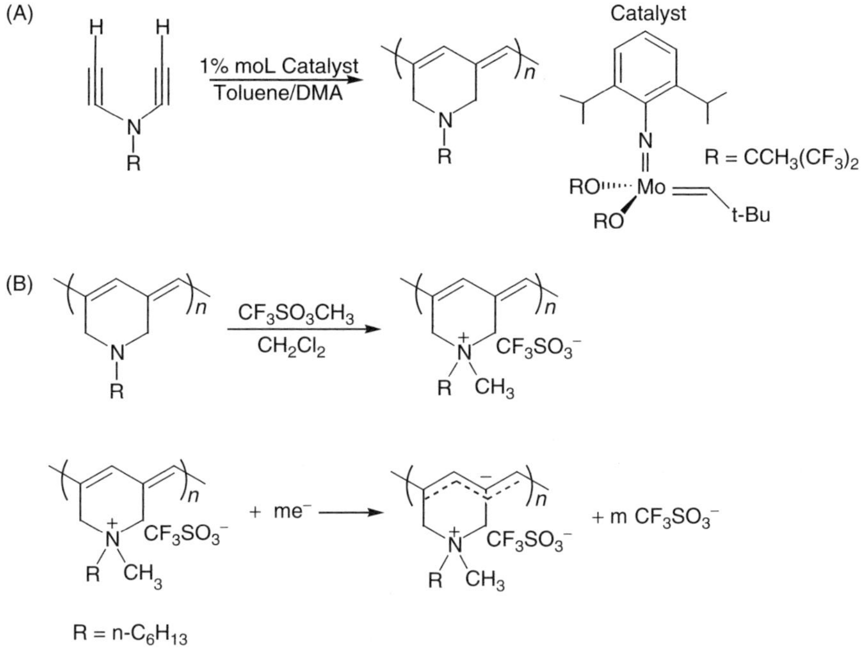

Figure 1.26 Synthesis of (A) poly(dipropargylhexylamine) and (B) poly(dipropargyl-N-hexyl-N-methyl ammonium triflate. (Reprinted with permission from *Chemistry of Materials*, 5, 1598. Copyright (1993) American Chemical Society.)

dope solely through the movement of anions. A comparison between a normal conducting polymer and a self-n-doped polymer shows that during the reduction (n-doping) process, normal polymers insert cations whereas self-n-doped polymers expel anions. Furthermore, this study was extended to a series of self-n-doped polymers derived from 3-(4-dimethyl dodecyl ammonium phenyl)thiophene triflate shown in Figure 1.27B [179].

Berlin *et al.* [166] reported the synthesis of n-doped polycationic polythiophenes using thiophene monomer functionalized with ammonium groups (see Figure 1.20 (7)). The polymer exhibits reversible p- and n-doping characteristics. The n-doping process and the associated expulsion of anions, is reportedly fast and independent of cation size, and the *in situ* conductivity is reportedly 2×10^{-2} S/cm. Kumar *et al.* synthesized a fully sulfonated n-doped polyaniline electrochemically using an acetonitrile–water (4:1) mixture [180]. The n-doping was confirmed

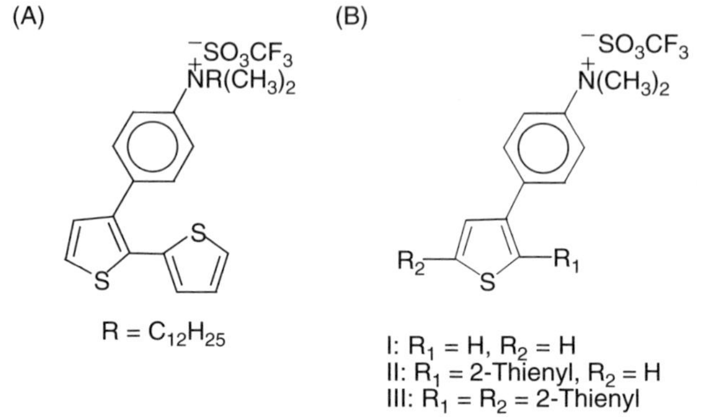

Figure 1.27 Structures of monomer based on phenylthiophene architecture. (Reproduced from *Polymer Preprints*, 1998, **39**, 137, 'Characterization of self-n-dopable polymers, A. A. Moxey, D. C. Loveday, I. D. Brotherson, J. P. Ferraris.)

by measuring *in situ* conductance of the polymer grown on a Pt twin wire electrode. Pei *et al.* [181] synthesized self-doped, n-type conducting polymers, poly(2-hydroxy-1-4-phenylene), sodium and tetrabutyl ammonium salts. Delocalization of the electrons associated with the side group oxygen anion onto the conjugated poly(*p*-phenylene) main chain occurs in the polymer generating intrinsic n-type carriers. This delocalization completely quenches the photoluminescence of the polymer, and increases the conductivity of the polymer by more than one order of magnitude. The conductivity of poly(2-hydroxy-1,4-phenylene) tetrabutyl ammonium salt is reportedly 5×10^{-9} S/cm. The low conductivity is attributed to the immobility of intrinsic charge carriers generated from side chain anion delocalization.

1.4.3 Auto Doping

Wudl and Heeger *et al.* prepared the self-dopable polymers poly(n-(3'-thienyl)alkanesulfonic acids) (P3TASH) and their sodium salts with alkane chain lengths ranging from 2 to 4 [127–129]. In an extension of this study, Ikenoue *et al.* found that poly(3-(3'-thienyl)propanesulfonic acid) (P3TPSH), which was obtained by exchanging the sodium ions in poly(sodium-3-(3'-thienyl)propanesulfonate) (P3TPSH) with protons using H^+-type ion exchange resin, was actually already self-doped, as evidenced by the presence of an additional optical absorption peak at 800 nm associated with the formation of polarons [132, 182, 183]. Since no electrical potential was applied to the system, 'ejection' or 'popping out' of protons was not expected to occur. Thus, the structure and doping mechanism of P3TASH was assumed to be different from that obtained by the electrochemical doping of P3TASH or its sodium salt. In order to distinguish between these two types of doping mechanism, the term 'auto-doping' or 'self-acid-doping' was first introduced by Ikenoue, Wudl and Heeger *et al*. They observed that P3TASH was extremely hygroscopic and hence the films of the polymer contained moisture. When the acid film was fully dehydrated, it turned greenish blue with a concomitant loss of conductivity; exposure to water vapor regenerated the original orange color. They therefore ascribed this aqueous solvatochromism to an acid–base reaction. In the presence of moisture, the acidic proton is solvated by water molecules. When water is removed by heating, the next most basic species able to accommodate the protons is the polythiophene backbone itself, as shown in Figure 1.28 [182]. Protonation of the backbone creates a polaronic residue on the polymer chain. This phenomenon is not observed for the sodium salts. Due to auto-doping, the sulfonic acid form of the polymer exhibits a higher conductivity in its electrochemically neutral state than the corresponding sodium salt.

The most detailed study of auto-doping in P3TASHs has been reported by Chen *et al*. [184]. In this report, they studied the structure and effect of the side chain length on the doping level of P3TASHs with alkanes containing two, six, and ten carbons, i.e., poly (2,3'-(thienyl)ethanesulfonic acid) (P3TESH), poly (2,3'-(thienyl)hexanesulfonic acid) (P3THSH), and poly (2,3'-(thienyl)decanesulfonic acid) (P3TDSH). The conductivities of the sulfonic acid derivatives are reported to be orders of magnitude higher than their sodium salts. The conductivity of P3TESH is reportedly two orders of magnitude higher than P3THSH. The important observation was that the spin density, as determined by electron spin resonance measurements, for the sulfonic acid forms is relatively high and comparable to that of poly(3-methylthiophene) doped by AsF_5^-. Electron spin resonance measurements of the aqueous solutions of P3TESH and P3THSH confirms the presence of large concentration of free spins, suggesting that the auto-doping or self-acid-doping is a redox process. Infrared spectra of P3TESH, P3THSH and their sodium

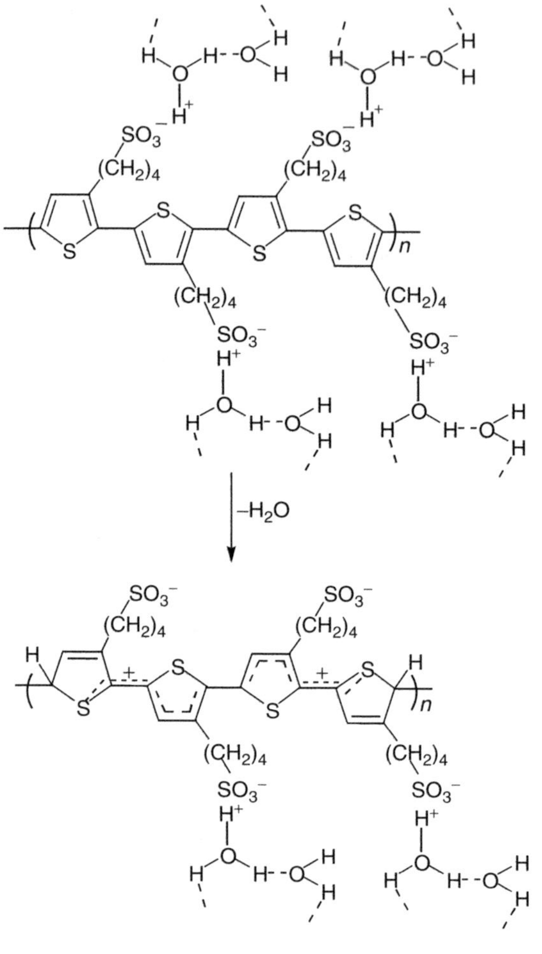

Figure 1.28 Auto-doping mechanism of P3TASH. (Reprinted from *Synthetic Metals*, 30, Y. Ikenoue, N. Uotani, A. O. Patil, F. Wudl, A. J. Heeger, 305. Copyright (1989), with permission from Elsevier.)

salts (Figure 1.29) suggest that the structure of auto-doped P3TESH and P3THSH involves polaron/bipolaron and proton addition on α-carbons similar to nonsulfonated polythiophenes.

An important study was performed by these authors to investigate the sulfonic acid derivatives P3TESH and P3THSH by titrating them with base while monitoring the pH of the solution. The titration curves shown in Figure 1.30 exhibit a two stage variation in pH values for both polymers. It is suggested that the first stage involves neutralization with free protons (H_f^+) while the second stage involves proton

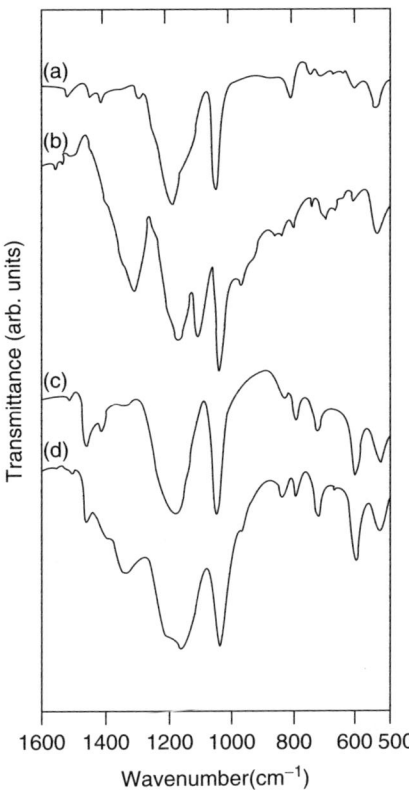

Figure 1.29 IR spectra of thin films at room temperature; (a) P3TESNa, (b) P3TESH, (c) P3THSNa and (d) P3THSH. (Reprinted with permission from *Macromolecules*, **26**, 7108. Copyright (1993) American Chemical Society.)

addition on α-carbons (H_α^+). The ratios of H_f^+ and H_α^+ are reported to be 48 % and 30 % for P3TESH and 61 % and 25 % for P3THSH, respectively. Surprisingly, 22 % and 14 % respectively of the anticipated protons (H_m^+) could not be detected by this method for P3TESH and P3THSH. The authors suggest that a percentage of the protons associated with the sulfonic acid derivatives oxidize the π-system of the polymer chain, which then either add to α-carbons or recombine to liberate hydrogen gas. The latter processes are considered responsible for the appearance of a large number of free spins. The doping level of P3TESH and P3THSH in aqueous solutions is reported to be sum of the fraction of H_α^+ and H_m^+ and is 52 % and 39 %, respectively. The doping level of a particular poly(n-(3'-(thienyl)alkanesulfonic acid) is thus a function of the degree of protonation and the extent of the

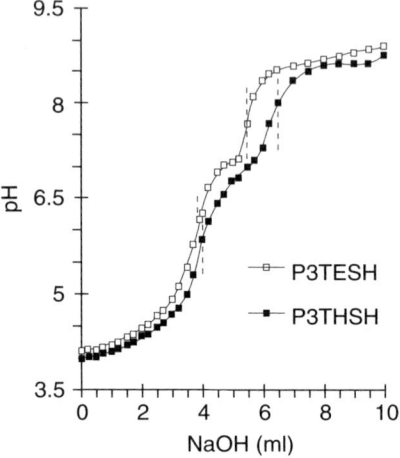

Figure 1.30 Titration curves of P3TESH and P3THSH with aqueous NaOH. (Reprinted with permission from *Macromolecules*, **26**, 7108. Copyright (1993) American Chemical Society.)

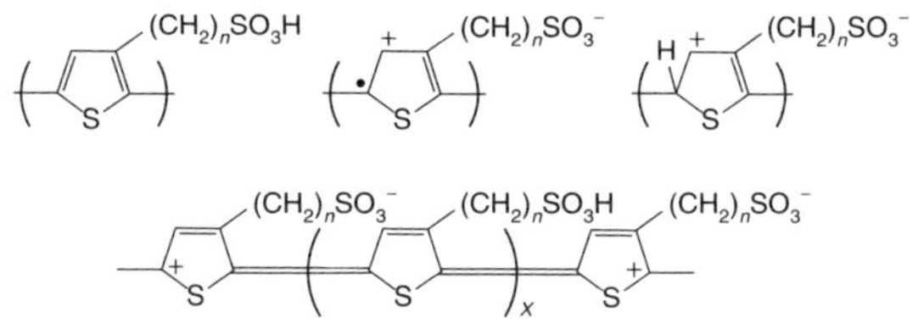

Figure 1.31 Basic units of poly(n-(3'-thienyl)alkanesulfonic acid). (Reprinted with permission from *Macromolecules*, **26**, 7108. Copyright (1993) American Chemical Society.)

redox process. It appears that the structures of auto-doped P3TASHs involve polaron/bipolaron with additional protons on some α-carbons of the rings resulting from redox processes and comprise four species as indicated in Figure 1.31.

1.5 EFFECT OF SUBSTITUENTS ON PROPERTIES OF POLYMERS

Functionalized conducting polymers were initially pursued as a means for overcoming issues of processability. For example, the first conducting polymers discovered were insoluble, intractable, nonmelting and thus not processible. However, over the past several years, remarkable progress has been made towards improving the processability of these materials. To improve melt and solution processability of the stiff backbone, the standard approach consisted of substitution of flexible side chains (alkyl, alkoxy, etc.) to the main chain. This approach was first applied to polythiophene [185]. However, as is well known, the presence of substituents on the backbone of the conducting polymer can not only modify the processability but can also modulate the electrical, electrochemical, and optical properties of the resulting polymer. Similarly, in self-doped conducting polymers, substitution of ionizable anionic functional groups has a significant impact on various properties. The distinctive properties of self-doped polymers that differ from the parent polymers are discussed below.

1.5.1 Solubility

The insolubility or aggregation of doped conducting polymers in solution is believed to be due to increased rigidity of the polymer chains on doping and the increased polar interactions between the polymeric chains. The manifestation of electronic delocalization throughout a π-system requires that adjacent monomer units along the backbone be coplanar. This coplanarity tends to make the polymers inflexible and insoluble. The low solubility is most likely due to lack of conformational mobility in solution and due to efficient molecular packing or crystallization forces in the solid state. However, this problem can be overcome by structural modifications. Substituents along the polymer backbone can significantly increase the solubility by disrupting the packing forces in the solid state, and by providing new polymer solvent interactions. The substitution of long, flexible, pendant, alkyl or alkoxy chains to a rigid polymer backbone has induced solubility in polythiophenes [113, 114, 186], polyanilines [115–118], poly(phenylene vinylenes) [187] and polyphenylenes [188, 189] in common organic solvents. The conformational mobility of the side chains, which act as 'bound solvent' [190], provides enough entropic driving force to carry the rigid polymer chain into solution.

A distinctive property of self-doped polymers is their water solubility in the neutral (insulating) and doped (conducting) states. This solubility is due to the covalently attached negatively charged groups on the polymer backbone. Solubility allows a deposition of conductive and electroactive layers onto any, even a nonconducting, surface by a simple casting of self-doped polymers. Such layers could find numerous applications in diverse areas of technology. For example, in self-doped sulfonated polyaniline, with increasing sulfur to nitrogen ratio from 0.65 to 1.3, the solubility in water increases from 22 to 88 g/L [191]. Zotti *et al.* [164] reported the solubility of self-doped polythiophene *ca* 10 g/L.

1.5.2 DC Conductivity

While substituents on conjugated polymer chains can increase their processability, it is recognized that there is a tradeoff in terms of conductivity. This is a direct result of the fact that the properties that strengthen chain–chain interactions are also responsible for charge transport along the backbone of the polymer. For example, the twisting of polymer backbone via steric interactions decreases π-orbital overlap resulting in decreased conductivity [192, 193]. Grubbs *et al.* were able to strike a balance between these two properties using ring opening metathesis polymerization of monosubstituted cyclooctatetraenes [194]. This approach resulted in a single substituent on every eighth carbon in polyacetylene, resulting in a soluble polymer that retained a significant amount of its conductivity.

In self-doped polymers, the same steric effects associated with covalently attached groups are largely responsible for the decrease in the conductivity as a result of decreased interchain transport or reduced conjugation of the π-system [153]. Self-doped polyanilines (ring substituted or through a alkyl or aryl linker to nitrogen sites) have lower conductivity in the range of 10^{-2} to 10^{-9} S/cm than parent polyaniline ~1–10 S/cm [136, 138, 146, 149, 153, 173, 195–202]. However, for ring sulfonated polyaniline with 50 % and 75 % degrees of sulfonation, conductivity is comparatively higher than with other self-doped polyanilines, around 0.1 and 1 S/cm, respectively [153, 173]. Commercially available ring sulfonated polyaniline with a sulfonation degree of 100 % has been reported to have a conductivity of 0.02 S/cm [202], which is lower than for 50 % and 75 % degrees of sulfonation. This decrease in conductivity with increasing degree of sulfonation has been attributed to the higher degree of twisting of the phenyl rings relative to one another and increased interchain separation due to increasing density of sulfonate groups [202]. Unlike polyaniline, the conductivity of self-doped polyaniline is independent of pH within a broad range of 0–12 [173].

The highest conductivity of 65 S/cm for electrochemically prepared self-doped ring-sulfonated polypyrrole was reported by Sahin *et al.* [203]. However, conductivity decreases from 65 S/cm to 6.5 S/cm with increase in sulfonation (or S/N ratio) from 0.29 to 0.53. The conductivity of ring sulfonated polypyrrole is higher than alkylsulfonate ($0.5-10^{-3}$ S/cm) [135] and N-substituted self-doped polypyrroles ($10^{-4}-10^{-6}$ S/cm) [133, 134, 204, 205]. The spacer alkyl chains of sulfonic acid and N-substitution leads to a large decrease in the conjugation along the polypyrrole chain, and in turn conductivity decreases. The conductivities found for chemically prepared self-doped polypyrrole are lower than for the polymer prepared electrochemically. The conductivity of sulfonic-acid-substituted self-doped polythiophene and polypyrrole is higher than with carboxylic acid substituents, likely due to its relatively weak acidity [206]. The structure and effect of the side chain length on the doping level of poly(n-(3'-thienyl)-alkanesulfonic acid) with alkanes of carbon numbers two, six and ten has been reported and it was found that the doping level decreases as a function of chain length [184]. The lower doping level of poly(n-(3'-thienyl)-alkanesulfonic acid) with higher chain length was attributed to its higher oxidation potential due to shorter conjugation length.

1.5.3 Molecular Weight

The molecular weights of substituted polymers are typically low compared with parent polymers due to the bulky nature of the ring substituents, which can hinder polymerization; however, solubility in common organic solvent is increased. Dao *et al.* [193] have studied the effect of alkyl substituents on the molecular weight of polyaniline. The molecular weight of chemically prepared poly(2-methyl aniline) and poly(2-ethyl aniline) was reportedly 7000 and 5000 g/mol, respectively, which is 15-times lower than polyaniline (80 000) prepared under identical conditions. A similar effect of alkyl substituents was observed on molecular weights of polythiophenes [207, 208].

Although self-doped polymers are typically processable and retain most of their electronic and redox properties, the insolubility of the products under the polymerization conditions leads to premature termination of the polymerization reaction. This results in low molecular weight polymers. Post polymerization modification, in the case of sulfonated polyaniline, can get around this problem to some extent; however, there is a lack of control over the degree of sulfonation as well as the distribution along polymer chains. A breakthrough was achieved with a self-doped form of poly(anilineboronic acid), which is soluble under polymerization conditions and resulted in a high molecular weight polymer [151]. Gel permeation chromatography results indicated a number average molecular weight of 1 676 000 g/mol, a weight average molecular weight of 1 760 000 g/mol and a polydispersity of approximately 1.05.

1.5.4 Redox Properties

The structure and conformational changes of a conducting polymer backbone in the presence of substituents such as alkyl, alkoxy and halogens, etc. do not show a significant impact on the redox properties (redox switching, cycleability and stability at high pH) except for shifts in redox potentials. However, in self-doped polymers, ionizable, negatively charged, covalently bonded, functional groups act as intramolecular dopant anions that are able to compensate positive charges in the polymer backbone, thus replacing auxiliary solution dopant anions. This 'inner' anion doping determines many distinctive properties of self-doped polymers, setting them apart from their parent polymers. Among them, the electrochemical redox activity in pH neutral solutions seems to be of great interest. In contrast to polyaniline, which shows its redox activity at solution pH not exceeding 3 or 4, self-doped derivatives are active in a broad range of solution pH, extended to higher values [209], in some cases up to pH 12 [210, 211]. Therefore, the use of conducting polymer coated electrodes in such fields as electrocatalytic conversion of solution species [212] sensors and biosensors could be possible, even in weakly neutral solutions, with the use of self-doped polyaniline derivatives.

1.5.5 Electronic and Spectroscopic Properties

The addition of substituents on the backbone of conjugated polymers decreases the degree of coplanarity. The steric repulsions between adjacent side groups result in twisting of the chain and hence yield a reduction in conjugation of the rings in the polymer. The optical absorption spectroscopic studies of the substituted polymer have given an indication of the effective conjugation length of the double bonds in the main chain. There are many reports in the literature showing the shift in absorbance with the addition of substituents on the conjugated polymer backbone associated with decreasing chain length [192, 194]. Similar structural impact is expected with self-doped polymers in addition to the impact of the charge on polaron/bipolaron structures.

In self-doped polyaniline, sulfonic groups induce changes in the geometry of the polyaniline backbone [152], affecting the physicochemical properties of the polymer. Comparative electronic absorption spectroscopic studies have shown a hypsochromic shift of the $\pi-\pi^*$ transition on going from emeraldine salt (ES) to sulfonated polyaniline salt, and a bathochromic shift of the polaron band transition. This is consistent with a decreased extent of conjugation, because of an increase in torsion angle between adjacent phenyl rings with respect to the plane of the nitrogen atoms, caused by the repulsion between the sulfonic group and hydrogen atoms on the adjacent phenyl rings [152, 153]. Based on X-ray photoelectron spectroscopic investigations [154] indicating a higher proportion of positively charged nitrogens, it was concluded that the polaron of sulfonated polyaniline is more localized as compared with the conductive form of polyaniline.

1.5.6 Mechanical and Thermal Properties

Mechanical properties of conducting polymers are limited in general due to low molecular weight, limited branching and crosslinking. Thermal properties are largely limited by the volatility of the dopants. For example, polyaniline is known for its limited thermal stability and loss of conductivity at high temperature. From a chemical point of view, several processes take place in polyaniline upon heating: loss of dopant, oxidation of chains by oxygen, and cross-linking. Since the protonated polymer undergoes irreversible chemical modifications upon heating in air, its conductivity cannot be recovered by redoping [199, 213–217]. The thermal stability, together with conductivity at high temperature, are important properties for many commercial applications. For example, thermal stability is important for conductive coatings and electronic circuits as well as polymer electrolyte membranes for fuel cells. Attempts to improve thermal stability have focused on stabilizing the dopant, which is typically volatile. Thermal stability of self-doped polymers is expected to be better since the dopant moiety is covalently bonded to the polymer backbone. Han et al. [218] have reported that the thermal stability of self-doped propylthiosulfonated polyaniline is significantly higher than sulfonated polyaniline [219] and unsubstituted polyaniline. The covalently bonded electron donating mercaptopropanesulfonic acid constituent group on the polymer backbone reportedly stabilizes the aromatic benzenoid ring, thereby reducing the decomposition of the backbone. The propylthiosulfonated polyaniline retains its sulfonic acid groups up to $\sim 260\,°C$ [218], whereas sulfonated polyaniline begins to lose sulfonic acid groups at a much lower temperature of about $185\,°C$ [219].

According to a recent report, the thermal stability of self-doped poly(anilineboronic acid) [220] is greater than that of HCl-doped polyaniline and other self-doped forms of polyaniline. This is due to a self-doped structure where the anion is immobilized in the form of a crosslink, resulting in conductivity up to 500 °C. This self-doped crosslinked structure is also responsible for enhanced mechanical properties including a hardness of 0.5 GPa [221].

1.6 APPLICATIONS OF SELF-DOPED POLYMERS

Conducting polymers are attractive materials for use in existing and emerging technologies due to their light weight, low cost and versatility compared with other standard conductors and semiconductors. Some examples of potential applications include rechargeable batteries, electrochromic displays and smart windows, light emitting diodes, toxic waste cleanup, sensors, field effect transistors and electromagnetic interference shielding, etc. Self-doped polymers are expected to play an important role in the technical implementation of conducting polymers since they can overcome many of the limitations associated with conducting polymers. For example, these polymers have proven to be an enabling step in the development of plastic electronic devices and biosensors.

1.6.1 Molecular Level Processing

The processing of ICPs into thin films is oftentimes extremely challenging due to their insolubility, intractability and infusibility. Therefore, the development and utilization of these polymers as active elements of thin film electronic and optical devices continues to be an intense area of research. These materials are generally manipulated into thin films *via* simple spin-casting techniques. However, it is becoming more apparent that better control over the molecular and supramolecular organizations of these materials is needed to exploit fully their novel optical and electrical properties. Recently, a new layer-by-layer processing technique involving the alternate deposition of adsorbed layers of oppositely charged polyelectrolytes from dilute solutions has emerged, as a variable means to manipulate polymers into multilayer thin films [222, 223]. This approach provides molecular level control over thickness and architecture of multilayer thin films, is remarkably simple to use, and is readily extended to a wide variety of polymers, including many different electroactive polymers.

Electronically conducting multilayers have been prepared by using oxidized polyaniline and polypyrrole as polycations [224–228]. Owing to their interesting electrical and optical properties, mono- and multilayers prepared from conducting polyelectrolytes should be interesting materials for potential application in the field of sensors, electrooptics and LED technology [162, 222, 229–231]. Water soluble self-doped conducting polymers have the attractive feature of producing uniform conducting coatings on any charged substrate with precise control of thickness, irrespective of form or size of the substrate, by simply dipping it into an aqueous polymer solution. There are only a few papers dealing with the fabrication and properties of multilayers prepared from self-doped conducting polymers. Rubner *et al.* were first to demonstrate that ultrathin conducting layers can be fabricated using carboxylic-acid-derivatized polythiophenes [223, 232–234]. Light-emitting diodes have been prepared by sequential adsorption of anionic and cationic poly(*p*-phenylene)s [235]. Lukkari *et al.* have shown that multilayers can be prepared from sulfonated poly(alkoxythiophene) and that the linear charge density of the conducting polymer, (i.e., its oxidation state) affects the multilayer formation [236]. They were the first to report on the all thiophene polyelectrolyte multilayers containing poly(3-(3'-thienyloxy)propanesulfonate) (P3TOPS) as polyanion and poly(3-(3'-thienyloxy)propyl triethyl ammonium) (P3TOPA) as polycation [237] as shown in Figure 1.32.

Sulfonated polyaniline has been used in the fabrication of multilayer heterostructures in light emitting diodes, the electrochemical control of electrolyte activity [238] and biosensors [239]. Recently,

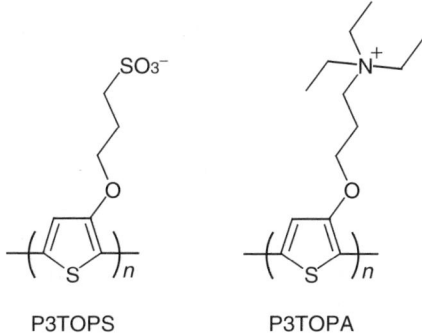

Figure 1.32 Structures of P3TOPS and P3TOPA. (Reprinted with permission from *Journal of the American Chemical Society*, **123**, 6083. Copyright (2001) American Chemical Society.)

nanoassemblies of sulfonated polyaniline multilayers have been formed [240]. Cao et al. [241] have shown the photovoltaic properties of a multilayer film based on highly sulfonated polyaniline and diazoresin. Torresi et al. [242] have used self-doped polyaniline multilayer films to prepare oxalate biosensors. Chan et al. [243] fabricated multilayer photovoltaic devices based on ruthenium containing poly(p-phenylenevinylene) and sulfonated polyaniline.

1.6.2 Transistors

The metal oxide semiconductor field effect transistor (MOSFET) is the most important device for very large scale integrated (VLSI) circuits such as microprocessors and memory devices [244]. Much research has explored the use of conducting polymers as an active material in field effect transistor devices; dedoped or doped conjugated polymers such as *trans*-polyacetylene [101], polythiophene [245], thiophene oligomers [246], poly(3-alkylthiophenes) [247, 248], polythienylene vinylene [249], poly(N-methylpyrrole) [250], polyaniline [251] and polynaphthalene vinylene or polyphenylene vinylene [252] have all been demonstrated as suitable materials for field effect transistor architecture. However, the performance of these organic devices has not been comparable with that of their inorganic counterparts due to low charge carrier mobility and relatively poor stability. Experiments have shown that the field effect mobility can be increased by two to three orders of magnitude by additional doping, confirming the fact that the carrier mobility in conjugated polymers depends strongly on doping [253-256]. Although the increased mobility is favorable when applications are considered, the threshold voltage V_t is also increased upon doping, and therefore the field effect transistor characteristics become more of the 'normally on' type [257]. To improve both carrier mobility and stability of conjugated polymer FET devices, Kuo et al. have fabricated water soluble, self-acid-doped, conducting polyanilines, poly(aniline-co-N-propanesulfonic acid aniline) (PAPSAH) and sulfonic acid ring substituted polyaniline MOSFETs (SPAN) [257]. The schematic cross sectional view of fabricated MOSFETs is shown in Figure 1.33. These field effect transistors have ideal current drain source (ds) voltage characteristics and their field effect mobilities can reach 2.14 (PAPSAH) and 0.33 cm^2 V^{-1}s^{-1} (SPAN) (Figure 1.34). These values are reportedly close to those of amorphous silicon inorganic transistors (0.1–1.0 cm^2 V^{-1}s^{-1}). Also, these field effect transistors are found to be environmentally more stable than those of other polyaniline field effect transistors.

APPLICATIONS OF SELF-DOPED POLYMERS 51

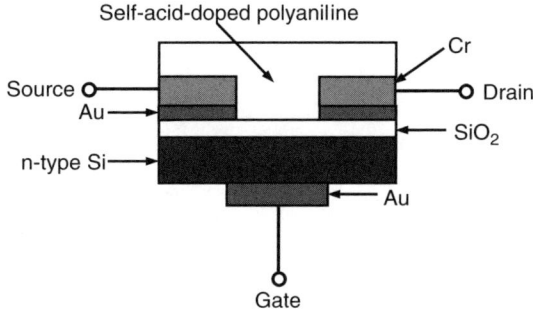

Figure 1.33 Cross sectional view of fabricated MOSFETs.

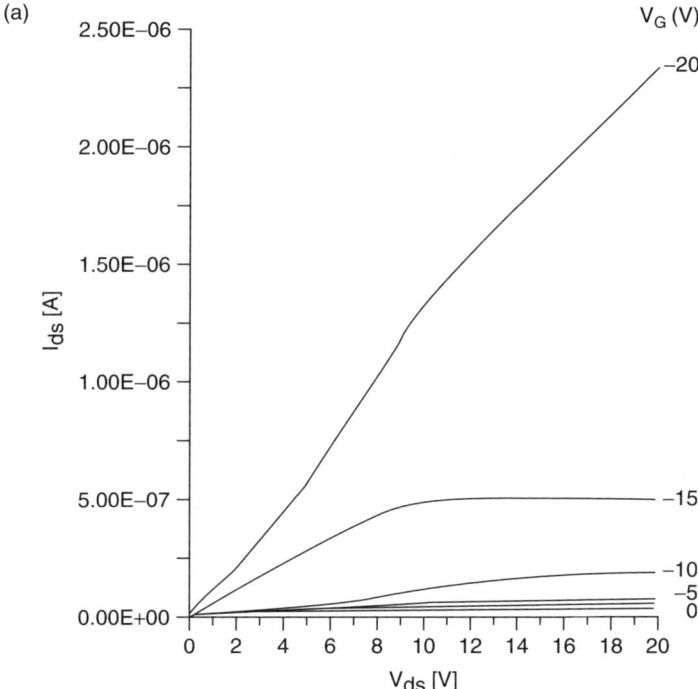

Figure 1.34 Drain-source current (I_{ds}) vs drain-source voltage (V_{ds}) characteristics of (a) SPAN (0.327 μm) and (b) PAPSAH (0.812 μm) MOSFETs at various V_G. The devices were exposed to air for two days prior to measurement. (Reprinted from *Synthetic Metals*, 93, C. T. Kuo, S. A. Chen, G. W. Hwang, H. H. Kuo, 155. Copyright (1998), with permission from Elsevier.)

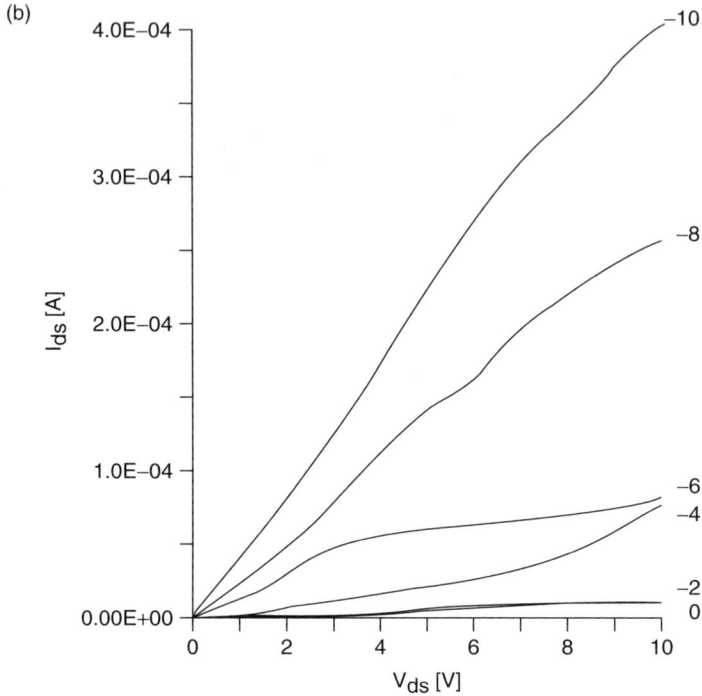

Figure 1.34 (*continued*).

1.6.3 Biosensors

Polyaniline is the conducting polymer most commonly used as an electrocatalyst and immobilizer for biomolecules [258–260]. However, for biosensor applications, a nearly neutral pH environment is required, since most biocatalysts (enzymes) operate only in neutral or slightly acidic or alkaline solutions. Therefore, it has been difficult or impossible to couple enzyme catalyzed electron transfer processes involving solution species with electron transport or electrochemical redox reactions of mostly polyaniline and its derivatives. Polyaniline is conducting and electroactive only in its protonated (proton doped) form i.e., at low pH values. At pH values above 3 or 4, polyaniline is insulating and electrochemically inactive. Self-doped polyaniline exhibits redox activity and electronic conductivity over an extended pH range, which greatly expands its applicability toward biosensors [209, 210, 261]. Therefore, the use of self-doped polyaniline and its derivatives could, in principle, enable the direct or mediated electron transfer between the polymer matrix and active centres of biocatalysts.

Karyakin *et al.* have developed a potentiometric glucose biosensor using self-doped polyaniline [262]. Self-doped polyaniline acts as a pH transducer exhibiting a stable potentiometric response of 70 mV/pH. Also, the self-doped polyaniline biosensor reportedly shows a three- to fourfold higher response compared with a glucose sensitive field effect transistor [263]. Chen *et al.* have used self-doped polyaniline as an ascorbate sensor [264]. Self-doped polyaniline has also been successfully utilized for enzyme immobilization in multilayers [265]. Rusling *et al.* [266] have shown enhanced electrochemical sensitivity to hydrogen peroxide by efficient wiring of the active enzymes with sulfonated polyaniline.

1.6.4 e-Beam Lithography

Electron-beam (e-beam) lithography allows the production of features much smaller than standard lithographic techniques, and as a result is expected to allow the production of increased memory capacity and more advanced large-scale integrated circuits. However, e-beam lithography has the serious disadvantage of positional errors arising from electrons accumulating in the resist and substrate during the e-beam writing process. Conducting polymers have been found to be useful in several lithographic applications [267, 268]; however, they are not sufficiently soluble in the conducting state, thus restricting their widespread use in lithographic applications. The processability of conducting self-doped polymers makes the material more convenient for lithographic applications. Ikenoue *et al.* have proposed a useful method for the coating of water-soluble, self-doped, polymers e.g., polythienylalkanesulfonate, as a charge dissipating top coat on the resist [269, 270]. This polymer has a variety of advantages including: (i) an electron conduction mechanism capable of giving an excellent shielding effect allowing fast e-beam writing speeds, (ii) good processability for coating and removal of the polymer, and (iii) good film formation owing to high molecular weight. They suggest that the self-doped polymer is not only useful for the charge dissipation but is also useful for stabilizing the sensitivity of chemically amplified resist as an acidic polymer film. Watanabe and Shimizu *et al.* [196] have reported the use of self-doped polyaniline films to prevent charging during e-beam lithography and to improve post-exposure delay latitude for chemically amplified resists, and as a conductive bottom layer in multilevel resists. Charge buildup on the resist layer during plasma etching causes gate oxide leakage or breakdown due to plasma nonuniformities. This damage was reportedly reduced by using self-doped polyaniline.

1.6.5 Electrochromic Devices

Chromatic changes caused by electrochemical processes were originally described in the literature in 1876 for the product of the anodic deposition of aniline [271]. However, the electrochromism was defined as an electrochemically induced phenomenon in 1969, when Deb observed its occurrence in films of some transition metal oxides [272]. Electrochromism in polypyrrole was first reported by Diaz *et al.* in 1981 [273]. Electrochromism is defined as the persistent change of optical properties of a material induced by reversible redox processes. Electronic conducting polymers have been known and studied as electrochromic materials since the initial systematic studies of their electronic properties.

Electronic conducting polymer-based electrochromics have received increasing attention due to their potential for facile production, structurally controllable states, high contrast, fast switching speeds and the ability to be applied to flexible electrochromic devices [274–277]. In conducting polymers, the electrochromism, as well as the conductivity, is explained using band theory [276]. The doping of a conducting polymer modifies its electronic structure, i.e., introduces charge carriers, producing new electronic states in the band gap, thus allowing new electronic transitions resulting in color changes.

Self-doped polymers exhibit faster electronic and optical responses to changes in electrochemical potential relative to the parent polymers, and thus they are useful for fabricating electrochemical chromic displays. Ikenoue *et al.* reported the electrochromic fast switching behavior of self-doped poly(3-(3′-thienyl)-propanesulfonic acid). The polymer shows a considerably faster optical switching response (saturated within 50 ms) than does the well known polyisothianaphthene (saturated within 500 ms), and long term stability over 10 000 cycles [183]. Self-doped polyaniline derivatives gave reversible color changes in all-solid-state electrochromic windows assembled from an electrochromic poly(aniline-N-butyl sulfonate) and an ion conducting polymer electrolyte membrane. The all-solid-state electrochromic window responded to potential steps between $+2.3$ and -1.5 V by changing its color from green-blue to transparent yellow within 60 s. The color contrast and optical response of poly(aniline-N-butyl sulfonate) in contact with different polymer/electrolyte systems is reportedly a function of electrolyte composition [278].

Reynolds et al. [279] reported electrochromic behavior of self-doped propanesulfonated poly(3,4-propylenedioxypyrrole). The polymer has not only shown interesting electrochromic properties in the visible, but, upon doping, also exhibits a very strong absorption in the near infrared with changes in transmittance up to 97%, extending the use of the polymer as the active layer in a visible/near infrared switchable device. Viinikanoja et al. [280] reported the electrochromism and pH-induced halochromism of self-doped poly(3-(3'-thienyloxy)propanesulfonate) multilayers.

1.6.6 Ion Exchangers

A unique property of self-doped polymers is their ability to act as a 'charge controllable membrane' with cation exchange abilities [133, 163, 281, 282], in contrast to the anion exchange properties of classical polymers, i.e., doped with small or free moving anions. Reynolds et al. [283] have reported the ion exchange properties of self-doped poly(3,6-(carbaz-9-yl)propanesulfonate). According to electrochemical analysis, the polymer has two separate oxidation states involving interactions of the polymer backbone with the covalently bound, self-doping, ion and counterions from the electrolyte, indicating that the polymer has the ability to act as a potential dependent, charge controllable, membrane with both cation exchange and anion exchange properties.

1.6.7 Rechargeable Batteries

Lithium secondary batteries are one of the most important applications of electronically conducting polymers. Conducting polymers such as polyaniline, polypyrrole and polythiophene are expected to be promising materials for the electrodes of secondary batteries because they are relatively stable in air and have good electrochemical properties. In the discharging cycle of these batteries, the electrons flowing from the lithium anode (negative pole) through an external electric circuit must be consumed at the cathode (positive pole). When used as the active mass of a positive electrode, a conducting polymer ensures efficient utilization of electrons by converting the oxidized form of the conducting polymer into the reduced form. Conversely, electrochemical oxidation of the reduced form takes place in the reverse charging cycle. Most of the common conducting polymers exchange anions during the charging and discharging processes. The reversible exchange of anions between active polymer mass and electrolyte limits many important characteristics such as the electric charge density (relative to the mass of a polymer), the maximum electric current available (due to limited rate of electrochemical processes including anion mobility), cycleability of the battery, storage characteristics, etc. The use of self-doped conducting polymers excludes the participation of anions in the charging–discharging process. Instead, the charge compensation occurs at the expense of lithium cation insertion or expulsion, and the increased efficiency of this process leads to a significant improvement of some characteristics of rechargeable lithium batteries [201].

Sulfonated polyanilines are typically used as the cathode in rechargeable batteries. They exhibit better chemical stability due to steric protection by the sulfonic group [145]. This fact reduces the degradation of the quinoid structure during oxidation, increasing the cycleability of the electrode. Also, the electrostatic environment provided by sulfonic groups decreases the participation of anions during the charge compensation process [284]. In lithium batteries, it is desirable that only lithium cations intercalate into the cathode, because this leads to the use of small amounts of electrolyte that only serves as a carrier for the cations to migrate/diffuse from the anode to the cathode, and this permits the fabrication of higher specific capacity batteries. A sulfonated polyaniline prepared by copolymerization of o-aminobenzenesulfonic acid and a relatively small amount of aniline reportedly exhibits a specific charge of approximately 47 Ah kg^{-1} in aqueous solutions (first redox process only) and over 80 Ah kg^{-1} in nonaqueous solutions (both redox steps) [285]. For a complete Li/sulfonated polyaniline cell with nonaqueous LiClO$_4$ electrolyte, specific charge of 52 Ah kg^{-1} and specific energy of 130 Wh kg^{-1} were calculated on the basis of experimental results. The specific energy of a sulfonated polyaniline–Li battery was ca 50 % higher than a PANi–Li battery [286]. Mousavi *et al.* [287] reported that the maximum capacity of a self-doped polyaniline–Zn battery is 146.4 Ah kg^{-1} and the specific energy 172.8 Wh kg^{-1} with coulumbic efficiency of 97–100 % over at least 200 cycles between 0.8 to 1.6 V.

1.6.8 Dip-Pen Nanolithography

Controlled patterning of conducting polymers at a micro- or nanoscale is the first step towards the fabrication of miniaturized functional devices.

APPLICATIONS OF SELF-DOPED POLYMERS

Dip-pen nanolithography [288] is a promising new nanofabrication tool, which allows one to pattern molecules on a variety of surfaces with a coated AFM tip in a controlled fashion on a sub-100 nm to many micrometer scale [289, 290]. Recently, Mirkin *et al.* [291] fabricated nanopatterns of self-doped conducting polyaniline by dip-pen nanolithography using electrostatic interactions as a driving force (Figure 1.35). The water soluble self-doped polyaniline 'ink' is converted to its solid state after patterning. The smallest feature size generated with these polymer inks had a diameter of 130 nm. In this case, the electrostatic interaction between the charged polymer chains and the oppositely charged substrate acts as a primary driving force in the dip-pen nanolithography process. Others have utilized electrically biased electrodes to polymerize monomers on conducting substrates [292]. Mirkin's approach differs from the literature approach in that no bias is required, and one can write on nonconducting substrates.

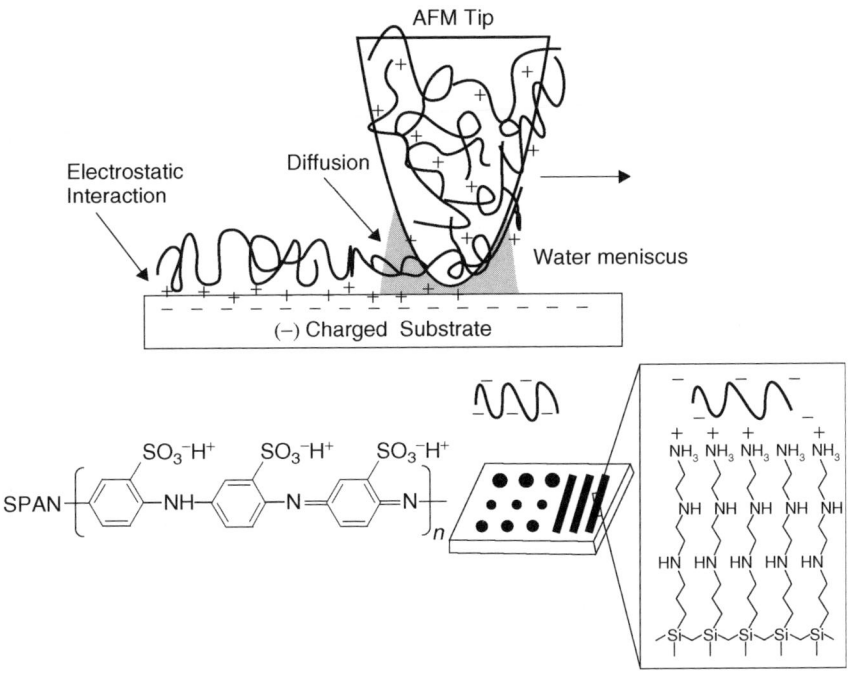

Figure 1.35 Schematic representation of dip-pen nanolithography for charged self-doped polyaniline. (Reproduced from *Advanced Materials*, 2002, **14**, 1474, J. H. Lim, C. A. Mirkin, with permission from Wiley-VCH.)

REFERENCES

[1] C. K. Chiang, C. R. J. Fincher, Jr., Y. W. Park, A. J. Heeger, H. Shirakawa, E. J. Louis, S. C. Gau, A. G. MacDiarmid, 'Electrical conductivity in doped polyacetylene,' *Physical Review Letters* **1977**, *39*, 1098.

[2] H. Shirakawa, E. J. Louis, A. G. MacDiarmid, C. K. Chiang, A. J. Heeger, 'Synthesis of electrically conducting organic polymers: halogen derivatives of polyacetylene, (CH),' *Chemical Communications* **1977**, 578.

[3] H. Shirakawa, 'The discovery of polyacetylene film: the dawning of an era of conducting polymers (Nobel lecture),' *Angewandte Chemie – International Edition* **2001**, *40*, 2575.

[4] A. J. Heeger, 'Semiconducting and metallic polymers: the fourth generation of polymeric materials (Nobel lecture),' *Angewandte Chemie – International Edition* **2001**, *40*, 2591.

[5] A. G. MacDiarmid, '"Synthetic metals": a novel role for organic polymers (Nobel lecture),' *Angewandte Chemie – International Edition* **2001**, *40*, 2581.

[6] H. Letheby, 'On the production of a blue substance by the electrolysis of sulphate of aniline,' *Journal of the Chemical Society* **1862**, *15*, 161.

[7] J. D. Rose, F. S. Statham, 'Acetylene reactions. VI. Trimerization of ethynyl compounds,' *Chemical Communications* **1950**, 69.

[8] M. Hatano, K. Kambara, K. O. Kamoto, 'Paramagnetic and electric properties of polyacetylene,' *Journal of Polymer Science.* **1961**, *51*, S26.

[9] E. Menefee, Y. H. Pao, 'Electron conduction in charge-transfer molecular crystals,' *Journal of Chemical Physics* **1962**, *36*, 3472.

[10] V. V. J. Walatka, M. M. Labes, J. H. Perlstein, 'Poly(sulfide nitride), a one-dimensional chain with a metallic ground state,' *Physical Review Letters* **1973**, *31*, 1139.

[11] R. L. Greene, G. B. Street, L. J. Suter, 'Superconductivity in poly(sulfur nitride) $(SN)_x$,' *Physical Review Letters* **1975**, *34*, 577.

[12] W. D. Gill, W. Bludau, R. H. Geiss, P. M. Grant, R. L. Greene, J. J. Mayerle, G. B. Street, 'Structure and electronic properties of polymeric sulfur nitride $(SN)_x$ modified by bromine,' *Physical Review Letters* **1977**, *38*, 1305.

[13] G. Natta, G. Mazzanti, P. Corradini, 'Stereospecific polymerization of acetylene,' *Atti accad. nazl. Lincei Rend.* **1958**, *25*, 3.

[14] H. Shirakawa, S. Ikeda, 'Infrared spectra of polyacetylene,' *Polymer Journal* **1971**, *2*, 231.

[15] T. Ito, H. Shirakawa, S. Ikeda, 'Simultaneous polymerization and formation of polyacetylene film on the surface of a concentrated soluble Ziegler-type catalyst solution,' *Journal of Polymer Science, Polymer Chemistry Edition* **1974**, *12*, 11.

[16] T. Ito, H. Shirakawa, S. Ikeda, 'Thermal *cis–trans* isomerization and decomposition of polyacetylene,' *Journal of Polymer Science, Polymer Chemistry Edition* **1975**, *13*, 1943.

[17] C. K. Chiang, S. C. Gau, C. R. J. Fincher, Y. W. Park, A. G. MacDiarmid, A. J. Heeger, 'Polyacetylene, $(CH)_x$: n-type and p-type doping and compensation,' *Applied Physics Letters* **1978**, *33*, 18.

[18] A. O. Patil, A. J. Heeger, F. Wudl, 'Optical-properties of conducting polymers,' *Chemical Reviews* **1988**, *88*, 183.

REFERENCES

[19] R. B. Seymour, *Conductive Polymers; Polymer Science and Technology*, 1st edn, Plenum Press, New York, **1981**.

[20] K. M. Maness, R. H. Terrill, T. J. Meyer, R. W. Murray, R. M. Wightman, 'Solid-state diode-like chemiluminescence based on serial, immobilized concentration gradients in mixed-valent poly[Ru(vbpy)(3)](PF$_6$)(2) films,' *Journal of the American Chemical Society* **1996**, *118*, 10609.

[21] R. Dupon, D. H. Whitmore, D. F. Shriver, 'Transference number measurements for the polymer electrolyte poly(ethylene oxide)' *Journal of the Electrochemical Society* **1981**, *128*, 715.

[22] J. C. Huang, 'Carbon black filled conducting polymers and polymer blends,' *Advances in Polymer Technology* **2002**, *21*, 299.

[23] E. K. Sichel, *Carbon Black-Polymer Composites: the Physics of Electrically Conducting Composites (Plastics engineering)*, Marcel Dekker Inc. (Aug1/82), **1982**.

[24] N. C. Billingham, P. D. Calvert, 'Electrically conducting polymers – a polymer science viewpoint,' *Advances in Polymer Science* **1989**, *90*, 1.

[25] D. M. Kelly, J. G. Vos, *Electroactive Polymer Electrochemistry: Part 2: Methods and Applications*, 1st edn, Plenum Press, New York, **1996**.

[26] H. D. Abruna, P. Denisevich, M. Umana, T. J. Meyer, R. W. Murray, 'Rectifying interfaces using 2-layer films of electrochemically polymerized vinylpyridine and vinylbipyridine complexes of ruthenium and iron on electrodes,' *Journal of the American Chemical Society* **1981**, *103*, 1.

[27] A. Deronzier, J. C. Moutet, 'Polypyrrole films containing metal complexes: syntheses and applications,' *Coordination Chemistry Reviews* **1996**, *147*, 339.

[28] F. Bedioui, J. Devynck, C. Biedcharreton, 'Immobilization of metalloporphyrins in electropolymerized films – Design and applications,' *Accounts of Chemical Research* **1995**, *28*, 30.

[29] A. R. Hillman, *Electrochemical Science and Technology of Polymers*, Vol. 1, Elsevier Applied Science, London, **1987**.

[30] P. G. Pickup, 'Conjugated metallopolymers. Redox polymers with interacting metal based redox sites,' *Journal of Materials Chemistry* **1999**, *9*, 1641.

[31] B. J. Holliday, T. M. Swager, 'Conducting metallopolymers: the roles of molecular architecture and redox matching,' *Chemical Communications* **2005**, 23.

[32] D. E. Fenton, J. M. Parker, P. V. Wright, 'Complexes of alkali metal ions with poly(ethylene oxide),' *Polymer* **1973**, *14*, 589.

[33] P. G. Bruce, *Solid State Electrochemistry*, new edn, Cambridge University Press, Cambridge, **1997**.

[34] F. M. Gray, *Polymer Electrolytes; RSC Materials Monographs*, 1st edn, The Royal Society of Chemistry, Cambridge, **1997**.

[35] B. Scrosati, *Applications of Electroactive Polymers*, Chapman and Hall, London, **1993**.

[36] M. B. Armand, J. M. Chabagno, M. J. Duclot, 'Polyethers as solid electrolytes,' *Fast Ion Transparent Solids: Electrodes Electrolytes, Proceedings of the International Conference* **1979**, 131.

[37] F. M. Gray, *Solid Polymer Electrolytes: Fundamentals and Technological Applications*, new edn, Wiley-VCH Verlag GmbH, New York, **1991**.

[38] P. G. Bruce, ' Rechargeable lithium batteries,' *Philosophical Transactions of the Royal Society of London* **1996**, *354*, 1577.
[39] P. Lightfoot, M. A. Mehta, P. G. Bruce, 'Crystal-structure of the polymer electrolyte poly(ethylene Oxide)$_3$LiCF$_3$SO$_3$,' *Science* **1993**, *262*, 883.
[40] G. S. MacGlashan, Y. G. Andreev, P. G. Bruce, 'Structure of the polymer electrolyte poly(ethylene oxide)(6): LiAsF$_6$,' *Nature* **1999**, *398*, 792.
[41] T. A. Skotheim, *Handbook of Conducting Polymers*, 2nd edn, CRC, **1986**.
[42] H. Shirakawa, 'Synthesis and characterization of highly conducting polyacetylene,' *Synthetic Metals* **1995**, *69*, 3.
[43] J. Heinze, 'Electrochemistry of conducting polymers,' *Synthetic Metals* **1991**, *43*, 2805.
[44] R. D. McCullough, 'The chemistry of conducting polythiophenes,' *Advanced Materials* **1998**, *10*, 93.
[45] A. G. MacDiarmid, 'Polyaniline and polypyrrole: where are we headed?,' *Synthetic Metals* **1997**, *84*, 27.
[46] A. G. MacDiarmid, A. J. Epstein, 'Conducting polymers: past, present and future,' *Materials Research Society Symposium Proceedings* **1994**, *328*, 133.
[47] S. Holdcroft, 'Patterning pi-conjugated polymers,' *Advanced Materials* **2001**, *13*, 1753.
[48] P. Novak, K. Muller, K. S. V. Santhanam, O. Haas, 'Electrochemically active polymers for rechargeable batteries,' *Chemical Reviews* **1997**, *97*, 207.
[49] F. Garnier, 'Functionalized conducting polymers – towards intelligent materials,' *Angewandte Chemie – International Edition in English* **1989**, *28*, 513.
[50] J. L. Bredas, D. Beljonne, V. Coropceanu, J. Cornil, 'Charge-transfer and energy-transfer processes in pi-conjugated oligomers and polymers: a molecular picture,' *Chemical Reviews* **2004**, *104*, 4971.
[51] A. Moliton, R. C. Hiorns, 'Review of electronic and optical properties of semiconducting pi-conjugated polymers: applications in optoelectronics,' *Polymer International* **2004**, *53*, 1397.
[52] J. L. Bredas, G. B. Street, 'Polarons, bipolarons, and solitons in conducting polymers,' *Accounts of Chemical Research* **1985**, *18*, 309.
[53] C. K. Chiang, M. A. Druy, S. C. Gau, A. J. Heeger, E. J. Louis, A. G. MacDiarmid, Y. W. Park, H. Shirakawa, 'Synthesis of highly conducting films of derivatives of polyacetylene, (CH)$_x$,' *Journal of the American Chemical Society* **1978**, *100*, 1013.
[54] Y. Cao, J. J. Qiu, P. Smith, 'Effect of solvents and cosolvents on the processibility or polyaniline. 1. Solubility and conductivity studies,' *Synthetic Metals* **1995**, *69*, 187.
[55] Y. Cao, P. Smith, A. J. Heeger, 'Counterion induced processibility of conducting polyaniline and of conducting polyblends of polyaniline in bulk polymers,' *Synthetic Metals* **1992**, *48*, 91.
[56] W. J. Moore, *Seven Solid States: An Introduction to the Chemistry and Physics of Solids*, Benjamin, New York, **1967**.
[57] W. P. Su, J. R. Schrieffer, A. J. Heeger, 'Solitons in polyacetylene,' *Physical Review Letters* **1979**, *42*, 1698.
[58] W. P. Su, J. R. Schrieffer, A. J. Heeger, 'Soliton excitations in polyacetylene,' *Condensed Matter and Materials Physics* **1980**, *22*, 2099.

REFERENCES

[59] W. P. Su, J. R. Schrieffer, 'Soliton dynamics in polyacetylene,' *Proceedings of the National Academy of Sciences of the United States of America* **1980**, *77*, 5626.

[60] A. R. Bishop, D. K. Campbell, K. Fesser, 'Polyacetylene and relativistic field-theory models,' *Molecular Crystals and Liquid Crystals* **1981**, *77*, 253.

[61] J. L. Bredas, R. R. Chance, R. Silbey, 'Theoretical-studies of charged defect states in doped polyacetylene and polyparaphenylene,' *Molecular Crystals and Liquid Crystals* **1981**, *77*, 319.

[62] A. Pron, P. Rannou, 'Processible conjugated polymers: from organic semiconductors to organic metals and superconductors,' *Progress in Polymer Science* **2002**, *27*, 135.

[63] C. Rebbi, 'Solitons,' *Scientific American* **1979**, *240*, 92.

[64] D. S. Boudreaux, R. R. Chance, J. L. Bredas, R. Silbey, 'Solitons and polarons in polyacetylene – self-consistent-field calculations of the effect of neutral and charged defects on molecular-geometry,' *Physical Review B* **1983**, *28*, 6927.

[65] H. Thomann, L. R. Dalton, Y. Tomkiewicz, N. S. Shiren, T. C. Clarke, 'Electron-nuclear double-resonance determination of the C-13 and H-1 hyperfine tensors for polyacetylene,' *Physical Review Letters* **1983**, *50*, 533.

[66] D. Ofer, R. M. Crooks, M. S. Wrighton, 'Potential dependence of the conductivity of highly oxidized polythiophenes, polypyrroles and polyaniline: finite windows of high conductivity,' *Journal of the American Chemical Society* **1990**, *112*, 7869.

[67] D. Moses, A. Denenstein, J. Chen, A. J. Heeger, P. McAndrew, T. Woerner, A. G. MacDiarmid, Y. W. Park, 'Effect of nonuniform doping on electrical transport in trans-$(CH)_x$: studies of the semiconductor-metal transition,' *Physical Review B* **1982**, *25*, 7652.

[68] E. J. Mele, M. J. Rice, 'Semiconductor–metal transition in doped polyacetylene,' *Physical Review B* **1981**, *23*, 5397.

[69] A. J. Epstein, H. Rommelmann, R. Bigelow, H. W. Gibson, D. M. Hoffmann, D. B. Tanner, 'Role of solitons in nearly metallic polyacetylene – reply,' *Physical Review Letters* **1983**, *51*, 2020.

[70] A. J. Epstein, R. W. Bigelow, A. Feldblum, H. W. Gibson, D. M. Hoffman, D. B. Tanner, 'Solitons, disorder and charge conduction in nearly metallic polyacetylene,' *Synthetic Metals* **1984**, *9*, 155.

[71] J. C. Scott, P. Pfluger, M. T. Krounb, G. B. Street, 'Electron-spin-resonance studies of pyrrole polymers – evidence for bipolarons,' *Physical Review B* **1983**, *28*, 2140.

[72] J. C. Scott, J. L. Bredas, K. Yakushi, P. Pfluger, G. B. Street, 'The evidence for bipolarons in pyrrole polymers,' *Synthetic Metals* **1984**, *9*, 165.

[73] J. L. Bredas, J. C. Scott, K. Yakushi, G. B. Street, 'Polarons and bipolarons in polypyrrole: evolution of the band structure and optical spectrum upon doping,' *Physical Review B* **1984**, *30*, 1023.

[74] J. L. Bredas, B. Themans, J. G. Fripiat, J. M. Andre, R. R. Chance, 'Highly conducting polyparaphenylene, polypyrrole and polythiophene chains – an ab initio study of the geometry and electronic structure modifications upon doping,' *Physical Review B* **1984**, *29*, 6761.

[75] T. C. Chung, J. H. Kaufman, A. J. Heeger, F. Wudl, 'Charge storage in doped poly(thiophene): optical and electrochemical studies,' *Physical Review B* **1984**, *30*, 702.

[76] G. Crecelius, M. Stamm, J. Fink, J. J. Ritsko, 'AsF$_5$-doped polyparaphenylene: evidence for polaron and bipolaron formation,' *Physical Review Letters* **1983**, *50*, 1498.
[77] J. C. Scott, P. Pfluger, M. T. Krounbi, G. B. Street, 'Electron-spin-resonance studies of pyrrole polymers – evidence for bipolarons,' *Physical Review B* **1983**, *28*, 2140.
[78] J. C. Chiang, A. G. MacDiarmid, 'Polyaniline – protonic acid doping of the emeraldine form to the metallic regime,' *Synthetic Metals* **1986**, *13*, 193.
[79] A. G. MacDiarmid, J. C. Chiang, A. F. Richter, A. J. Epstein, 'Polyaniline – a new concept in conducting polymers,' *Synthetic Metals* **1987**, *18*, 285.
[80] A. G. MacDiarmid, C. K. Chiang, A. F. Richter, N. L. D. Somasiri, A. J. Epstein, *Conducting Polymers: Special Applications*, 1st edn, D. Reidel Publishing Co., Dordrecht, **1987**.
[81] C. Menardo, M. Nechtschein, A. Rousseau, J. P. Travers, P. Hany, 'Investigation on the structure of polyaniline – C-13 NMR and titration studies,' *Synthetic Metals* **1988**, *25*, 311.
[82] C. C. Han, R. L. Elsenbaumer, 'Protonic acids – generally applicable dopants for conducting polymers,' *Synthetic Metals* **1989**, *30*, 123.
[83] A. J. Heeger, 'Polyaniline with surfactant counterions – conducting polymer materials which are processable in the conducting form,' *Synthetic Metals* **1993**, *57*, 3471.
[84] Y. Xia, J. M. Wiesinger, A. G. MacDiarmid, A. J. Epstein, 'Camphorsulfonic acid fully doped polyaniline emeraldine salt: conformations in different solvents studied by an ultraviolet/visible/near-infrared spectroscopic method,' *Chemistry of Materials* **1995**, *7*, 443
[85] J. H. Hwang, S. C. Yang, 'Morphological modification of polyaniline using poly-electrolyte template molecules,' *Synthetic Metals* **1989**, *29*, E271.
[86] J. M. Liu, S. C. Yang, 'Novel colloidal polyaniline fibrils made by template guided chemical polymerization,' *Chemical Communications* **1991**, 1529.
[87] S. H. Li, H. Dong, Y. Cao, 'Synthesis and characterization of soluble polyaniline,' *Synthetic Metals* **1989**, *29*, E329.
[88] B. D. Malhotra, S. Ghosh, R. Chandra, 'Polyaniline polymeric acid composite, a novel conducting rubber,' *Journal of Applied Polymer Science* **1990**, *40*, 1049.
[89] Y. K. Kang, M. H. Lee, S. B. Rhee, 'Electrochemical properties of polyaniline doped with poly(styrenesulfonic acid),' *Synthetic Metals* **1992**, *52*, 319.
[90] R. Nagarajan, S. Tripathy, J. Kumar, F. F. Bruno, L. Samuelson, 'An enzymatically synthesized conducting molecular complex of polyaniline and poly(vinylphosphonic acid)' *Macromolecules* **2000**, *33*, 9542.
[91] R. Nagarajan, W. Liu, J. Kumar, S. K. Tripathy, F. F. Bruno, L. A. Samuelson, 'Manipulating DNA conformation using intertwined conducting polymer chains,' *Macromolecules* **2001**, *34*, 3921.
[92] N. A. Lokshin, V. G. Sergeyev, A. B. Zezin, V. B. Golubev, K. Levon, V. A. Kabanov, 'Polyaniline-containing interpolymer complexes synthesized in low-polar organic media,' *Langmuir* **2003**, *19*, 7564.
[93] B. Deore, Z. D. Chen, T. Nagaoka, 'Potential-induced enantioselective uptake of amino acid into molecularly imprinted overoxidized polypyrrole,' *Analytical Chemistry* **2000**, *72*, 3989.

REFERENCES

[94] K. Gurunathan, A. V. Murugan, R. Marimuthu, U. P. Mulik, D. P. Amalnerkar, 'Electrochemically synthesized conducting polymeric materials for applications towards technology in electronics, optoelectronics and energy storage devices,' *Materials Chemistry and Physics* **1999**, *61*, 173.

[95] R. Cabala, V. Meister, K. PotjeKamloth, 'Effect of competitive doping on sensing properties of polypyrrole,' *Journal of the Chemical Society, Faraday Transactions* **1997**, *93*.

[96] M. X. Wan, J. C. Li, 'Formation mechanism of polyaniline microtubules synthesized by a template-free method,' *Journal of Polymer Science. Part A, Polymer Chemistry* **2000**, *38*, 2359

[97] P. A. Hassan, S. N. Sawant, N. C. Bagkar, J. V. Yakhmi, 'Polyaniline nanoparticles prepared in rodlike micelles,' *Langmuir* **2004**, *20*, 4874.

[98] Z. M. Zhang, Z. X. Wei, M. X. Wan, 'Nanostructures of polyaniline doped with inorganic acids,' *Macromolecules* **2002**, *35*, 5937

[99] A. J. Heeger, S. Kivelson, J. R. Schrieffer, W. P. Su, 'Solitons in conducting polymers,' *Reviews of Modern Physics* **1988**, *60*, 781.

[100] K. E. Ziemelis, A. T. Hussain, D. D. C. Bradley, R. H. Friend, J. Ruehe, G. Wegner, 'Optical spectroscopy of field-induced charge in poly(3-hexylthienylene) metal-insulator-semiconductor structures: evidence for polarons,' *Physical Review Letters* **1991**, *66*, 2231.

[101] J. H. Burroughes, C. A. Jones, R. H. Friend, 'New semiconductor-device physics in polymer diodes and transistors,' *Nature* **1988**, *335*, 137.

[102] J. H. Burroughes, D. D. C. Bradley, A. R. Brown, R. N. Marks, K. Mackay, R. H. Friend, P. L. Burns, A. B. Holmes, 'Light-emitting diodes based on conjugated polymers,' *Nature* **1990**, *347*, 539.

[103] A. F. Diaz, K. K. Kanazawa, G. P. Gardini, 'Electrochemical polymerization of pyrrole,' *Chemical Communications* **1979**, *14*, 635.

[104] M. Aizawa, S. Watanabe, H. Shinohara, H. Shirakawa, 'Electrochemical cation doping of a polythienylene film,' *Chemical Communications* **1985**, *5*, 264.

[105] J. M. Kern, J. P. Sauvage, 'Photochemical deposition of electrically conducting polypyrrole,' *Chemical Communications* **1989**, *10*, 657.

[106] K. Teshima, K. Yamada, N. Kobayashi, R. Hirohashi, 'Photopolymerization of aniline with a tris(2,2'-bipyridyl)ruthenium complex–methylviologen polymer bilayer electrode system,' *Chemical Communications* **1996**, *7*, 829.

[107] W. Liu, A. L. Cholli, R. Nagarajan, J. Kumar, S. Tripathy, F. F. Bruno, L. Samuelson, 'The role of template in the enzymatic synthesis of conducting polyaniline,' *Journal of the American Chemical Society* **1999**, *121*, 11345.

[108] M. Angelopoulos, G. E. Asturias, S. P. Ermer, A. Ray, E. M. Scherr, A. G. MacDiarmid, M. Akhtar, Z. Kiss, A. J. Epstein, 'Polyaniline – solutions, films and oxidation-state,' *Molecular Crystals and Liquid Crystals* **1988**, *160*, 151.

[109] K. G. Neoh, E. T. Kang, K. L. Tan, 'Structural study of polyaniline films in reprotonation/deprotonation cycles,' *Journal of Physical Chemistry* **1991**, *95*, 10151.

[110] A. R. Hopkins, P. G. Rasmussen, R. A. Basheer, 'Characterization of solution and solid-state properties of undoped and doped polyanilines processed from hexafluoro-2-propanol,' *Macromolecules* **1996**, *29*, 7838.

[111] A. Andreatta, Y. Cao, J. C. Chiang, A. J. Heeger, P. Smith, 'Electrically-conductive fibers of polyaniline spun from solutions in concentrated sulfuric acid,' *Synthetic Metals* **1988**, *26*, 383.

[112] Y. Cao, A. J. Smith, A. J. Heeger, *Conjugated Polymeric Materials: Opportunities in Electronics, Optoelectronics, and Molecular Electronics*, 1st edn, Kluwe, Dordrecht, **1990**.

[113] K. Y. Jen, R. Oboodi, R. L. Elsenbaumer, 'Processible and environmentally stable conducting polymers,' *Polymeric Materials Science and Engineering* **1985**, *53*, 79.

[114] M. Sato, S. Tanaka, K. Kaeriyama, 'Soluble conducting polythiophenes,' *Chemical Communications* **1986**, 873.

[115] J. Yano, 'The transformation of electroinactive polymers derived from aniline derivatives into electroactive and functional polymers. 2. Making poly(N,N-di-N-butylaniline) films have anion exchangeability and selective potential response to dissolved iodide ions,' *Journal of the Electrochemical Society* **1991**, *138*, 455.

[116] Y. Wei, R. Hariharan, S. A. Patel, 'Chemical and electrochemical copolymerization of aniline with alkyl ring-substituted anilines,' *Macromolecules* **1990**, *23*, 758.

[117] G. Daprano, M. Leclerc, G. Zotti, G. Schiavon, 'Synthesis and characterization of polyaniline derivatives – poly(2-alkoxyanilines) and poly(2,5-dialkoxyanilines),' *Chemistry of Materials* **1995**, *7*, 33.

[118] W. A. Gazotti, M. A. DePaoli, 'High yield preparation of a soluble polyaniline derivative,' *Synthetic Metals* **1996**, *80*, 263.

[119] L. A. Samuelson, A. Anagnostopoulos, K. S. Alva, J. Kumar, S. K. Tripathy, 'Biologically derived conducting and water soluble polyaniline,' *Macromolecules* **1998**, *31*, 4376.

[120] W. Liu, J. Kumar, S. Tripathy, K. J. Senecal, L. Samuelson, 'Enzymically synthesized conducting polyaniline,' *Journal of the American Chemical Society* **1999**, *121*, 71.

[121] M. R. Nabid, A. A. Entezami, 'A novel method for synthesis of water-soluble polypyrrole with horseradish peroxidase enzyme,' *Journal of Applied Polymer Science* **2004**, *94*, 254.

[122] B. Lee, V. Seshadri, G. A. Sotzing, 'Poly(thieno 3,4-b thiophene)-poly(styrene sulfonate): a low band gap, water dispersible conjugated polymer,' *Langmuir* **2005**, *21*, 10797.

[123] J. Stejskal, 'Colloidal dispersions of conducting polymers,' *Journal of Polymer Materials* **2001**, *18*, 225.

[124] B. Wessling, 'Dispersion – the key tool for understanding, improving and using conductive polymers and organic metals,' *Synthetic Metals* **2005**, *152*, 5.

[125] M. S. Freund, C. Karp, N. S. Lewis, 'Growth of thin processable films of poly(pyrrole) using phosphomolybdate clusters,' *Inorganica Chimica Acta* **1995**, *240*, 447.

[126] M. C. Lonergan, 'A tunable diode based on an inorganic semiconductor vertical bar conjugated polymer interface,' *Science* **1997**, *278*, 2103.

[127] A. O. Patil, Y. Ikenoue, F. Wudl, A. J. Heeger, 'Water-soluble conducting polymers,' *Journal of the American Chemical Society* **1987**, *109*, 1858.

REFERENCES

[128] A. O. Patil, Y. Ikenoue, N. Basescu, N. Colaneri, J. Chen, F. Wudl, A. J. Heeger, 'Self-doped conducting polymers,' *Synthetic Metals* **1987**, *20*, 151.

[129] Y. Ikenoue, J. Chiang, A. O. Patil, F. Wudl, A. J. Heeger, 'Verification of the cation-popping doping mechanism of self-doped polymers,' *Journal of the American Chemical Society* **1988**, *110*, 2983.

[130] E. E. Havinga, L. W. Vanhorssen, W. Tenhoeve, H. Wynberg, E. W. Meijer, 'Self-doped water-soluble conducting polymers,' *Polymer Bulletin* **1987**, *18*, 277.

[131] E. E. Havinga, L. W. Van Horssen, 'Electrochemical polymerization of substituted thiophenes,' *Makromolekulare Chemie, Macromolecular Symposia* **1989**, 67.

[132] Y. Ikenoue, Y. Saida, M. Kira, H. Tomozawa, H. Yashima, M. Kobayashi, 'A facile preparation of a self-doped conducting polymer,' *Chemical Communications* **1990**, 1694.

[133] N. S. Sundaresan, S. Basak, M. Pomerantz, J. R. Reynolds, 'Electroactive copolymers of pyrrole containing covalently bound dopant ions – poly(pyrrole-Co-3-(pyrrol-1-yl)propanesulphonate),' *Chemical Communications* **1987**, 621.

[134] J. R. Reynolds, N. S. Sundaresan, M. Pomerantz, S. Basak, C. K. Baker, 'Self-doped conducting copolymers – a charge and mass-transport study of poly (pyrrole-co-3-(pyrrol-1-yl)propanesulfonate),' *Journal of Electroanalytical Chemistry* **1988**, *250*, 355.

[135] E. E. Havinga, W. Ten Hoeve, E. W. Meijer, H. Wynberg, 'Water-soluble self-doped 3-substituted polypyrroles,' *Chemistry of Materials* **1989**, *1*, 650.

[136] P. Hany, E. M. Genies, C. Santier, 'Polyanilines with covalently bonded alkyl sulfonates as doping agent – Synthesis and properties,' *Synthetic Metals* **1989**, *31*, 369.

[137] J. Yue, A. J. Epstein, 'Synthesis of self-doped conducting polyaniline,' *Journal of the American Chemical Society* **1990**, *112*, 2800.

[138] J. Y. Bergeron, J. W. Chevalier, L. H. Dao, 'Water-soluble conducting poly (aniline) polymer,' *Chemical Communications* **1990**, 180.

[139] T. I. Wallow, B. M. Novak, '*In aqua* synthesis of water-soluble poly(paraphenylene) derivatives,' *Journal of the American Chemical Society* **1991**, *113*, 7411.

[140] V. Chaturvedi, S. Tanaka, K. Kaeriyama, 'Preparation of poly(*p*-phenylene) via a new precursor route,' *Macromolecules* **1993**, *26*, 2607.

[141] I. U. Rau, M. Rehahn, 'Rigid-rod polyelectrolytes: carboxylated poly(paraphenylenes) via a novel precursor route,' *Polymer* **1993**, *34*, 2889.

[142] A. D. Child, J. R. Reynolds, 'Water-soluble rigid-rod polyelectrolytes – a new self-doped, electroactive sulfonatoalkoxy-substituted poly(*p*-phenylene),' *Macromolecules* **1994**, *27*, 1975.

[143] S. Q. Shi, F. Wudl, 'Synthesis and characterization of a water-soluble poly (para-phenylenevinylene) derivative,' *Macromolecules* **1990**, *23*, 2119.

[144] J. Yue, G. Gordon, A. J. Epstein, 'Comparison of different synthetic routes for sulfonation of polyaniline,' *Polymer* **1992**, *33*, 4410.

[145] E. Kim, M. H. Lee, B. S. Moon, C. Lee, S. B. Rhee, 'Redox cycleability of a self-doped polyaniline,' *Journal of the Electrochemical Society* **1994**, *141*, L26.

[146] C. Dearmitt, S. P. Armes, J. Winter, F. A. Uribe, S. Gottesfeld, C. Mombourquette, 'A novel N-substituted polyaniline derivative,' *Polymer* **1993**, *34*, 158.

[147] M. T. Nguyen, P. Kasai, J. L. Miller, A. F. Diaz, 'Synthesis and properties of novel water-soluble conducting polyaniline copolymers,' *Macromolecules* **1994**, *27*, 3625.

[148] H. S. O. Chan, S. C. Ng, W. S. Sim, K. L. Tan, B. T. G. Tan, 'Preparation and characterization of electrically conducting copolymers of aniline and anthranilic acid – evidence for self-doping by X-ray photoelectron spectroscopy,' *Macromolecules* **1992**, *25*, 6029.

[149] S. C. Ng, H. S. O. Chan, H. H. Huang, P. K. H. Ho, 'Poly(o-aminobenzylphosphonic acid) – a novel water-soluble, self-doped functionalized polyaniline,' *Journal of the Chemical Society-Chemical Communications* **1995**, 1327.

[150] B. Deore, M. S. Freund, 'Saccharide imprinting of poly(aniline boronic acid) in the presence of fluoride,' *Analyst* **2003**, *128*, 803.

[151] B. A. Deore, I. Yu, M. S. Freund, 'A switchable self-doped polyaniline: interconversion between self-doped and non-self-doped forms,' *Journal of the American Chemical Society* **2004**, *126*, 52.

[152] J. Yue, A. J. Epstein, A. G. MacDiarmid, 'Sulfonic acid ring-substituted polyaniline, a self-doped conducting polymer,' *Molecular Crystals and Liquid Crystals* **1990**, *189*, 255.

[153] J. Yue, Z. H. Wang, K. R. Cromack, A. J. Epstein, A. G. MacDiarmid, 'Effect of sulfonic-acid group on polyaniline backbone,' *Journal of the American Chemical Society* **1991**, *113*, 2665.

[154] J. Yue, A. J. Epstein, 'XPS study of self-doped conducting polyaniline and parent systems,' *Macromolecules* **1991**, *24*, 4441.

[155] T. Kawai, H. Mizobuchi, N. Yamasaki, H. Araki, K. Yoshino, 'Optical and magnetic-properties of polyaniline derivatives having ionic groups,' *Japanese Journal of Applied Physics Part 2 – Letters* **1994**, *33*, L357.

[156] R. S. Wang, L. M. Wang, Z. M. Su, Y. J. Fu, 'Study of self-doping conductive properties of different oxidized states of poly-3-(2-ethane carboxylate) pyrrole,' *Synthetic Metals* **1995**, *69*, 511.

[157] E. T. Kang, K. G. Neoh, Y. L. Woo, K. L. Tan, 'Self-doped polyaniline and polypyrrole – a comparative study by X-ray photoelectron spectroscopy,' *Polymer Communications* **1991**, *32*, 412.

[158] D. Delabouglise, F. Garnier, 'Poly (3-carboxymethyl pyrrole), a pH sensitive, self-doped conducting polymer,' *New Journal of Chemistry* **1991**, *15*, 233.

[159] M. D. Ingram, H. Staesche, K. S. Ryder, ' "Activated" polypyrrole electrodes for high-power supercapacitor applications,' *Solid State Ionics* **2004**, *169*, 51.

[160] D. J. Liaw, B. Y. Liaw, J. P. Gong, Y. Osada, 'Synthesis and properties of poly(3-thiopheneacetic acid) and its networks via electropolymerization,' *Synthetic Metals* **1999**, *99*, 53.

[161] M. Chayer, K. Faid, M. Leclerc, 'Highly conducting water-soluble polythiophene derivatives,' *Chemistry of Materials* **1997**, *9*, 2902.

[162] K. Faid, M. Leclerc, 'Responsive supramolecular polythiophene assemblies,' *Journal of the American Chemical Society* **1998**, *120*, 5274.

REFERENCES

[163] F. Tran-Van, M. Carrier, C. Chevrot, 'Sulfonated polythiophene and poly (3,4-ethylenedioxythiophene) derivatives with cations exchange properties,' *Synthetic Metals* **2004**, *142*, 251.

[164] G. Zotti, S. Zecchin, G. Schiavon, A. Berlin, G. Pagani, A. Canavesi, 'Doping-induced ion-exchange in the highly conjugated self-doped polythiophene from anodic coupling of 4-(4*H*-cyclopentadithien-4-yl)butanesulfonate,' *Chemistry of Materials* **1997**, *9*, 2940.

[165] A. A. Moxey, D. C. Loveday, I. D. Brotherson, J. P. Ferraris, 'Synthesis and characterization of poly[2-(2-thienyl)3-(4-dimethyl dodecyl ammonium phenyl) thiophene triflate] a processable, self n-dopable polymer,' *Polymer Preprints (American Chemical Society, Division of Polymer Chemistry)* **1998**, *39*, 137.

[166] A. Berlin, G. Schiavon, S. Zecchin, G. Zotti, 'New highly conjugated self-doped polythiophenes functionalized with alkyl ammonium groups,' *Synthetic Metals* **2001**, *119*, 153.

[167] Y. Cao, P. Smith, A. J. Heeger, 'Counterion induced processibility of conducting polyaniline,' *Synthetic Metals* **1993**, *57*, 3514.

[168] M. Lapkowski, 'Electrochemical synthesis of polyaniline poly(2-acryl-amido-2-methyl-1-propane-sulfonic acid) composite,' *Synthetic Metals* **1993**, *55*, 1558.

[169] M. Angelopoulos, N. Patel, R. Saraf, 'Amic acid doping of polyaniline – characterization and resulting blends,' *Synthetic Metals* **1993**, *55*, 1552.

[170] S. K. Sahoo, R. Nagarajan, S. Roy, L. A. Samuelson, J. Kumar, A. L. Cholli, 'An enzymatically synthesized polyaniline: a solid-state NMR study,' *Macromolecules* **2004**, *37*, 4130.

[171] W. J. Bae, K. H. Kim, Y. H. Park, W. H. Jo, 'A novel water-soluble and self-doped conducting polyaniline graft copolymer,' *Chemical Communications* **2003**, 2768.

[172] W. J. Bae, K. H. Kim, W. H. Jo, Y. H. Park, 'A water-soluble and self-doped conducting polypyrrole graft copolymer,' *Macromolecules* **2005**, *38*, 1044.

[173] X. L. Wei, Y. Z. Wang, S. M. Long, C. Bobeczko, A. J. Epstein, 'Synthesis and physical properties of highly sulfonated polyaniline,' *Journal of the American Chemical Society* **1996**, *118*, 2545.

[174] H. K. Lin, S. A. Chen, 'Synthesis of new water-soluble self-doped polyaniline,' *Macromolecules* **2000**, *33*, 8117.

[175] K. Takahashi, K. Nakamura, T. Yamaguchi, T. Komura, S. Ito, R. Aizawa, K. Murata, 'Characterization of water-soluble externally HCl-doped conducting polyaniline,' *Synthetic Metals* **2002**, *128*, 27.

[176] M. Leclerc, K. Faid, 'Electrical and optical properties of processable polythiophene derivatives: structure–property relationships,' *Advanced Materials* **1997**, *9*, 1087.

[177] M. T. Nguyen, M. Leclere, A. F. Diaz, 'Water-soluble conductive-electroactive polymers,' *Trends in Polymer Science* **1995**, *3*, 186.

[178] N. Zhang, R. Wu, Q. Li, K. Pakbaz, C. O. Yoon, F. Wudl, 'Synthesis and properties of an n-self-doped conducting polymer,' *Chemistry of Materials* **1993**, *5*, 1598.

[179] D. C. Loveday, A. A. Moxey, M. Hmyene, X. Ren, D. Guerrero, J. P. Ferraris, 'Characterization of self n-dopable polymers,' *Polymer Preprints (American Chemical Society, Division of Polymer Chemistry)* **1998**, *39*, 145.

[180] K. Krishnamoorthy, A. Q. Contractor, A. Kumar, 'Electrochemical synthesis of fully sulfonated n-dopable polyaniline: poly(metanilic acid),' *Chemical Communications* **2002**, 240.
[181] Q. Pei, F. Klavetter, Y. Yang, 'Self-doped n-type conducting polymers through side-group anion charge delocalization,' *Polymer Preprints (American Chemical Society, Division of Polymer Chemistry)* **1995**, *36*, 213.
[182] Y. Ikenoue, N. Uotani, A. O. Patil, F. Wudl, A. J. Heeger, 'Electrochemical studies of self-doped conducting polymers – verification of the cation-popping doping mechanism,' *Synthetic Metals* **1989**, *30*, 305.
[183] Y. Ikenoue, H. Tomozawa, Y. Saida, M. Kira, H. Yashima, 'Evaluation of electrochromic fast-switching behaviour of self-doped conducting polymer,' *Synthetic Metals* **1991**, *40*, 333.
[184] S. A. Chen, M. Y. Hua, 'Structure and doping level of the self-acid-doped conjugated conducting polymers – poly(n-(3'-thienyl)alkanesulfonic acid)s,' *Macromolecules* **1993**, *26*, 7108.
[185] K. Y. Jen, G. G. Miller, R. L. Elsenbaumer, 'Highly conducting, soluble, and environmentally-stable poly(3-alkylthiophenes),' *Chemical Communications* **1986**, 1346.
[186] S. Hotta, S. D. D. V. Rughooputh, A. J. Heeger, F. Wudl, 'Spectroscopic studies of soluble poly(3-alkylthienylenes)' *Macromolecules* **1987**, *20*, 212.
[187] R. L. Elsenbaumer, K. Jen, G. G. Miller, H. Eckhardt, L. W. Shacklette, R. Jow, *Electronic Properties of Conjugated Polymers*, Vol. 76, Springer-Verlag, Berlin, **1987**.
[188] M. Rehahn, A. D. Schlueter, G. Wegner, W. J. Feast, 'Soluble poly(paraphenylenes). 1. Extension of the Yamamoto synthesis to dibromobenzenes substituted with flexible side chains,' *Polymer* **1989**, *30*, 1054.
[189] M. Rehahn, A. D. Schlueter, G. Wegner, W. J. Feast, 'Soluble poly(paraphenylenes). 2. Improved synthesis of poly(para-2,5-di-n-hexylphenylene) via palladium-catalyzed coupling of 4-bromo-2,5-di-n-hexylbenzeneboronic acid,' *Polymer* **1989**, *30*, 1060.
[190] M. Ballauff, 'Stiff-chain polymers–structure, phase behavior and properties,' *Angewandte Chemie, International Edition in English* **1989**, *28*, 253.
[191] S. Ito, K. Murata, S. Teshima, R. Aizawa, Y. Asako, K. Takahashi, B. M. Hoffman, 'Simple synthesis of water-soluble conducting polyaniline,' *Synthetic Metals* **1998**, *96*, 161.
[192] Y. Wei, W. W. Focke, G. E. Wnek, A. Ray, A. G. MacDiarmid, 'Synthesis and electrochemistry of alkyl ring-substituted polyanilines,' *Journal of Physical Chemistry* **1989**, *93*, 495
[193] M. Leclerc, J. Guay, L. H. Dao, 'Synthesis and characterization of poly(alkylanilines)' *Macromolecules* **1989**, *22*, 649.
[194] C. B. Gorman, E. J. Ginsburg, R. H. Grubbs, 'Soluble, highly conjugated derivatives of polyacetylene from the ring-opening metathesis polymerization of monosubstituted cyclooctatetraenes: synthesis and the relationship between polymer structure and physical properties,' *Journal of the American Chemical Society* **1993**, *115*, 1397
[195] A. Kitani, K. Satoguchi, H. Q. Tang, S. Ito, K. Sasaki, 'Electrosynthesis and properties of self-doped polyaniline,' *Synthetic Metals* **1995**, *69*, 129.

REFERENCES

[196] S. Shimizu, T. Saitoh, M. Uzawa, M. Yuasa, K. Yano, T. Maruyama, K. Watanabe, 'Synthesis and applications of sulfonated polyaniline,' *Synthetic Metals* **1997**, *85*, 1337.
[197] H. S. O. Chan, A. J. Neuendorf, S. C. Ng, P. M. L. Wong, D. J. Young, 'Synthesis of fully sulfonated polyaniline: a novel approach using oxidative polymerisation under high pressure in the liquid phase,' *Chemical Communications* **1998**, 1327.
[198] E. Kim, M. Lee, M. H. Lee, S. B. Rhee, 'Liquid-crystalline assemblies from self-doped polyanilines,' *Synthetic Metals* **1995**, *69*, 101.
[199] S. A. Chen, G. W. Hwang, 'Water-soluble self-acid-doped conducting polyaniline – Structure and properties,' *Journal of the American Chemical Society* **1995**, *117*, 10055.
[200] M. Y. Hua, Y. N. Su, S. A. Chen, 'Water-soluble self-acid-doped conducting polyaniline: poly(aniline-co-N-propylbenzenesulfonic acid-aniline),' *Polymer* **2000**, *41*, 813.
[201] A. Malinauskas, 'Self-doped polyanilines,' *Journal of Power Sources* **2004**, *126*, 214.
[202] W. Lee, G. Du, S. M. Long, A. J. Epstein, S. Shimizu, T. Saitoh, M. Uzawa, 'Charge transport properties of fully-sulfonated polyaniline,' *Synthetic Metals* **1997**, *84*, 807.
[203] Y. Sahin, A. Aydin, Y. A. Udum, K. Pekmez, A. Yildiz, 'Electrochemical synthesis of sulfonated polypyrrole in FSO_3H/acetonitrile solution,' *Journal of Applied Polymer Science* **2004**, *93*, 526.
[204] P. Udebert, G. Bidan, M. Lapkowaki, D. Limosin, *Electronic Properties of Conjugated Polymers*, Springer-Verlag, Berlin, **1987**.
[205] G. Bidan, B. Ehui, M. Lapkowski, 'Conductive polymers with immobilised dopants: ionomer composites and auto-doped polymers-a review and recent advances,' *Journal of Physics. D, Applied Physics* **1988**, *21*, 1043.
[206] R. S. Wang, L. M. Wang, Y. J. Fu, Z. M. Su, 'The influence of different substituent on polymer self-doping conductive property,' *Synthetic Metals* **1995**, *69*, 713.
[207] S. Oztemiz, G. Beaucage, O. Ceylan, H. B. Mark, 'Synthesis, characterization and molecular weight studies of certain soluble poly(3-alkylthiophene) conducting polymers,' *Journal of Solid State Electrochemistry* **2004**, *8*, 928.
[208] I. T. Kim, R. L. Elsenbaumer, 'Synthesis, characterization, and electrical properties of poly(1-alkyl-2, 5-pyrrylene vinylenes): New low band gap conducting polymers,' *Macromolecules* **2000**, *33*, 6407
[209] L. V. Lukachova, E. A. Shkerin, E. A. Puganova, E. E. Karyakina, S. G. Kiseleva, A. V. Orlov, G. P. Karpacheva, A. A. Karyakin, 'Electroactivity of chemically synthesized polyaniline in neutral and alkaline aqueous solutions – role of self-doping and external doping,' *Journal of Electroanalytical Chemistry* **2003**, *544*, 59.
[210] A. A. Karyakin, A. K. Strakhova, A. K. Yatsimirsky, 'Self-doped polyanilines electrochemically active in neutral and basic aqueous-solutions – electropolymerization of substituted anilines,' *Journal of Electroanalytical Chemistry* **1994**, *371*, 259.
[211] A. A. Karyakin, I. A. Maltsev, L. V. Lukachova, 'The influence of defects in polyaniline structure on its electroactivity: optimization of "self-doped" polyaniline synthesis,' *Journal of Electroanalytical Chemistry* **1996**, *402*, 217.

[212] A. Malinauskas, 'Electrocatalysis at conducting polymers,' *Synthetic Metals* **1999**, *107*, 75.
[213] Y. Wei, K. F. Hsueh, 'Thermal analysis of chemically synthesized polyaniline and effects of thermal aging on conductivity,' *Journal of Polymer Science, Part A: Polymer Chemistry* **1989**, *27*, 4351.
[214] Y. D. Wang, M. F. Rubner, 'An investigation of the conductivity stability of acid-doped polyanilines,' *Synthetic Metals* **1992**, *47*, 255.
[215] S. Kim, I. J. Chung, 'Annealing effect on the electrochemical property of polyaniline complexed with various acids,' *Synthetic Metals* **1998**, *97*, 127.
[216] A. Gok, B. Sari, M. Talu, 'Synthesis and characterization of conducting substituted polyanilines,' *Synthetic Metals* **2004**, *142*, 41.
[217] N. A. Zaidi, J. P. Foreman, G. Tzamalis, S. C. Monkman, A. P. Monkman, 'Alkyl substituent effects on the conductivity of polyaniline,' *Advanced Functional Materials* **2004**, *14*, 479.
[218] C. C. Han, C. H. Lu, S. P. Hong, K. F. Yang, 'Highly conductive and thermally stable self-doping propylthiosulfonated polyanilines,' *Macromolecules* **2003**, *36*, 7908.
[219] J. Yue, A. J. Epstein, Z. Zhong, P. K. Gallagher, A. G. MacDiarmid, 'Thermal stabilities of polyanilines,' *Synthetic Metals* **1991**, *41*, 765.
[220] I. Yu, B. A. Deore, C. L. Recksiedler, T. C. Corkery, A. S. Abd-El-Aziz, M. S. Freund, 'Thermal stability of high molecular weight self-doped poly(aniline-boronic acid),' *Macromolecules* **2005**, *38*, 10022.
[221] B. A. Deore, I. S. Yu, P. M. Aguiar, C. Recksiedler, S. Kroeker, M. S. Freund, 'Highly cross-linked, self-doped polyaniline exhibiting unprecedented hardness,' *Chemistry of Materials* **2005**, *17*, 3803.
[222] G. Decher, J. D. Hong, J. Schmitt, 'Buildup of ultrathin multilayer films by a self-assembly process. 3. Consecutively alternating adsorption of anionic and cationic polyelectrolytes on charged surfaces,' *Thin Solid Films* **1992**, *210*, 831.
[223] M. Ferreira, M. F. Rubner, 'Molecular-level processing of conjugated polymers. 1. Layer-by-layer manipulation of conjugated polyions,' *Macromolecules* **1995**, *28*, 7107.
[224] J. H. Cheung, A. F. Fou, M. F. Rubner, 'Molecular self-assembly of conducting polymers,' *Thin Solid Films* **1994**, *244*, 985.
[225] A. C. Fou, M. F. Rubner, 'Molecular-level processing of conjugated polymers. 2. Layer-by-layer manipulation of *in situ* polymerized p-type doped conducting polymers,' *Macromolecules* **1995**, *28*, 7115.
[226] J. H. Cheung, W. B. Stockton, M. F. Rubner, 'Molecular-level processing of conjugated polymers. 3. Layer-by-layer manipulation of polyaniline via electrostatic interactions,' *Macromolecules* **1997**, *30*, 2712.
[227] W. B. Stockton, M. F. Rubner, 'Molecular-level processing of conjugated polymers. 4. Layer-by-layer manipulation of polyaniline via hydrogen-bonding interactions,' *Macromolecules* **1997**, *30*, 2717.
[228] M. K. Ram, M. Salerno, M. Adami, P. Faraci, C. Nicolini, 'Physical properties of polyaniline films: assembled by the layer-by-layer technique,' *Langmuir* **1999**, *15*, 1252.
[229] L. Kumpumbu-Kalemba, M. Leclerc, 'Electrochemical characterization of monolayers of a biotinylated polythiophene: towards the development of polymeric biosensors,' *Chemical Communications* **2000**, 1847.

REFERENCES

[230] P. K. H. Ho, J. S. Kim, J. H. Burroughes, H. Becker, S. F. Y. Li, T. M. Brown, F. Cacialli, R. H. Friend, 'Molecular-scale interface engineering for polymer light-emitting diodes,' *Nature* 2000, *404*, 481.

[231] P. K. H. Ho, M. Granstroem, R. H. Friend, N. C. Greenham, 'Ultrathin self-assembled layers at the ITO interface to control charge injection and electroluminescence efficiency in polymer light-emitting diodes,' *Advanced Materials* 1998, *10*, 769.

[232] J. H. Cheung, A. F. Fou, M. Ferreira, M. F. Rubner, 'Molecular self-assembly of conducting polymers: a new layer-by-layer thin film deposition process,' *Polymer Preprints (American Chemical Society, Division of Polymer Chemistry)* 1993, *34*, 757.

[233] M. Ferreira, J. H. Cheung, M. F. Rubner, 'Molecular self-assembly of conjugated polyions – a new process for fabricating multilayer thin-film heterostructures,' *Thin Solid Films* 1994, *244*, 806.

[234] A. C. Fou, O. Onitsuka, M. Ferreira, M. F. Rubner, B. R. Hseih, 'Interlayer interactions in self-assembled poly(phenylene vinylene) multilayer heterostructures, implications for light-emitting and photorectifying diodes,' *Materials Research Society Symposium Proceedings* 1995, *575*.

[235] J. W. Baur, S. Kim, P. B. Balanda, J. R. Reynolds, M. F. Rubner, 'Thin-film light-emitting devices based on sequentially adsorbed multilayers of water-soluble poly(p-phenylene)s,' *Advanced Materials* 1998, *10*, 1452.

[236] J. Lukkari, A. Viinikanoja, J. Paukkunen, M. Salomaki, M. Janhonen, T. Aaritalo, J. Kankare, 'Oxidation induced variation in polyelectrolyte multilayers prepared from sulfonated self-dopable poly(alkoxythiophene),' *Chemical Communications* 2000, 571.

[237] J. Lukkari, M. Salomaki, A. Viinikanoja, T. Aaritalo, J. Paukkunen, N. Kocharova, J. Kankare, 'Polyelectrolyte multilayers prepared from water-soluble poly(alkoxythiophene) derivatives,' *Journal of the American Chemical Society* 2001, *123*, 6083.

[238] M. Onoda, K. Yoshino, 'Heterostructure electroluminescent diodes prepared from self-assembled multilayers of poly(p-phenylene vinylene) and sulfonated polyaniline,' *Japanese Journal of Applied Physics Part 2-Letters* 1995, *34*, L260.

[239] C. Li, K. Mitamura, T. Imae, 'Electrostatic layer-by-layer assembly of poly (amido amine) dendrimer/conducting sulfonated polyaniline: structure and properties of multilayer films,' *Macromolecules* 2003, *36*, 9957.

[240] N. Sarkar, M. K. Ram, A. Sarkar, R. Narizzano, S. Paddeu, C. Nicolini, 'Nanoassemblies of sulfonated polyaniline multilayers,' *Nanotechnology* 2000, *11*, 30.

[241] T. B. Cao, L. H. Wei, S. M. Yang, M. F. Zhang, C. H. Huang, W. X. Cao, 'Self-assembly and photovoltaic property of covalent-attached multilayer film based on highly sulfonated polyaniline and diazoresin,' *Langmuir* 2002, *18*, 750.

[242] P. A. Fiorito, S. I. C. de Torresi, 'Optimized multilayer oxalate biosensor,' *Talanta* 2004, *62*, 649.

[243] K. Y. K. Man, H. L. Wong, W. K. Chan, C. Y. Kwong, A. B. Djurisic, 'Efficient photodetectors fabricated from a metal-containing conjugated polymer by a multilayer deposition process,' *Chemistry of Materials* 2004, *16*, 365.

[244] S. M. Sze, *Semiconductor Devices Physics and Technology*, 2nd edn, John Wiley & Sons, Inc., Hoboken, NJ, **1985**.
[245] A. Tsumura, H. Koezuka, T. Ando, 'Macromolecular electronic device – field-effect transistor with a polythiophene thin-film,' *Applied Physics Letters* **1986**, *49*, 1210.
[246] G. Horowitz, D. Fichou, X. Z. Peng, Z. G. Xu, F. Garnier, 'A field-effect transistor based on conjugated alpha-sexithienyl,' *Solid State Communications* **1989**, *72*, 381.
[247] J. Paloheimo, H. Stubb, P. Ylilahti, P. Kuivalainen, 'Field-effect conduction in polyalkylthiophenes,' *Synthetic Metals* **1991**, *41*, 563.
[248] A. Assadi, C. Svensson, M. Willander, O. Inganas, 'Field-effect mobility of poly(3-hexylthiophene),' *Applied Physics Letters* **1988**, *53*, 195.
[249] H. Fuchigami, A. Tsumura, H. Koezuka, *Ext. Abstr., [99] Int. ConI Solid State Devices and Materials*, 27–29, 596.
[250] A. Tsumura, H. Koezuka, S. Tsunoda, T. Ando, 'Chemically prepared poly (N-methylpyrrole) thin-film – its application to the field-effect transistor,' *Chemistry Letters* **1986**, 863.
[251] R. K. Yuan, S. C. Yang, H. Yuan, R. L. Jiang, H. Z. Qian, D. C. Gui, 'Surface field-effect of polyaniline film,' *Synthetic Metals* **1991**, *41*, 727.
[252] Y. Ohmori, K. Muro, M. Onoda, K. Yoshino, 'Fabrication and characteristics of Schottky gated field-effect transistors utilizing poly(1,4-naphthalene vinylene) and poly(*para*-phenylene vinylene),' *Japanese Journal of Applied Physics Part 2 – Letters* **1992**, *31*, L646.
[253] E. Punkka, M. F. Rubner, J. D. Hettinger, J. S. Brooks, S. T. Hannahs, 'Tunneling and hopping conduction in Langmuir–Blodgett thin films of poly(3-hexylthiophene),' *Physical Review B* **1991**, *43*, 9076.
[254] A. Tsumura, H. Koezuka, T. Ando, 'Polythiophene field-effect transistor – its characteristics and operation mechanism,' *Synthetic Metals* **1988**, *25*, 11.
[255] A. Tsumura, H. Fuchigami, H. Koezuka, 'Field-effect transistor with a conducting polymer film,' *Synthetic Metals* **1991**, *41*, 1181.
[256] H. Stubb, E. Punkka, J. Paloheimo, 'Electronic and optical properties of conducting polymer thin films,' *Materials Science & Engineering R-Reports* **1993**, *10*, 85.
[257] C.-T. Kuo, S.-A. Chen, G.-W. Hwang, H.-H. Kuo, 'Field-effect transistor with the water-soluble self-acid-doped polyaniline thin films as semiconductor,' *Synthetic Metals* **1998**, *93*, 155.
[258] S. Cosnier, 'Biomolecule immobilization on electrode surfaces by entrapment or attachment to electrochemically polymerized films. A review,' *Biosensors & Bioelectronics* **1999**, *14*, 443.
[259] A. Q. Contractor, T. N. Sureshkumar, R. Narayanan, S. Sukeerthi, L. Rakesh, R. S. Srinivasa, 'Conducting polymer-based biosensors,' *Electrochimica Acta* **1994**, *39*, 1321.
[260] R. J. Geise, J. M. Adams, N. J. Barone, A. M. Yacynych, 'Electropolymerized films to prevent interferences and electrode fouling in biosensors,' *Biosensors and Bioelectronics* **1991**, *6*, 151.
[261] H. Tang, A. Kitani, T. Yamashita, S. Ito, 'Highly sulfonated polyaniline electrochemically synthesized by polymerizing aniline-2,5-disulfonic acid and copolymerizing it with aniline,' *Synthetic Metals* **1998**, *96*, 43.

REFERENCES

[262] E. E. Karyakina, L. V. Neftyakova, A. A. Karyakin, 'A novel potentiometric glucose biosensor based on polyaniline semiconductor films,' *Analytical Letters* **1994**, *27*, 2871.
[263] S. D. Caras, J. Janata, D. Saupe, K. Schmitt, 'PH-based enzyme potentiometric sensors. 1. Theory,' *Analytical Chemistry* **1985**, *57*, 1917.
[264] D. M. Zhou, J. J. Xu, H. Y. Chen, H. Q. Fang, 'Ascorbate sensor based on 'self-doped' polyaniline,' *Electroanalysis (New York)* **1997**, *9*, 1185.
[265] O. Ngamna, A. Morrin, S. E. Moulton, A. J. Killard, M. R. Smyth, G. G. Wallace, 'An HRP based biosensor using sulphonated polyaniline,' *Synthetic Metals* **2005**, *153*, 185.
[266] X. Yu, G. A. Sotzing, F. Papadimitrakopoulos, J. F. Rusling, 'Wiring of enzymes to electrodes by ultrathin conductive polyion underlayers: enhanced catalytic response to hydrogen peroxide,' *Analytical Chemistry* **2003**, *75*, 4565.
[267] M. Angelopoulos, J. M. Shaw, K. L. Lee, W. S. Huang, M. A. Lecorre, M. Tissier, 'Lithographic applications of conducting polymers,' *Journal of Vacuum Science and Technology B* **1991**, *9*, 3428.
[268] M. Angelopoulos, J. M. Shaw, R. D. Kaplan, S. Perreault, 'Conducting polyanilines: discharge layers for electron-beam lithography,' *Journal of Vacuum Science and Technology, B:* **1989**, *7*, 1519.
[269] F. Murai, H. Tomozawa, Y. Ikenoue, 'Characteristics and application of electric charging preventing film forming material for electronic rays lithography use,' *Denshi Zairyou (Japanese)* **1990**, *29*, 48.
[270] H. Tomozawa, Y. Saida, Y. Ikenoue, F. Murai, Y. Suzuki, T. Tawa, Y. Ohta, 'Properties and application of conducting polymer on electron-beam lithography,' *Journal of Photopolymer Science and Technology* **1996**, *9*, 707.
[271] M. F. Goppelsröder, *Compt. Rend.* **1876**, *82*, 331.
[272] S. K. Deb, 'Novel electrophotographic system,' *Applied Optics, Supplement* **1969**, 192.
[273] A. F. Diaz, J. I. Castillo, J. A. Logan, W. Y. Lee, 'Electrochemistry of conducting polypyrrole films,' *Journal of Electroanalytical Chemistry* **1981**, *129*, 115.
[274] R. J. Mortimer, 'Organic electrochromic materials,' *Electrochimica Acta* **1999**, *44*, 2971.
[275] M. A. De Paoli, G. Casalbore-Miceli, E. M. Girotto, W. A. Gazotti, 'All polymeric solid state electrochromic devices,' *Electrochimica Acta* **1999**, *44*, 2983.
[276] M. Mastrogostino, *Applications of Electroactive Polymers.*, 1st edn, Chapman & Hall, London, **1993**.
[277] M. A. De Paoli, W. A. Gazotti, 'Electrochemistry, polymers and optoelectronic devices: a combination with a future,' *Journal of the Brazilian Chemical Society* **2002**, *13*, 410.
[278] H. J. Kim, Y. B. Han, W. N. Kim, E. Kim, 'Electrochronic properties of poly(aniline-N-butylsulfonate)s in contact with solid polymer electrolyte membranes,' *Journal of the Japan Society of Colour Material* **1999**, *72*, 11.
[279] G. Sonmez, I. Schwendeman, P. Schottland, K. W. Zong, J. R. Reynolds, 'N-substituted poly(3,4-propylenedioxypyrrole)s: high gap and low redox potential switching electroactive and electrochromic polymers,' *Macromolecules* **2003**, *36*, 639.

[280] A. Viinikanoja, J. Lukkari, T. Aaritalo, T. Laiho, J. Kankare, 'Phosphonic acid derivatized polythiophene: a building block for metal phosphonate and polyelectrolyte multilayers,' *Langmuir* **2003**, *19*, 2768.
[281] O. Stephan, P. Schottland, P. Y. Le Gall, C. Chevrot, C. Mariet, M. Carrier, 'Electrochemical behaviour of 3,4-ethylenedioxythiophene functionalized by a sulphonate group. Application to the preparation of poly(3,4-ethylenedioxythiophene) having permanent cation-exchange properties,' *Journal of Electroanalytical Chemistry* **1998**, *443*, 217.
[282] S. Basak, C. S. C. Bose, K. Rajeshwar, 'Electrochemical quartz crystal microgravimetry of poly(pyrrole-Co-3-(pyrrol-1-yl)propanesulfonate) films – electrosynthesis, ion-transport, and ion assay,' *Analytical Chemistry* **1992**, *64*, 1813.
[283] Y. J. Qiu, J. R. Reynolds, 'Poly 3,6-(carbaz-9-yl)propanesulfonate – a self-doped polymer with both cation and anion-exchange properties,' *Journal of the Electrochemical Society* **1990**, *137*, 900.
[284] H. Varela, R. M. Torresi, D. A. Buttry, 'Mixed cation and anion transport during redox cycling of a self-doped polyaniline derivative in nonaqueous media,' *Journal of the Electrochemical Society* **2000**, *147*, 4217.
[285] C. Barbero, R. Koetz, 'Electrochemical formation of a self-doped conductive polymer in the absence of a supporting electrolyte. The copolymerization of o-aminobenzenesulfonic acid and aniline,' *Advanced Materials* **1994**, *6*, 577.
[286] C. Barbero, M. C. Miras, B. Schnyder, O. Haas, R. Kotz, 'Sulfonated polyaniline films as cation insertion electrodes for battery applications. 1. Structural and electrochemical characterization,' *Journal of Materials Chemistry* **1994**, *4*, 1775.
[287] M. S. Rahmanifar, M. F. Mousavi, M. Shamsipur, 'Effect of self-doped polyaniline on performance of secondary Zn–polyaniline battery,' *Journal of Power Sources* **2002**, *110*, 229.
[288] R. D. Piner, J. Zhu, F. Xu, S. H. Hong, C. A. Mirkin, ' "Dip-pen" nanolithography,' *Science* **1999**, *283*, 661.
[289] S. H. Hong, J. Zhu, C. A. Mirkin, 'A new tool for studying the *in situ* growth processes for self-assembled monolayers under ambient conditions,' *Langmuir* **1999**, *15*, 7897.
[290] S. H. Hang, C. A. Mirkin, 'A nanoplotter with both parallel and serial writing capabilities,' *Science* **2000**, *288*, 1808.
[291] J. H. Lim, C. A. Mirkin, 'Electrostatically driven dip-pen nanolithography of conducting polymers,' *Advanced Materials* **2002**, *14*, 1474.
[292] B. W. Maynor, S. F. Filocamo, M. W. Grinstaff, J. Liu, 'Direct-writing of polymer nanostructures: poly(thiophene) nanowires on semiconducting and insulating surfaces,' *Journal of the American Chemical Society* **2002**, *124*, 522.

2
Self-Doped Derivatives of Polyaniline

2.1 INTRODUCTION

While polyaniline was first described in the literature in the late nineteenth century [1], it was not until studies in 1910–1912 that the existence of different oxidation states of the polymer was established [2, 3]. Since the mid-1980s, polyaniline has been the most intensively studied among the conducting polymers due to its unique properties, and this focus has resulted in significant development. Polyaniline shows reversible insulator-to-metal transitions and electrochromic behavior (yellow–green–blue–violet), depending on its oxidation state and pH. It has good stability in the presence of air and humidity. The combination of these characteristics makes polyaniline useful for various applications including rechargeable batteries [4–7], light emitting diodes [8, 9], transistors [10, 11], molecular sensors [12], nonlinear optical devices [13], corrosion protection [14–17], electromagnetic interference shielding [18, 19], and electrochromic displays [20–22]. However, limitations such as poor solubility in common solvents and infusibility have been an impediment to its incorporation into industrial applications. Various approaches have been pursued to enhance solubility by the introduction of substituents onto its backbone. Examples include the addition of alkyl, alkoxy, aryl hydroxyl, amino or halogen groups [23–38]. This approach has resulted in improved solubility of the base form in common organic solvents; however, modification also results in lower conductivity and lower molecular weight due to steric effects. Other factors such as water insolubility, pH dependent conductivity and electroactivity have also limited applications of polyaniline.

In order to overcome these limitations, a design strategy to enhance solubility of polyaniline in aqueous solution without adversely impacting conductivity and electroactivity has been developed. In 1990, Yue and Epstein [39] and Dao *et al.* [40] reported the first water soluble conducting derivatives of polyaniline, i.e., 'self-doped sulfonated polyaniline.' In this form, negatively charged sulfonate groups, covalently attached to the polymer backbone, act as intramolecular dopant anions which are able to compensate positive charges at protonated nitrogen atoms on the polymer backbone (Figure 2.1, A), thus replacing auxiliary solution dopant anions. This self-doped polymer is regarded as being created via the initial formation of the strong acid as shown in Figure 2.1, B. Benzenesulfonic acid is a strong acid, which protonates ('dopes') the imine nitrogen atoms to give conducting self-doped polymer in a manner analogous to the protonation of the parent emeraldine base form of polyaniline by HCl. This inner anion doping determines many of the distinctive properties of self-doped polyaniline and distinguishes it from the parent polyaniline [41].

The solubility of polyaniline is greatly enhanced by the presence of $-SO_3^-$ groups along the backbone. In contrast to polyaniline, sulfonated polyaniline is highly soluble in basic aqueous solutions. The good environmental stability of the parent polyaniline is further improved by the presence of $-SO_3^-$ groups on the phenyl rings and the conductivity is independent of pH of values ≤ 7.5. The chemical, electrochemical, electrical and optical properties of the sulfonated polymer differ from those of the parent polyaniline, reflecting the crystallographic, steric and

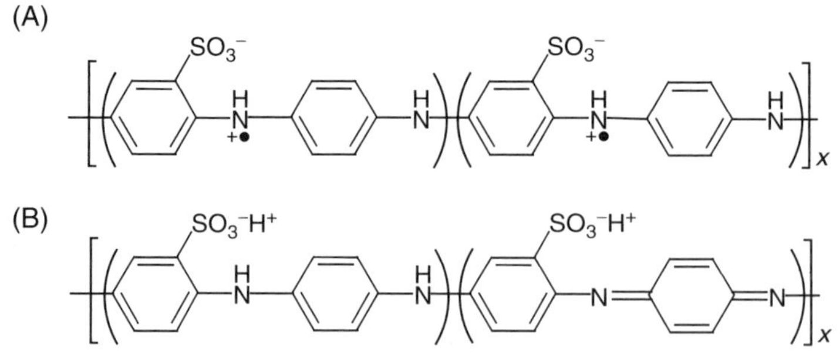

Figure 2.1 Structures of ring-sulfonated polyaniline. (Reprinted with permission from *Journal of the American Chemical Society*, 112, 2800. Copyright (1990) American Chemical Society.)

electronic effects of the $-SO_3^-$ groups. Transport studies have shown that the charge is more localized in sulfonated polyaniline (Figure 2.1, A) compared with the parent polyaniline due to the electron withdrawing effect of the $-SO_3^-$ groups. Since these initial reports, there has been significant research activity directed towards the synthesis, properties and applications of self-doped conducting polyanilines. This chapter summarizes some of the most important achievements in the field of self-doped polyanilines.

2.2 CHEMICAL SYNTHESIS OF SULFONIC ACID DERIVATIVES

Structural modifications of polyaniline have mainly been exploited to achieve improved processability and environmental stability. In general, the substituted polyanilines can be obtained via oxidative polymerization of the corresponding monomer. However, inductive and steric effects can make such monomers difficult to polymerize [42]. Several substituted polyanilines have been prepared by varying the nature (alkyl, alkoxy, halogen, etc.) and the position (2- vs 3-, 5-positions) of the substituent [24, 27–32, 34, 37, 43, 44]. These studies have shown that regardless of the nature and position of the substituent group, there is an adverse effect on polymerization and the properties of the polymer such as conductivity and electroactivity. To overcome these limitations, various synthetic methods have been developed to prepare self-doped sulfonated polyanilines. These methods involve controlled postpolymerization modifications by synthetic reactions on the whole polymer and copolymerization of less reactive monomers with aniline as described below.

2.2.1 Post-Polymerization Modification

2.2.1.1 Electrophilic Aromatic Substitution

Electrophilic substitution is a straightforward way to functionalize polyaniline. Substitution of sulfonic acid groups on the backbone of polyaniline, as shown in Figure 2.2, was first introduced by Epstein et al. [39] in the very first report of self-doped water soluble polyaniline. Their synthetic method involved the sulfonation of polyaniline using fuming sulfuric acid. The emeraldine base form of polyaniline (0.5 g) was dissolved in 40 mL of fuming sulfuric acid with constant stirring.

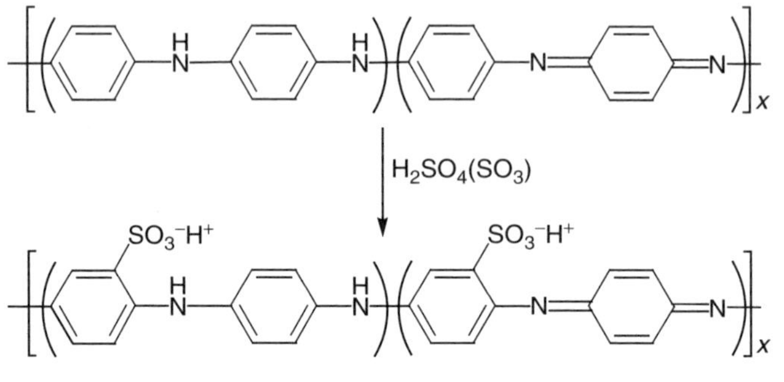

Figure 2.2 Electrophilic aromatic substitution. (Reprinted from *Polymer*, 33, J. Yue, G. Gordan, A. J. Epstein, 4410. Copyright (1992) with permission from Elsevier.)

The color of the solution changed from dark purple to dark blue over a period of 2 h at room temperature. The polymer was precipitated in methanol at a temperature in the range of 10 to 20 °C in an ice bath. Precipitation was completed by the addition of 100 mL of acetone. The green precipitate was then collected and was washed thoroughly with methanol until the filtrate had a pH value of 7 when tested with wet pH paper. The resulting precipitate, sulfonated polyaniline, was only slightly soluble in water. However, the polymer became highly soluble in water after its conversion to the salt form after dedoping upon exposure to a basic aqueous solution. Based on elemental analysis, approximately 50 % of the total number of phenyl rings in the polymer were substituted with sulfonic acid groups. This procedure has been extensively used by others to obtain sulfonated polyaniline [45–53]. A higher degree of sulfonation, up to 75 %, has been obtained for electrophilic substitution reactions carried out using the leucoemeraldine base form of the polymer [54]. The dried leucoemeraldine base was sulfonated in 10 mL of fuming sulfuric acid (precooled to 5 °C) for 1 h. The reaction mixture was subsequently introduced into 0.75 L of a 75:25 ice–water mixture to precipitate the sulfonated leucoemeraldine form of polyaniline. The polymer was then washed with cold water and dried at room temperature in a vacuum oven. A yield of 70 % was obtained using this method. Epstein *et al.* suggested that the higher electron density on the phenyl rings in the leucoemeraldine base form resulted in a higher sulfonation reaction rate, S/N ratio and yield. When the emeraldine base and pernigraniline base forms of polyaniline are dissolved in fuming sulfuric acid, the nitrogen atoms at quinoid sites are protonated, causing the positive charges to delocalize into the quinoid ring units due to conjugation of the nitrogen p_z-orbital with the C6 ring's π-orbitals. Therefore, the positive charge resonates between the protonated quinoid ring and nitrogen. The protonated imine repeat units are thus deactivated for the subsequent electrophilic aromatic substitution reaction, i.e., the sulfonation reaction. This may be the cause for the emeraldine base and pernigraniline base forms yielding S/N ratios of no more than 0.5 [54].

Unfortunately, the procedure for sulfonation of polyaniline using fuming sulfuric acid requires handling of toxic and corrosive reactants, making it less than ideal and difficult to scale up. In addition, slow dissolution of the polyaniline in the concentrated sulfuric acid is observed and likely results from concurrent reaction with SO_3, resulting in complicated reaction patterns including homogeneous and heterogeneous sulfonation, and multiple sulfonation. The process of sulfonation of polyaniline is highly sensitive to the reaction conditions, which can lead to a wide variety of structures. For example, different degrees of backbone and phenyl ring sulfonation as well as the distributions of sulfonic acid groups for a given average sulfonation degree are observed [55]. Finally, chain scission by hydrolysis occurs during sulfonation resulting in a decrease in the molecular weight of the polymer produced [55, 56]. Careful control of the reaction conditions such as sulfonation time, starting from of the polyaniline, and temperature was proposed by Epstein *et al.* [55] to minimize the hydrolysis. In order to overcome these limitations, alternative synthetic approaches have been pursued.

Sulfonated polyaniline has also been prepared via postpolymerization treatment with chlorosulfonic acid [57] as well as sulfur trioxide/triethyl phosphate complex [55]. The direct sulfonation of the emeraldine salt form with chlorosulfonic acid in 1,2-dichloroethane at 80 °C resulted in the production of HCl-doped sulfonated polyaniline as shown in Figure 2.3. In this case, the HCl dopant from hydrolysis of the chlorosulfonic group exchanges with the original dopant. Because the $-SO_3H$ group is not involved in self-doping, the polymer is soluble in pure water up to 88 mg/mL. This method results in a high degree of sulfonation (S/N ratio 1.3), however the externally (HCl) doped form of the sulfonated polymer exhibits very poor conductivity of 1.7×10^{-5} S/cm.

2.2.1.2 Nucleophilic Substitution

An alternative route to modify polyaniline involves nucleophilic substitution. In 1989, Genies *et al.* [58] synthesized self-doped polyaniline using the reaction of the emeraldine base with propanesultone as well as butanesultone. The proposed mechanism involves nucleophilic attack

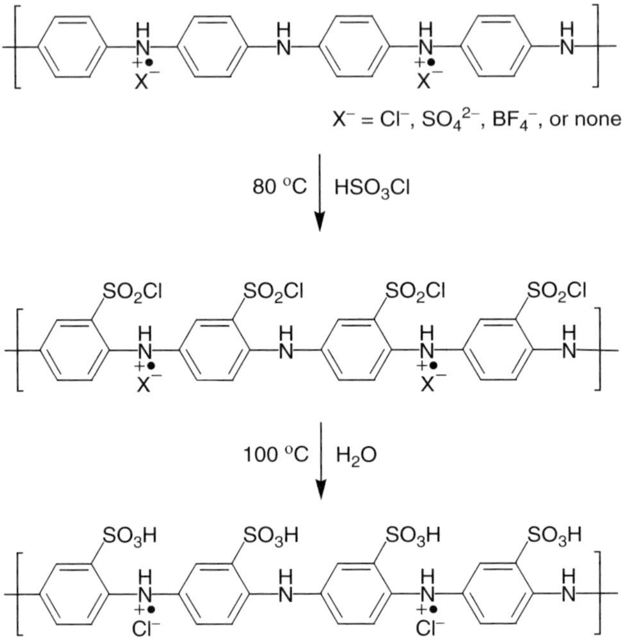

Figure 2.3 Synthesis of externally doped sulfonated polyaniline. (Reprinted from *Synthetic Metals*, 96, S. Ito, K. Murata, S. Teshima, R. Aizawa, Y. Asako, K. Takahaski, B. M. Hoffman 161. Copyright (1998), with permission from Elsevier.)

by the nitrogen lone pair electrons on the C–O bonding carbon of the sultone as shown in Figure 2.4. The product of this reaction exhibited very low water solubility and conductivity (10^{-9} S/cm). Following this report, Dao *et al.* [40] prepared a water-soluble conducting poly(aniline propanesulfonic acid) by derivatization of the leucoemeraldine form of polyaniline. The leucoemeraldine base was reacted with 1.5 equivalents per unit of NaH in dimethyl sulfoxide at 40–50 °C under argon for 4 h. The resulted green viscous solution was subsequently reacted with 1.5 equivalents of 1,3-propanesultone for 18 h at room temperature. The sodium salt of poly(aniline propanesulfonate) was isolated by precipitation into tetrahydrofuran, followed by extensive washing and vacuum drying. Poly(aniline propanesulfonic acid) was obtained by exchanging the sodium salt of the polymer through a Dowex 50 W-X8-H$^+$ column. Poly(aniline propanesulfonic acid) and its sodium salt are reportedly highly soluble in water. This was the first report where sulfonated polyaniline was soluble in the conducting form. The existence of the

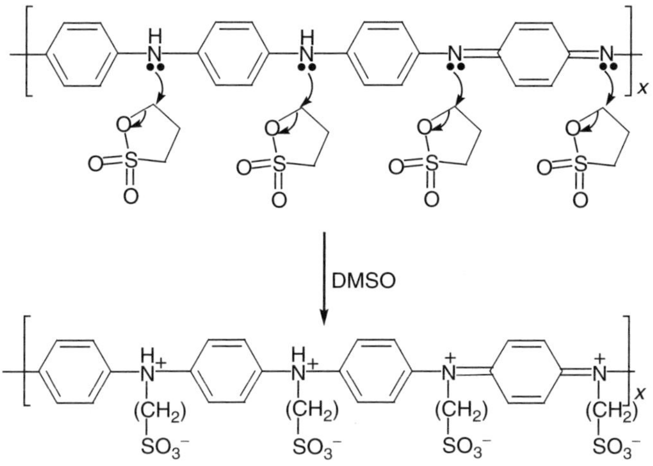

Figure 2.4 The mechanism of reaction of polyemeraldine with propanesultone. (Reprinted from *Synthetic Metals*, **31**, P. Hany, E. M. Genies, C. Santier, 369. Copyright (1989), with permission from Elsevier.)

conducting form of the polymer was inferred from the absorption band at about 900 nm, which appears for protonic-acid doped polyaniline.

In 1994, Chen *et al*. [59] synthesized a water-soluble self-acid-doped polyaniline, poly(aniline-co-N-propanesulfonic acid-aniline) and its sodium salt (Figure 2.5). In this case the dried emeraldine base form of polyaniline was reacted with NaH in dried dimethyl sulfoxide at around 45 °C under dry nitrogen for 6 h to obtain a green-black solution. The authors suggested that in this process the original emeraldine base was probably converted to an emeraldine base without H on nitrogen (anionic nitrogen). Subsequently, the solution was reacted with 1,3-propanesultone for 20 h at room temperature. The resulting solution was precipitated with 1 M HCl aqueous solution to obtain HCl doped poly(aniline-co-N-propanesulfonic acid-aniline) and was washed with a large amount of acetonitrile. The sodium salt of the polymer was prepared by dedoping it in an NaOH aqueous solution. The conductivity of the acid form of polymer was 1.5×10^{-2} S/cm without external doping. Electron spin resonance results confirmed the existence of polarons in both the aqueous solution and solid films of poly(aniline-co-N-propanesulfonic acid-aniline). It was suggested that the alkylsulfonic acid bound on the side chain, attaches to the amine nitrogen, protonates ('dopes') the imine nitrogen atom to make the

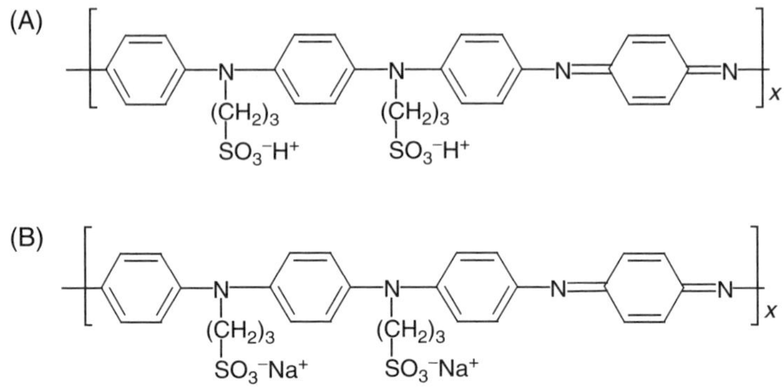

Figure 2.5 Structures of poly(aniline-co-N-propanesulfonic acid-aniline) (A) and its sodium salt (B) (Reprinted with permission from *Journal of the American Chemical Society*, **116**, 7939. Copyright (1994) American Chemical Society.)

polymer conducting. The drawback of the polymer was reportedly the insolubility in water after drying. In order to increase the thermal stability of poly(aniline-co-N-propanesulfonic acid-aniline), a benzene ring was introduced between the $-SO_3H$ and $-(CH_2)_3-$ groups attached to the amine nitrogens in the polyaniline to prepare poly(aniline-co-N-propylbenzenesulfonic acid-aniline) [60]. In a subsequent study, Chen *et al.* [61] prepared a water-soluble self-doped polyaniline derivative, poly(aniline-co-N-benzoylsulfonic aniline) by direct reaction of polyaniline with o-sulfobenzoic anhydride (Figure 2.6). This polymer is reportedly completely water soluble and can be redissolved after drying. The self-doped polymer had a conductivity of 4.7×10^{-4} S/cm.

The postpolymerization modification of polyaniline by addition of nucleophiles such as amines, thiols, cyanide, carbanions, arylsulfinate and sulfite has been successfully carried out by Han and Barbero *et al.* [42, 62–64]. The addition of nucleophiles to quinoimine units present in the oxidized polyaniline is shown in Figure 2.7 [42]. This nucleophilic substitution is proposed as a powerful way to obtain modified polyaniline linked with different moieties. The modification reportedly changes the properties of polyaniline and increases the solubility in common organic solvents. Sulfonated polyaniline was prepared by the reaction of SO_3^{2-}, a strong nucleophile, with the emeraldine or pernigraniline forms of polyaniline in aqueous media. The different forms of polyaniline were stirred with a saturated solution of SO_3^{2-} and HSO_3^- at 80 °C for 2 h, and the final product was washed with water and 1 M HCl. The degree of sulfonation was reported to be up to 68 % obtained using the highly oxidized pernigraniline form [65]. The polymer is reportedly

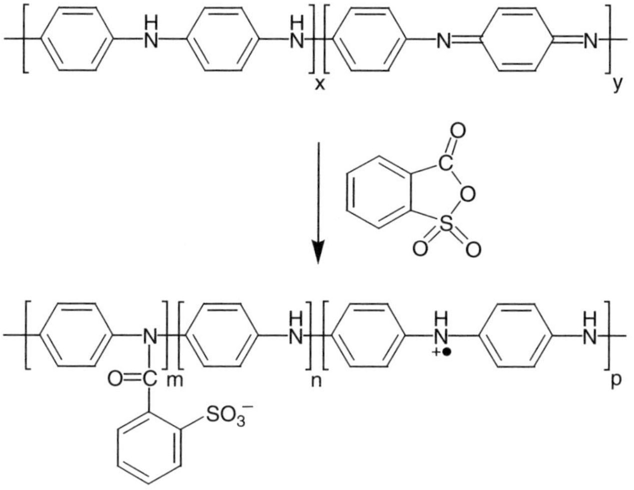

Figure 2.6 Reaction mechanism of polyaniline with o-sulfobenzoic anhydride (Reprinted with permission from *Macromolecules*, 33, 8117. Copyright (2000) American Chemical Society.)

soluble in aqueous base. According to Barbero *et al.*, the nucleophilic addition method resulted in minimal degradation in comparison with electrophilic substitution using fuming sulfuric acid [55]. A minimal chain length alteration was reported for reactions involving sulfite ($\eta = 1.14$) or bisulfite ($\eta = 1.1$) with emeraldine base compared to 50 % ring sulfonated polyaniline prepared via electrophilic substitution ($\eta = 0.9$).

Han *et al.* [64, 66] reported the synthesis of highly conductive and thermally stable self-doped mercaptopropanesulfonic-acid-substituted polyanilines by the concurrent reduction and substitution reaction between polyaniline and a nucleophile. These reactions were carried out on both electrochemically generated and free standing polyaniline films prepared from emeraldine base dissolved in N-methylpyrrolidinone. The electrochemically prepared films were dedoped with 5 % aqueous Na_2CO_3 to convert them the into the emeraldine base form. The sulfonated polyaniline was prepared by reaction of a polyaniline emeraldine base film with 0.1 M 3-mercapto-1-propanesulfonic acid sodium salt in methanol under nitrogen at room temperature for approximately 14 h [66]. A catalytic amount (0.01 M) of acetic acid was reported to accelerate the reaction. The resulting sulfonated polyaniline film was thoroughly rinsed with methanol, followed by 5 % aqueous Na_2CO_3 to remove reactants.

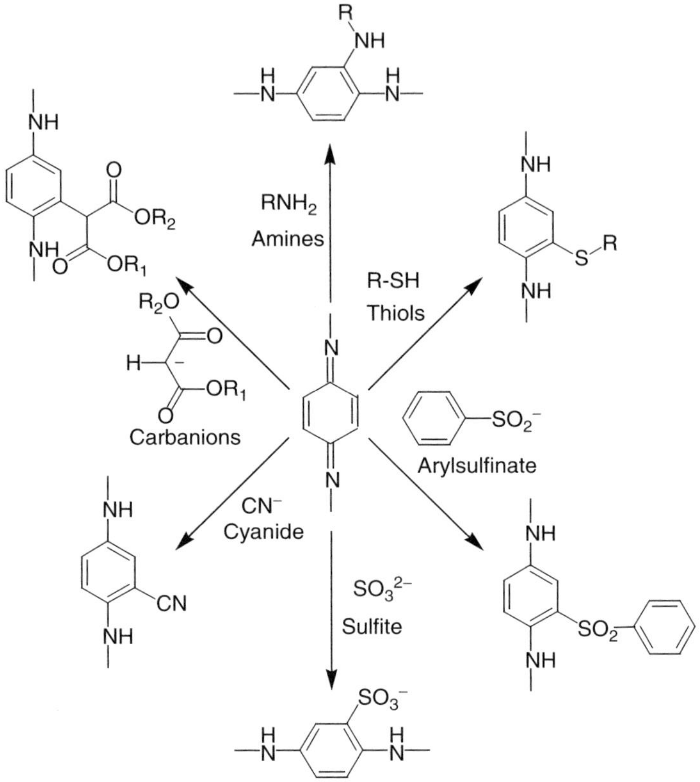

Figure 2.7 Reaction pathways of nucleophilic addition to polyaniline. (Reprinted from *Electrochimica Acta*, **49**, C. Barbero, H. J. Salavagione, D. F. Acevedo, D. E. Grumelli, F. Garay, G. A. Planes, G. M. Morales, M. C. Miras, 3671. Copyright (2004), with permission from Elsevier.)

2.2.1.3 Coupling of Diazonium Salts with Polyaniline

Reactions involving arenediazonium ions play an important role in synthetic organic chemistry [67]. Because of the reactivity of arene-diazonium ions with nucleophiles [67] and the high concentration of diarylamines in the polyaniline backbone, it is possible to introduce a broad range of functional groups into polyaniline using substituted arenediazonium ions. The coupling of diazonium salts with polyaniline was first reported by Freund *et al.* [68]. The reaction of diazonium salts with electrochemically reduced polyaniline was carried out in acidic media. Reduction of the polyaniline film was required since no reaction was reported with the oxidized emeraldine form under these conditions. In this reaction the diazonium ion loses molecular nitrogen and substitutes on the amine nitrogen of the polyaniline backbone, resulting in an electroinactive polymer. In a different approach, Barbero *et al.* [69] prepared electroactive poly(4-sulfobenzeneazo-(N-methylaniline) by reacting 4-sulfobenzenediazonium ion with poly(N-methylaniline) at low temperature in basic media (Figure 2.8). They suggested that electrophilic attack is favourable as poly(N-methylaniline) is less oxidized (14 % positive charge in emeraldine state) compared with polyaniline (50 % positive charge in emeraldine state). Also, the presence of the methyl group probably activates the ring for electrophilic attack. The polymer obtained was reported to be soluble in aqueous basic media, similar to other forms of sulfonated polyaniline. The reported degree of substitution was 26 %, however, the self-doping in this polymer is less favourable due to the relative remoteness of anionic -SO_3^- groups relative to the amino group.

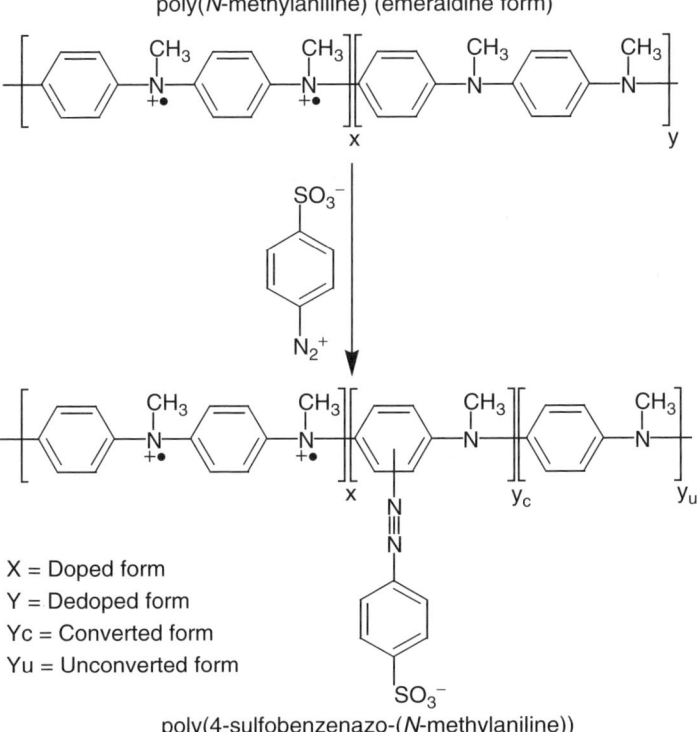

Figure 2.8 Reaction mechanism of diazonium coupling in poly(N-methylaniline). (Reprinted from *Synthetic Metals*, 97, G. A. Planes, G. M. Morales, M. C. Miras, C. Barbero, 223. Copyright (1998), with permission from Elsevier.)

2.2.2 Polymerization of Monomers

2.2.2.1 Homopolymerization

Sulfonated polyanilines are typically synthesized by postpolymerization treatment of polyaniline. The low reactivity of sulfonic acid substituted aniline to oxidation and polymerization is due to the presence of the electron withdrawing sulfonic acid group makes it less suitable for homopolymerization. However, Yano, Maruyama and Watanabe *et al.* [70] have prepared self-doped fully sulfonated (100 %) conducting polyaniline (Figure 2.9) using the monomer 2-methoxyaniline-5-sulfonic acid. The reactivity of aniline sulfonic acid was probably improved by the introduction of the electron donating methoxy group. The polymer was chemically prepared using ammonium peroxydisulfate under basic conditions in an aqueous pyridine solution at 4 °C with a yield of 96 %. While the polymer showed good water solubility, the conductivity and molecular weight were approximately 0.04 S/cm and 10 000 g/mol, respectively. The low conductivity of methoxy sulfonated polyaniline relative to postpolymerized sulfonated polyaniline has been attributed to the larger twist of the phenyl rings associated with steric effects and increased interchain separation [71]. Yong *et al.* [72] prepared self-doped, fully sulfonated water-soluble polyaniline at high pressure using *o*- and *m*-aminobenzesulfonic acid. These reactions were catalyzed with 5 mol% Co^{2+}. This reaction was not possible at ambient pressure, presumably because of the inhibiting influence of the electron-withdrawing sulfonic acid group on oxidation of the aromatic amine. At high pressure, polymerization was reported to be possible due to the enhancement of aniline oxidation rate. The low conductivity of the polymer (10^{-3}–10^{-4} S/cm) relative to methoxy substituted fully sulfonated polyaniline [70] was attributed to 'pressure induced defects,' although the nature of these defects was not explored.

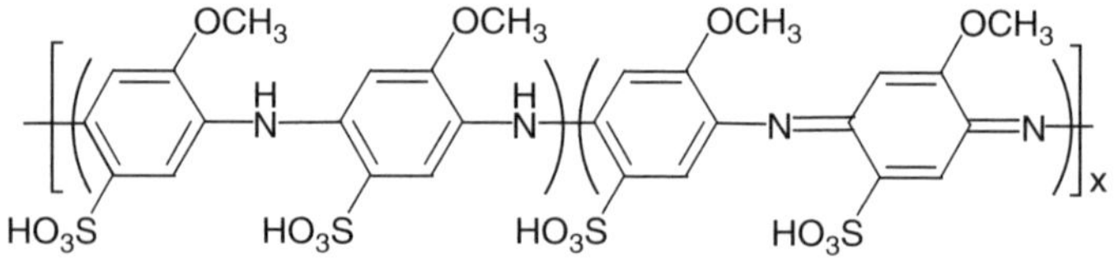

Figure 2.9 Structure of poly(2-methoxyaniline-5-sulfonic acid). (Reprinted from *Synthetic Metals*, 85, S. Shimizu, T. Saitoh, M. Uzawa, M. Yuasa, K. Yano, T. Maruyama, K. Watanabe, 1337. Copyright (1997), with permission from Elsevier.)

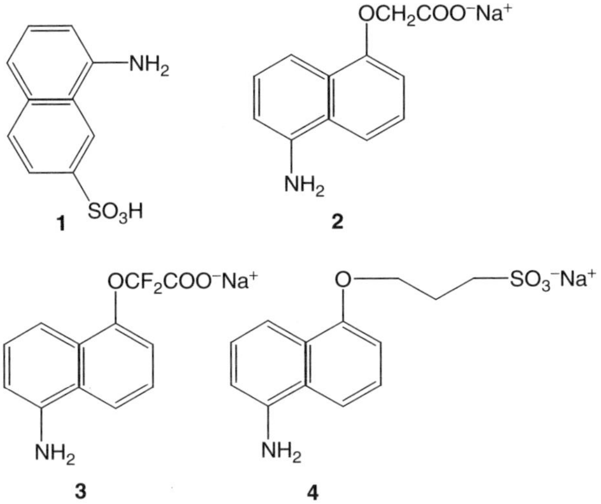

Figure 2.10 Structures of monomers: (1) 5-aminonaphthalene-2-sulfonic acid, (2) 5-amino-1-naphthyloxy acetic acid, (3) 5-amino-1-naphthyloxydifluro acetic acid, (4) 3-(5-aminonaphthyloxy)1-propanesulfonic acid). (Reprinted from *Polymer*, 43, V. George, D. J. Young, 4073. Copyright (2002), with permission from Elsevier.)

Armes et al. [73] chemically polymerized the water-soluble sodium salt of poly(diphenylamine-*p*-sulfonate). Half of the sulfonic acid groups in the polymer are reportedly neutralized to the Na$^+$ salt. The aqueous solution of this polymer undergoes a color change from red to green when the pH is varied from 6 to 4. The electrochemical response of the polymer in acetonitrile was poor and could not be measured in water. The chemical homopolymerization of substituted aminonaphthylenes (Figure 2.10) was carried at by Young et al. [74, 75]. These polymers are reported to be self-doped, water soluble and fluorescent. The yield of poly(5-aminonaphthalene-2-sulfonic acid) prepared using oxidant KIO$_4$ was high, approximately 95 %. In the case of poly (1–4) (Figure 2.10) prepared using ammonium persulfate, the yield was in the rage of 30–56 %. The electrical conductivity of polymer was low in the range of 10^{-4} to 10^{-5} S/cm.

2.2.2.2 Copolymerization

In general, copolymerization provides an additional variable by which the properties of a polymer can be tuned. Copolymerization of aniline with ring substituted aniline derivatives has been well studied as a means to improve conductivity and solubility of polyaniline for a variety of applications [76–85]. There are very few sulfonated monomers like 2-methoxyaniline-5-sulfonic acid which can be homopolymerized. For example, the homopolymerization of alkoxy-sulfonated aniline 1-(2-amino)-pronane-3-sulfonic acid or 1-(2-amino)-butane-3-sulfonic acid does not take place, possibly due to the bulkiness of the substituent [86]. As mentioned above, it is difficult to control the degree of sulfonation using postpolymerization methods; therefore, chemical copolymerization of aniline and sulfonated aniline derivatives such as alkoxy sulfonate [86], diaminodiphenylsulfone [87], sodium diphenylamine-4-sulfonate [88], amino-naphthalene sulfonic acids [89], aminobenzenesulfonic acid [90–94], and aminobenzoic acid [95, 96] have been carried out under mild conditions. Neoh *et al.* [97] copolymerized metanilic acid (aminobenzenesulfonic acid) and aniline, where the extent of self-doping was tuned through the choice of metanilic acid/aniline and oxidant/monomer ratios. Prévost *et al.* [86] prepared water soluble self-doped copolymers of *o*-alkoxysulfonated anilines, i.e., 1-(2-amino)-pronane-3-sulfonic acid or 1-(2-amino)-butane-3-sulfonic acid with aniline. The conductivity values were in the range of 10^{-3} to 1 S/cm depending on both self- and external doping.

2.2.2.3 Graft and Diblock Copolymerization

Graft and diblock copolymerization are other approaches that can be used to improve the physical and mechanical properties as well as the solubility in common organic solvents. Grafting of conducting polymer chains like polypyrrole [98–105] and polyaniline [106–109] onto a nonconductive polymer has been carried out. In 1992, Ikada *et al.* [110] first prepared a self-doped polyaniline film by graft copolymerization. In this study, surface modification of a polyaniline film was achieved by graft copolymerization with acrylamide, acrylic acid and 4-styrenesulfonic acid. The aim of this study was to improve hydrophilicity of the polyaniline film surface and provide appropriate functional groups for protein and enzyme immobilization. In this work, the authors suggested that a self-doped or self-protonated polyaniline film was formed as a result of grafting with acrylic acid and 4-styrenesulfonic acid due to the presence of carboxylic acid and sulfonic acid groups, respectively. However, this modification most likely would not result a self-doped polymer in the bulk. In 2000, Ruckenstein *et al.* [111] prepared

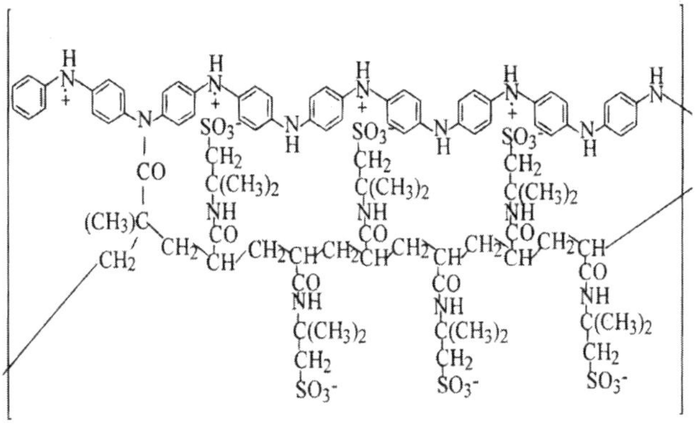

CHEMICAL SYNTHESIS OF SULFONIC ACID DERIVATIVES 89

Figure 2.11 Structure of poly(aniline-co-2-acrylamido-2-methyl-1-propanesulfonic acid). (Reprinted with permission from *Macromolecules*, 33, 1129. Copyright (2000) American Chemical Society.)

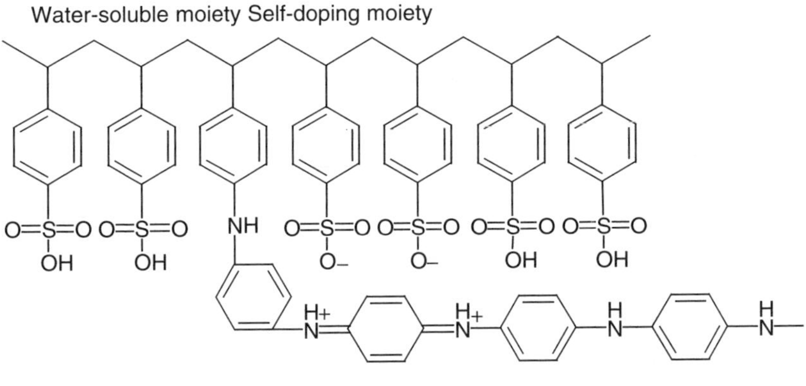

Figure 2.12 Structure of self-doped graft copolymer poly(styrenesulfonic acid-g-aniline). (*Chemical Communications*, 2003, 2768, W. J. Bae, K. H. Kim, Y. H. Park, W. H. Jo, reproduced by permission of the Royal Society of Chemistry.)

water-soluble self-doped conducting graft copolymer poly(aniline-co-2-acrylamido-2-methyl-1-propanesulfonic acid) (Figure 2.11) and its salt. The graft copolymerization of aniline onto poly(2-acrylamido-2-methyl-1-propanesulfonic acid) was done chemically, and the polymer showed a conductivity of around 0.8 S/cm following drying under vacuum at 40 °C. Park and Jo et al. [112] prepared a self-doped conducting polyaniline graft copolymer, poly(styrenesulfonic acid-g-aniline) (Figure 2.12), that

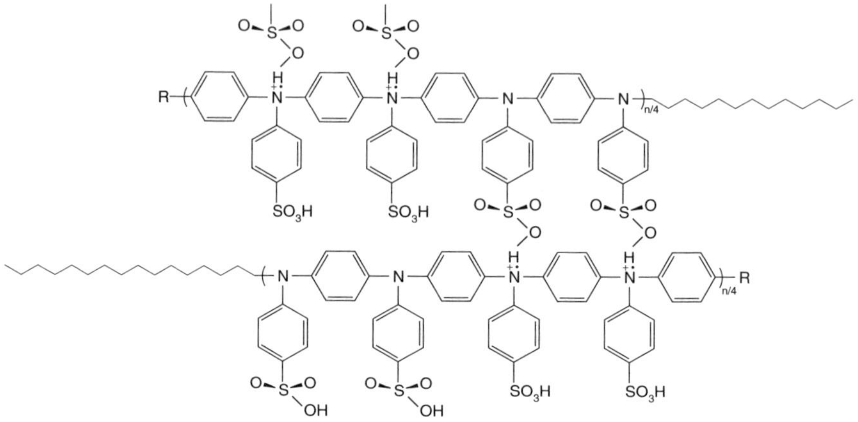

Figure 2.13 Proposed self-doping aggregation process for self-doped diblock copolymer. (Reproduced from *Journal of Polymer Science*, 2004, F. Hua, E. J. Ruckenstein, **42**, 2179, reprinted with permission of John Wiley & Sons, Inc.)

was soluble in both water and polar organic solvents. The conductivity of copolymer was approximately 10^{-1} S/cm.

Ruckenstein *et al.* also prepared a water-soluble diblock copolymer of polysulfonic diphenyl aniline and poly(ethylene oxide) [113] with various poly(ethylene oxide) segment lengths (number average molecular weight 350 and 2000 g/mol). These copolymers had low conductivities ranging from 10^{-6} to 10^{-3} S/cm, due to the electron withdrawing effect of the sulfonyl group as well as the steric effects of the bulky aromatic substituents at the N sites of the polyaniline backbone and of the poly(ethylene oxide) block. In these copolymers intermolecular self-doping was suggested. The aggregation in the polysulfonic diphenyl aniline backbone (Figure 2.13) in water due to self-doping resulted into formation of nanoscale rod-like aggregates. These diblock copolymers show temperature dependent conductivity, exhibiting an order of magnitude increase in conductivity upon heating from 32 to 57 °C.

In a subsequent study, Ruckenstein *et al.* prepared densely grafted copolymer poly(aniline-2-sulfonic acid-co-aniline) [114]. Water solubility of the copolymer increased with the molar ratio of aniline-2-sulfonic acid to aniline in the polymerization mixture. Also, in this study it was suggested that aggregation occurred due to intramolecular and intermolecular self-doping of the copolymer poly(aniline-2-sulfonic

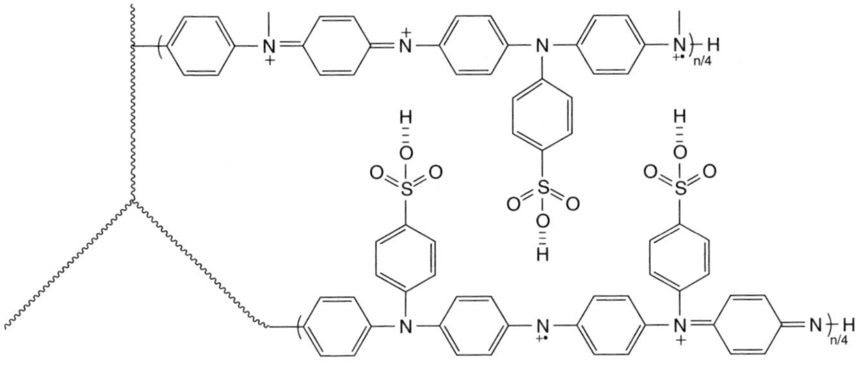

CHEMICAL SYNTHESIS OF SULFONIC ACID DERIVATIVES

Intramolecular self-doping

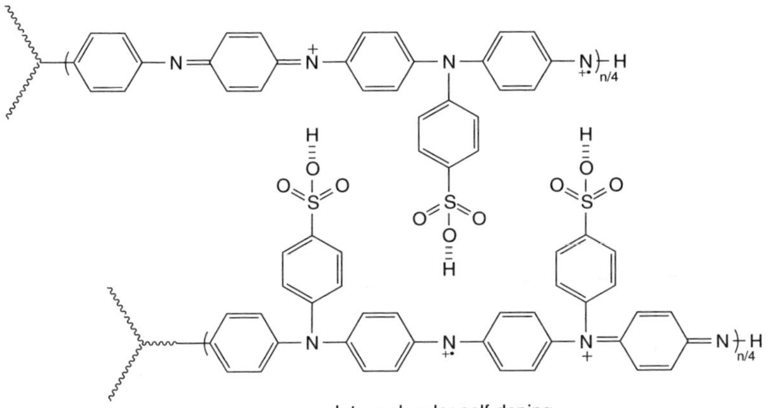

Intermolecular self-doping

Figure 2.14 Self-doped mechanism of hyperbranched sulfonated polydiphenylamine. (Reprinted with permission from *Macromolecules* 38, 888. Copyright (2005) American Chemical Society.)

acid-co-aniline) in water. Initially, spherical particles with sizes around 200 nm were reported to be formed. These particles combined into larger particles and finally formed numerous micrometer structures after 7 days. Ruckenstein *et al.* [115] have also prepared water-soluble self-doped hyperbranched sulfonated polydiphenylamine. The conductivity in the self-doped state was reportedly 1.2×10^{-2} S/cm. Similar to the above mentioned copolymers, intermolecular self-doping as shown in Figure 2.14 reportedly induces aggregation. The morphology of hyperbranched sulfonated polydiphenylamine is also reportedly dependent on pH (see section 2.5.7).

2.3 ELECTROCHEMICAL SYNTHESIS OF SULFONIC ACID DERIVATIVES

Apart from chemical synthesis, electrochemical synthesis is a versatile method for the preparation of self-doped sulfonated polyaniline in both soluble forms and thin films deposited on an electrode surface. Electropolymerization of sulfonated polyaniline homopolymers and copolymers has been carried out in aqueous and nonaqueous media. Similar to chemical synthesis, electrochemically controlled electrophilic and nucleophilic substitution reactions are also reported.

2.3.1 Aqueous Media

2.3.1.1 *Homopolymer*

Sulfonated anilines have for the most part been electropolymerized as copolymers with aniline. The low reactivity of sulfonic-acid-substituted aniline, due to the presence of electron attractive sulfonic acid, is less suitable for oxidative polymerization. There are very few reports on the homopolymerization of sulfonic aniline derivatives like aminobenzesulfonic acid [116], aniline-2,5-disulfonic acid [117, 118], 2-methoxyaniline-5-sulfonic acid [119, 120] and N-substituted aniline monomers such as 3-anilino-1-propanesulfonic acid [121]. Anodic electropolymerization of these monomers results in soluble self-doped polymers that do not form a stable polymer film at the electrode surface. In most of the cases, the polymer is precipitated from solution by adding alcohols and acetone. The isolated yield of *ortho-*, *meta-* and *para*-poly(aminobenzenesulfonic acid) prepared in 0.5 M H_2SO_4 was reported to be 92, 88 and 55 %, respectively, with molecular weights considerably higher than other homopolymers, of the order of 10 000 to 100 000 g/mol regardless of the position of sulfonic acid [116]. These polymers were also reportedly soluble in water (30–40 mg/mL) regardless of solution pH, and the self-doped conductivity of the polymers were in the region of 10^{-3} S/cm. Poly(aniline-2,5-disulfonic acid) [117, 118] was reported to be soluble in water up to 0.1 mg/mL. The polymer exhibited pH-independent conductivity (0.34 S/cm) up to pH 9. The molecular weight of the polymer was found to be distributed between 3000 to 22 000 g/mol. Wallace *et al.* electropolymerized poly(2-methoxyaniline-5-sulfonic acid) [120] using an electrohydrodynamic processing method. They have shown that both the applied potential and the flow rate employed have a significant impact on the molecular weight and the molecular weight distribution obtained. In addition, the solution pH was shown to affect the oxidation and subsequent polymerization of 2-methoxyaniline-5-sulfonic acid. The conductivity of the polymer was reported to be approximately 0.01 S/cm with or without controlling solution pH during polymerization [119]. A bimodal molecular weight distribution with a high molecular weight fraction (average M_w ~15 000 g/mol) and a low molecular weight (M_w ~2 200 g/mol) was reported.

Kane-Maguire and Wallace et al. [122] were first to report on self-doped, optically active poly(2-methoxyaniline-5-sulfonic acid). They electropolymerized 2-methoxyaniline-5-sulfonic acid in the presence of (R)-(+)- or (S)-(−)-1-phenylethylamine. The poly(2-methoxyaniline-5-sulfonic acid) films formed using enantiomers of amines exhibit intense mirror image circular dichroism spectra (Figure 2.15) in the visible region. Wallace et al. suggest that the optical activity in poly(2-methoxyaniline-5-sulfonic acid) is probably due to electrostatic binding

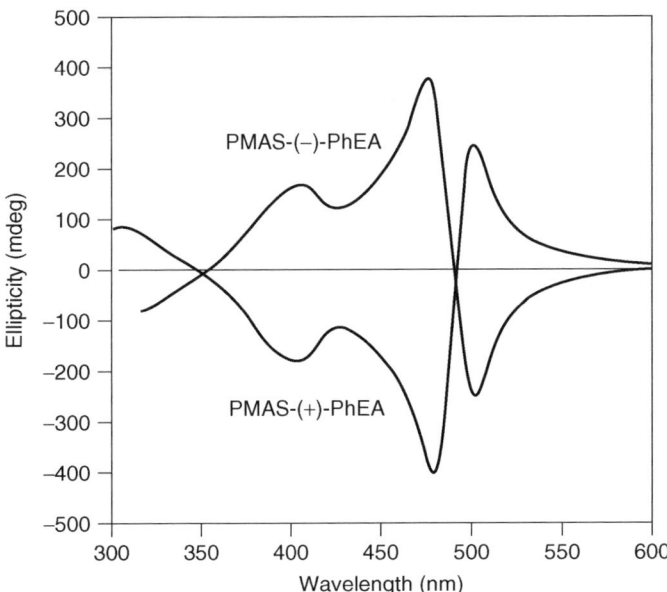

Figure 2.15 Circular dichroism spectra of poly(2-methoxyaniline-5-sulfonic acid) films potentiodynamically deposited on indium tin oxide glass from a 0.1 M aqueous solution of 2-methoxyaniline-5-sulfonic acid and 0.1 M (R)-(+)- or (S)-(−)-1-phenylethylamine, respectively. (Reprinted from *Synthetic Metals*, 106, E. V. Strounina, L. A. P. Kane-Maguire, G. G. Wallace, 129. Copyright (1999), with permission from Elsevier.)

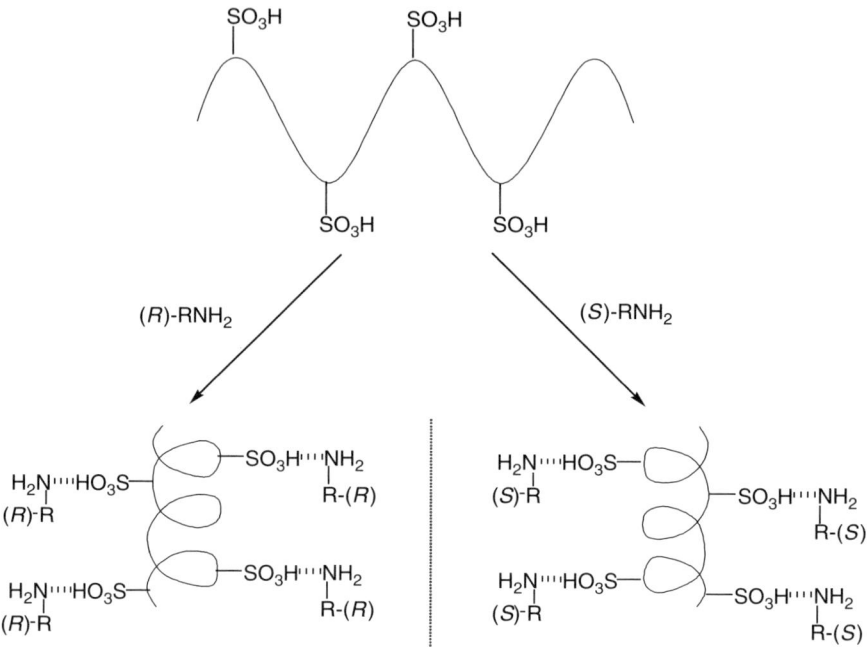

Figure 2.16 Electrostatic interaction of chiral ammonium ion of phenylethylamine to ionized sulfonate groups along the polymer backbone. (Reprinted from *Synthetic Metals*, 106, E. V. Strounina, L. A. P. Kane-Maguire, G. G. Wallace, 129. Copyright (1999), with permission from Elsevier.)

of the chiral ammonium ion of phenylethylamine to ionized sulfonate groups along the polymer backbone as shown in Figure 2.16. The chiral amines induced a preferred one-handed helical structure to the sulfonated polyaniline chains via acid-base interactions. The molecular weight of polymer after dissolution in water was reported to be 7400 g/mol and conductivity of dried pellet around 10^{-2} S/cm.

2.3.1.2 Copolymer

The method of electrochemical copolymerization has many advantages including:

(i) morphological modifications can be accomplished by adjusting electrodeposition conditions (method of deposition, substrate, scan rate, and potential range),
(ii) highly controlled deposition of polymer that results in well defined thin films ready for characterization and device development.

Electrochemical copolymerization of aniline with aminobenzenesulfonic acid (metanilic acid) was first carried out by Lee et al. [123] in HClO$_4$ using 0.01 M aniline and 0.1 M metanilic acid. It was found that oxidation of metanilic acid would not produce significant polymerization on its own, but rather only purple oligomers near the electrode surface. The addition of a small amount of aniline was necessary to obtain copolymer films. A 40 % level of self-doping was achieved using this approach. The level of self-doping could be increased up to 50 % by adjusting the ratio of aniline/metanilic acid in H$_2$SO$_4$ [124]. It has been reported that the growth rate of aniline-metanilic acid copolymer is much lower than conventional polyaniline due to presence of the electron withdrawing SO$_3$H group in metanilic acid, which greatly reduces its reactivity and increases steric hindrance. During the copolymerization, a higher number of aniline monomers were incorporated at the initial stage and the incorporation of metanilic acid monomer increased in the later stages of polymer growth [125]. The inhibition or steric effect of metanilic acid was observed even at an aniline/metanilic acid ratio of one. In contrast, for the copolymerization of 2,5-diaminobenzesulfonic acid with aniline, addition of 2,5-diaminobenzesulfonic acid increased the growth rate of the polymer [126], reportedly due to the autocatalytic effect of the two terminal aromatic amine groups as in *p*-phenylenediamine [127, 128]. However, the rate of copolymer growth decreased with increasing amounts of 2,5-diaminobenzesulfonic acid due to the inhibition or steric effects of sulfonate groups. There are several reports on the optimization of copolymerization of aniline with metanilic acid using different compositions and in different electrolytes as well as their characterization [129–134]. The self-doped copolymer obtained from the mixture of aniline and metanilic acid is electroactive up to pH 12 [129]. There are several reports on the copolymerization of aniline with various sulfonated aniline monomers. The conductivity of copolymer of aniline with 2,5-disulfonic acid is approximately 10^{-1} S/cm and independent of pH up to 9 [117, 118]. The copolymers of aminonaphthalene-sulfonic acid with aniline are redox active in neutral pH medium, however, conductivity of the copolymers was not reported [135–139]. Kotz et al. [140] suggested that a self-doped conducting polymer produced by electrochemical copolymerization of *o*-aminobenzenesulfonic acid and aniline could be prepared in the absence of a supporting electrolyte. The copolymer was prepared by potential cycling in the range of -0.2 to 0.8 V vs SCE at a glassy carbon electrode, using 0.05 M *o*-aminobenzenesulfonic acid and 0.001 M aniline solution in water. They suggest that in the presence of supporting electrolyte, the concentration of aniline cannot be reduced below 0.01 M. As per their proposed mechanism of polymerization without supporting electrolyte, radical cation of aniline is formed first and subsequently attacks the o-aminobenzenesulfonic acid to produce dimers. These dimers are then more easily oxidized than parent monomer and react with other monomers to form polymer.

2.3.1.3 Nucleophilic Substitution

Reaction of nucleophiles with quinoimine rings in polyaniline is discussed in section 2.2.1.2. Since these nucleophiles were found only to attack the quinoimine rings of emeraldine base, it should be possible to control the substitution reaction electrochemically by controlling the oxidation state of polyaniline. Recently, Calvo *et al.* [141] prepared sulfonated polyaniline films electrochemically by nucleophilic addition of sulfite ion controlled through the polymer oxidation state. The mechanism of nucleophilic addition of sulfite to polyaniline is shown in Figure 2.17. Sulfonation of electrochemically polymerized polyaniline film was performed by applying alternate potential steps of 0.20 and −0.15 V vs Ag/AgCl in 4M $NaHSO_3$. Nucleophilic substitution of sulfite to polyaniline only took place in the oxidized polymer. The degree of sulfonation was reportly as high as 50 %.

2.3.1.4 Composites

Kitani *et al.* have prepared polyaniline composites using poly(aniline-2-sulfonic acid) [142] and poly(2,5-disulfonic acid) [143] instead of conventional electroinactive sulfonated polymers like polystyrene sulfonic acid. These composites were prepared by electrochemical dopant exchange. Electrochemically prepared polyaniline films were dedoped by treating 1 M NaOH solution. Furthermore, these dedoped films were

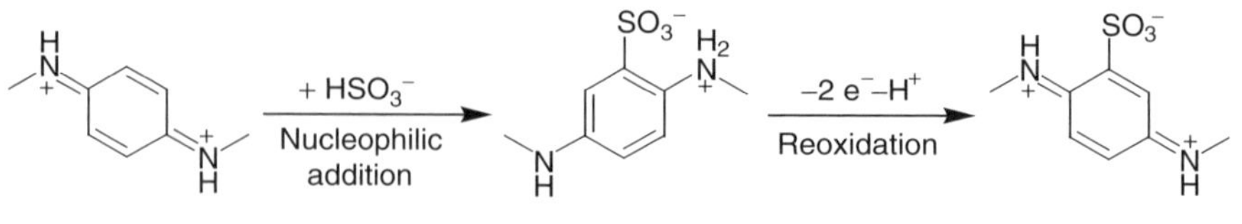

Figure 2.17 Mechanism of nucleophilic addition of sulfite to polyaniline. (Reprinted with permission from *Langmuir*, **20**, 2349. Copyright (2004) American Chemical Society.)

doped in poly(aniline-2-sulfonic acid) and poly(2,5-disulfonic acid) by potential cycling between -0.2 to 0.95 V (vs SCE) at 5 mV/s. These composites were redox active and conducting up to pH 7. The redox charge of composites increased up to 120 % of the value of unsubstituted polyaniline and the specific capacity was determined to be 117 Ah/kg, which is higher than that of unsubstituted polyaniline (108 Ah/kg).

2.3.2 Non-Aqueous Media

2.3.2.1 Homo- and Copolymer

Electrochemical homopolymerization of poly(aniline-N-alkylsulfonates) (alkyl = propyl, butyl and pentyl) in acetonitrile containing 0.1 M $NaClO_4$ and 5 % (v/v) 0.3 M $HClO_4$ was carried out by Rhee et al. [144]. The polymers were prepared on a platinum electrode by cyclic voltammetry (0.0 to 1.0 V vs Ag/AgCl) or potentiostatic techniques (1.0 V). These polymers were found to form liquid crystalline solutions in water. The conductivity of poly(aniline-N-propanesulfonic acid) and poly(aniline-N-butanesulfonic acid) was reportly 9×10^{-5} and 6×10^{-5} S/cm, respectively. Electrochemical polymerization of orthanilic acid, metanilic acid and sulfonic acid and their copolymerization with aniline in dimethyl sulfoxide containing tetrabutyl ammonium perchlorate were carried out by Sahin et al. [145]. These polymers and copolymers were found to be soluble in water, dimethyl sulfoxide and N-methylpyrrolidinone. The conductivity of orthanilic acid, metanilic acid and sulfonic acid was reportly 0.052, 0.087 and 0.009 S/cm, respectively. The conductivity of copolymers for these three isomers of aminobenzenesulfonic acid was reported as 0.094, 0.26 and 0.033 S/cm, respectively. Sahin et al. [146] have also prepared the copolymers of these three isomers with aniline in acetonitrile containing fluorosulfonic acid (FSO_3H). The copolymers were found to be soluble in water, dimethyl sulfoxide and N-methylpyrrolidinone.

2.3.2.2 Electrophilic Substitution

Electrophilic substitution of sulfonic acid on the polyaniline backbone in nonaqueous media has been carried out electrochemically by Sahin et al. [147]. They prepared self-doped sulfonated polyaniline electrochemically in acetonitrile containing anhydrous FSO_3H. FSO_3H was used as both sulfonation reagent for aniline and also as supporting electrolyte. *In situ* sulfonation occurred during polymerization under these conditions. The

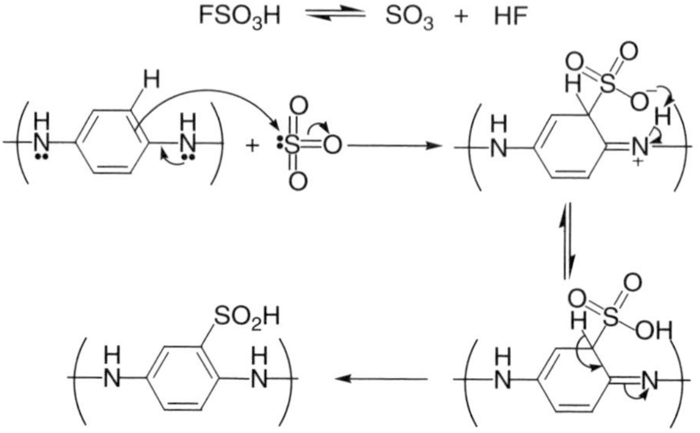

Figure 2.18 Proposed mechanism for the electrochemical polymerization and *in situ* sulfonation of aniline. (Reprinted from *Synthetic Metals*, **129**, Y. Sahin, K. Pekmez, A. Yildiz, 107. Copyright (2002), with permission from Elsevier.)

degree of sulfonation of the polymer was controlled by varying FSO_3H and aniline concentrations in the solutions used for polymerization. Electrodeposition was performed by potential cycling in the range of −0.3 to 1.9 V (vs Ag/AgCl) at a sweep rate of 100 mV/s. It was suggested that the sulfonation reaction took place as a result of the attack of the strong electrophile SO_3^- on the benzenoid ring before the formation of the quinoid ring. The protonated imine repeat units are deactivated for the subsequent electrophilic aromatic substitution reaction, i.e., the sulfonation reaction. The electron density on the benzenoid ring is higher so that the electrophilic substitution (i.e., the sulfonation reaction) occurs more readily. A proposed reaction mechanism for the electrochemical polymerization and *in situ* sulfonation of aniline is shown in Figure 2.18. The maximum S/N ratio reported was 0.89 and the conductivity values of the films were shown to increase from 1.24 to 14.6 S/cm with decreasing S/N ratio. In addition, the polymer was found to be soluble in basic aqueous solution. An *in situ* sulfonation reaction was also carried out in water and water–acetonitrile mixtures by changing both aniline and FSO_3H concentrations [148]. Sulfonated polyaniline was shown to be resistant to electro-oxidative degradation as evidenced from UV-Vis data.

2.4 ENZYMATIC SYNTHESIS OF SULFONIC ACID DERIVATIVES

Recently, the use of enzymes as biological catalysts in the synthesis of polyanilines has attracted great interest due to their environmental

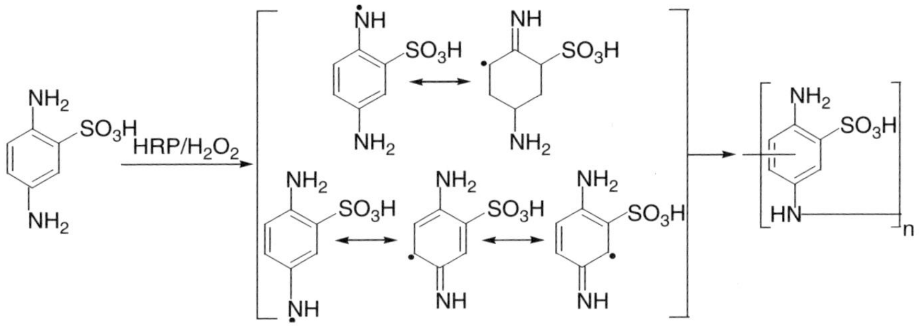

Figure 2.19 Proposed mechanism of enzyme catalyzed polymerization. (Reprinted with permission from *Macromolecules*, 30, 4024. Copyright (1997) American Chemical Society.)

compatibility. The enzymes offer a higher degree of control over the kinetics of the reaction conditions and a higher yield of product [149–152]. The synthesis of a self-doped, water-soluble polyaniline by enzyme catalyzed polymerization was first reported by Tripathy et al. [149, 150]. They synthesized self-doped poly(p-aminobenzoic acid) [149] and poly(2,5-diaminobenzenesulfonate) [150] using horseradish peroxidase (HRP) and hydrogen peroxide through oxidative free radical coupling. The mechanism for the enzyme-catalyzed polymerization of 2,5-diaminobenzene sulfonic acid is shown in Figure 2.19. In the presence of hydrogen peroxide, horseradish peroxidase catalyzes the oxidation of 2,5-diaminobenzene sulfonic acid and generates free radicals. These free radicals undergo coupling to produce dimers. Successive oxidation and coupling reactions result in the fully sulfonated polyaniline [150]. The polymerization of 2,5-diaminobenzene sulfonic acid was carried out in 0.1 M sodium phosphate buffered solution at pH 6.0. Monomer, 2,5-diaminobenzene sulfonic acid, was dissolved in sodium phosphate buffered solution containing the enzyme. The reaction was initiated with the addition of 100 μL of 30 % hydrogen peroxide solution with stirring. The reaction was carried out at room temperature for a minimum of 3 h with constant stirring. The reaction medium was then dialyzed (cutoff 2000 MW) overnight against water to remove the unbound buffer, and the water was evaporated off to get a dark brown solid. The polymer was then extracted with methanol, which was later evaporated off to obtain dark brown colored polymer with 80 % yield. The polymer showed pH dependent (pH 1.2 to 12.8) optical properties. The absorption peak at around 540 nm assigned to the doped form of the polymer decreased with increase in pH of solution. A new peak at

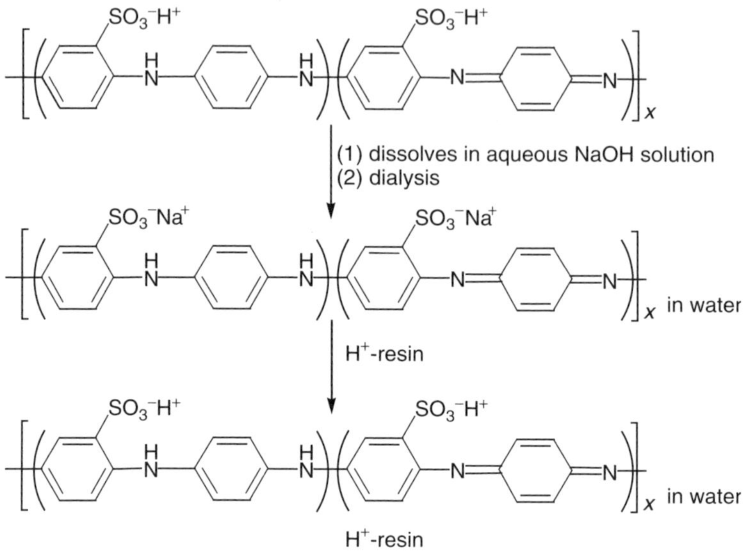

Figure 2.20 Synthesis mechanism of water soluble self-acid-doped sulfonated polyaniline. (Reprinted with permission from *Macromolecules*, 29, 3950. Copyright (1996) American Chemical Society.)

445 nm appeared at higher pH (above pH 9) reportedly due to dedoping of the polymer. This polymer was shown to be soluble in water at all pH values, which makes it suitable material for self-assembly into organized structures with biological macromolecules such as enzymes for fabrication of biosensors. The conductivity of polymer, however, was very low (10^{-5} S/cm) and the average molecular weight was only approximately 18 000 g/mol. Poly(p-aminobenzoic acid) was also soluble in neutral and alkaline conditions and exhibited a conductivity around 10^{-5} S/cm.

2.5 PROPERTIES OF SULFONIC ACID DERIVATIVES

The discovery of self-doped polyanilines was a major breakthrough in the field of conducting polymers due to their desirable properties such as water solubility, pH and temperature independent conductivity, redox activity over a wider pH range and thermal and environmental stability. These polymers can be prepared chemically and electrochemically by various methods discussed in the above sections 2.2–2.4. The distinctive properties of self-doped sulfonated polyanilines are discussed in this section.

2.5.1 Solubility

The insolubility of polyaniline in aqueous solvent and in most common solvents is well known. However, the high solubility of polyaniline is essential for postsynthetic processing. The high solubility of self-doped sulfonated polyanilines in basic aqueous solution, pure water and, in some cases, across a broad range of pH values is due to the presence of anionic -SO_3^- groups. The ring sulfonated polyanilines are highly soluble in aqueous NH_4OH and $NaOH$ solution and slightly soluble in water. In basic aqueous solution, a solubility of 23 mg/mL was observed for 50 % ring sulfonated polyanilines [41]. For higher degrees of sulfonation, e.g., 75 % ring sulfonated polyanilines, a 1.5 fold increase in solubility of about 38 mg/mL has been observed [54]. In basic aqueous solution, the sulfonated polyaniline backbone is in the emeraldine base form. Chen *et al.* [59] synthesized the first water-soluble self-acid-doped (autodoped) poly(aniline-co-*N*-propanesulfonic acid aniline) (Figure 2.5). They have also prepared water-soluble self-acid-doped ring sulfonated polyaniline by a two step procedure shown in Figure 2.20 [47]. Sulfonated polyaniline was prepared following Yue and Epstein's method [39]. In the first step, the polymer was dissolved in aqueous NaOH solution and then excess NaOH was removed by dialysis with a semipermeable membrane in water. In the second step, the self-acid-doped sulfonated polyaniline solution was obtained by exchanging Na^+ of the sulfonated polyaniline sodium salt aqueous solution to H^+ using H^+-type ion exchange resin. The aqueous solution of the conducting form of the sulfonated polyaniline was then concentrated to a solid content of 50 % by weight and was stable for more than one year.

An increased water solubility was observed for externally HCl doped fully sulfonated polyaniline (Figure 2.3) prepared by reacting polyaniline with chlorosulfonic acid in dichloroethane at elevated temperature [57]. With an increase in sulfur to nitrogen ratio from 0.65 to 1.3, the water solubility increased from 22 to 88 mg/mL. It was suggested that the externally doped state of sulfonated polyaniline resulted in an increased free proton concentration (from the -SO_3H group) and that this was responsible for the increased water solubility. Electrochemically prepared poly(*o*- and *m*-aminobenzenesulfonic acid) homopolymers were soluble in water in the range of 30–40 mg/mL regardless of the solution pH [116]. Similarly, enzymatically synthesized self-doped sulfonated polyaniline showed water solubility under all pH conditions [150]. Electrochemically prepared sulfonated polyaniline by *in situ* sulfonation reaction using FSO_3H was found to be soluble in 0.1 M KOH, dimethyl sulfoxide and N-methylpyrrolidinone at about 20–35 mg/mL based on degree of sulfonation [147, 148].

2.5.2 Conductivity

Self-doped sulfonated polyanilines have lower conductivity than externally doped polyanilines due to steric effects of sulfonic acid substituents associated with the presence of a bulky substituent on the phenyl ring. Substituent groups induce additional ring twisting along the polymer backbone due to the increased steric interactions and as a result lower the crystallographic order of the polymer chains. Such induced ring twisting increases the energy barrier for charge transport and reduces the extent of polaron delocalization along the chain [47]. In sulfonated polyaniline, the conductivity is dependent on the degree of sulfonation, and the position and nature of the substituent. The conductivity of various sulfonated polyanilines is summarized in Table 2.1.

2.5.2.1 pH Dependence

In unsubstituted polyaniline, degree of protonation and resulting conductivity are controlled by changing the pH of a Brønsted acid doping solution [153]. At low pH (less than 1), polyaniline is highly protonated and conductive. In the protonation process, the insertion of anions to maintain electrical neutrality occurs. However, at pH values greater than 4, the polyaniline is dedoped and insulating. Unlike polyaniline, the conductivity of self-doped sulfonated polyaniline is independent of pH within a broad range of values, due to self-protonation by the presence of intrinsic acid on the polymer backbone. Due to the presence of sulfonic acid, the internal acid–base equilibrium of sulfonated polyaniline is not affected by the external medium over wider pH range, which in most cases ranges from 0 to 7 [41, 90, 154, 155]. For poly(aniline disulfonic acid), the pH independence of conductivity is reported up to 9 [117, 118]. The conductivity of leucoemeraldine-base sulfonated polyaniline (degree of sulfonation 75 %) was reportedly stable up to pH 12 [54]. Epstein et al. have shown the comparison of pH dependent conductivity of leucoemeraldine-base sulfonated polyaniline (degree of sulfonation 75 %) with emeraldine-base sulfonated polyani-

PROPERTIES OF SULFONIC ACID DERIVATIVES

Table 2.1 Conductivity of various sulfonated polyanilines

Structure	Degree of Sulfonation (%)	Conductivity (S/cm)	References
(SO$_3$H on ring, NH)	50	0.1	[41]
	75	1.0	[54]
	100	10^{-2}–10^{-3}	[55]
N–(CH$_2$)$_2$–SO$_3$H	50	4×10^{-6}	[88]
	100	10^{-5}	[40]
N–(CH$_2$)$_3$–SO$_3$H	50	10^{-2}	[59]
	100	10^{-8}	[58]
N–phenyl–SO$_3$H	50	10^{-2}	[88]
	100	10^{-3}	[73]
N–(CH$_2$)$_3$–phenyl–SO$_3$H	47	8.5×10^{-5}	[60]
N–C(=O)–phenyl–SO$_3$H	copolymer	5×10^{-4}	[61]

Table 2.1 Conductivity of various sulfonated polyanilines

Structure	Degree of Sulfonation (%)	Conductivity (S/cm)	References
(structure 1: polyaniline with -S-(CH$_2$)$_2$-SO$_3$H substituent)	15 20 26	0.017 0.42 0.77	[66]
(structure 2: polyaniline with SO$_3$H and HO$_3$S substituents)	50 100	0.1–0.5 0.04–0.3	[118] [118]
(structure 3: polyaniline with OCH$_3$ and HO$_3$S substituents)	100	10^{-2}	[70]
(structure 4: graft copolymer with SO$_3$H groups)	Graft copolymer	10^{-1}	[112]
(structure 5: graft copolymer with SO$_3$H and amide linkages)	Graft copolymer	1	[111]

line (degree of sulfonation 50 %) and HCl-doped polyaniline as shown in Figure 2.21. The conductivity of leucoemeraldine-base sulfonated polyaniline is pH independent over the range of pH 0 to 12. However, the conductivity of emeraldine-base sulfonated polyaniline decreases by

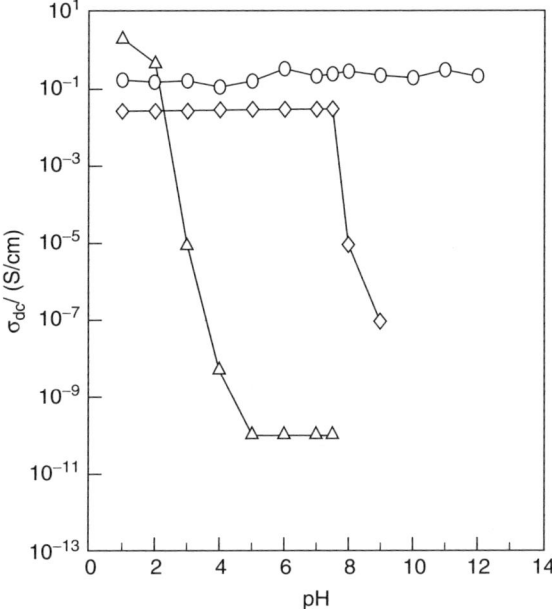

Figure 2.21 pH dependence of DC conductivity at room temperature for leucoemeraldine-base sulfonated polyaniline (O), emeraldine-base sulfonated polyaniline (◊) and polyaniline-HCl (△). (Reprinted with permission from *Journal of the American Chemical Society*, 118, 2545. Copyright (1996) American Chemical Society.)

five orders of magnitude from pH 7.5 to 9 and for HCl-doped polyaniline conductivity decreases by more than ten orders of magnitude in the pH range 2–4. The leucoemeraldine-base form of sulfonated polyaniline has higher pH independent conductivity than emeraldine-base sulfonated and HCl-doped polyaniline due to a higher concentration of sulfonic groups attached to the polyaniline backbone and formation of six member ring complexes (for details see Chapter 1, Section 1.4.1).

2.5.2.2 Temperature Dependence

Epstein *et al*. [41, 54, 71] suggested that the sulfonated polyanilines have much stronger temperature dependence conductivity than emeraldine hydrochloride due to greater electron localization. The temperature dependence of the conductivity of 50 % [41], 75 % [54] and fully ring sulfonated polyanilines [71] was best fit by the quasi one-dimensional variable range hopping model described by Equation (2.1):

$$\sigma(T) = \sigma_0 \exp[-(T_0/T)^{1/2}] \qquad (2.1)$$

where $T_0 = 8\alpha/(zN(\varepsilon_F)k_B)$, α^{-1} is the localization length, $N(\varepsilon_F)$ is the density of states at the Fermi level, k_B is the Boltzmann constant, and z is the number of nearest neighbor chains. The temperature dependent conductivity of the 50 % ring sulfonated polyaniline was best fit with $T_0 = 39\,000\,K$ (Figure 2.22), much larger than the 6000 K for polyaniline hydrochloride. The larger T_0 suggests a much greater localization of charge carriers in sulfonated polyaniline compared with that in polyaniline hydrochloride. In the case of 75 % ring sulfonated polyaniline, T_0 was smaller (25 000 K), consistent with the greater room temperature conductivity (Table 2.1). However, in the case of fully ring sulfonated polyaniline, a lower T_0 (15 000 K), a lower room temperature conductivity and a smaller apparent density of states at the Fermi level were observed due to the presence of an excess of -SO_3^- groups. These groups probably force the chain out of planarity by twisting the phenyl rings relative to one another to lower the overlap of orbitals along the conjugated system. As a consequence, the conduction electron wave functions in sulfonated polyanilines are expected to be substantially more localized than those in polyaniline hydrochloride, leading to a lower mobility for the charge carriers both along the polymer chain and between polymer chains.

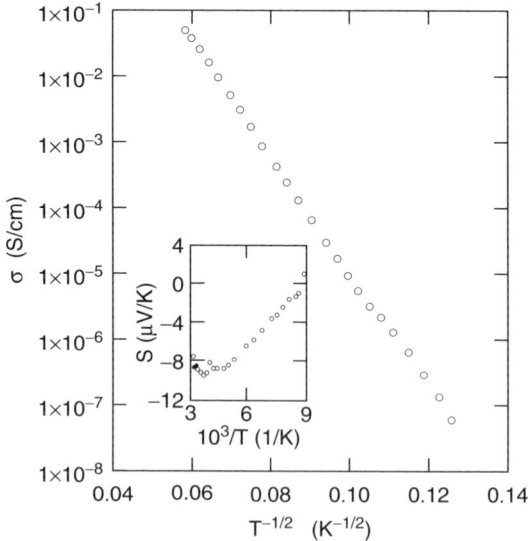

Figure 2.22 Temperature dependence conductivity of 50 % ring sulfonated polyaniline. The inset shows the thermopower vs reciprocal temperature for 50 % ring sulfonated polyaniline. (Reprinted with permission from *Journal of the American Chemical Society*, **113**, 2665. Copyright (1991) American Chemical Society.)

2.5.3 pH Dependent Redox Behavior

The redox activity of polyaniline (evaluated by the positions of the oxidation and reduction peaks, anodic and cathodic charge and shape of the peak) depends mainly on the pH of the medium and is ultimately lost in neutral and alkaline media. However, the redox activity of self-doped sulfonated polyanilines is independent of pH over a broad range of values, due to the enhancement of doping strength of protons from the sulfonic acid groups onto the imine nitrogen atoms of the polyaniline chain. The redox behaviour of sulfonated polyanilines in acidic pH solution is similar to conventional polyaniline. A cyclic voltammogram of 50 % ring sulfonated polyaniline in acidic solutions (1 M HCl) exhibits two redox couples ($E_{1/2}$, 0.28 V and 0.77 V vs Ag/AgCl) associated with the conversion of leucoemeraldine to emeraldine and subsequent conversion to the pernigraniline oxidation state (Figure 2.23) [41].

However, the redox couples are more closely spaced than observed for polyaniline. The first redox process occurs at a higher potential and the second redox process occurs at a lower potential relative to polyaniline and they are similar to alkyl substituted polyaniline [156]. These results are explained by the electronic and steric effects of SO_3^- groups on the backbone of polymer as follows. The bulky substituent on the ring will cause ring twisting and hence a reduction in chain conjugation, which will in turn result in reduced stability of the half oxidized semiquinone radical cation (emeraldine) formed during the first redox process (Figure 2.23). Since the half oxidized semiquinone radical cation is less stable than in the parent polyaniline, it will require a higher potential for its formation. In a second redox process, further oxidation of the half oxidized semiquinone to quinonediimine (pernigraniline) (Figure 2.23), with simultaneous deprotonation, occurs with the formation of imine nitrogens with a wider angle. Some of the steric strain may be relieved by the wider C–N=C angle at the quinoid group in comparison with the benzenoid group. On the other hand, the better conjugation of lone-pair electrons at the nitrogen atoms with π electrons on the phenyl rings as well as with the electron withdrawing groups, -SO_3^-, lowers the electron density on the nitrogen atoms and therefore raises the oxidation potential. These two effects partially counteract each other and lead to a net lower oxidation potential with a higher substitution level. In the case of 75 % ring sulfonated polyaniline in acidic solution, two sets of redox couples are observed at 0.32 and 0.67 V vs Ag/AgCl, corresponding to 0.35 V wide potential window for the conducting emeraldine form. This window is narrower than 50 % ring

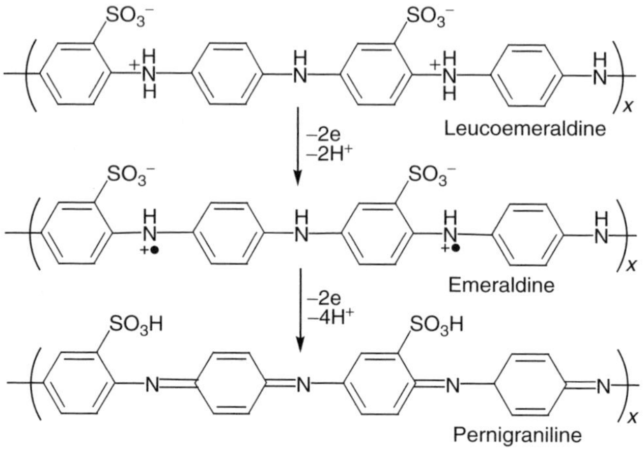

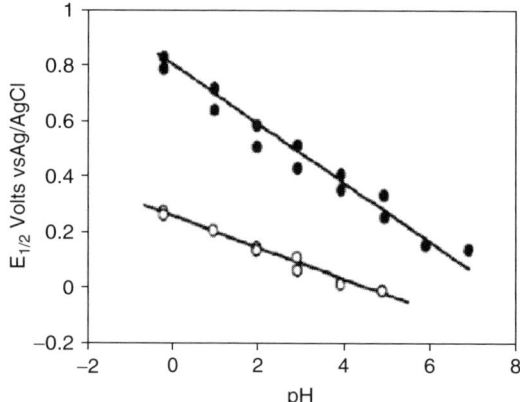

Figure 2.23 Interconversion of oxidation states of sulfonated polyaniline. (Reprinted with permission from *Journal of the American Chemical Society*, **113**, 2665. Copyright (1991) American Chemical Society.)

Figure 2.24 $E_{1/2}$ (50 mV/s) of first (open circle) and second (closed circle) redox processes of 50 % ring-sulfonated polyaniline as a function of pH range. (Reprinted with permission from *Journal of the American Chemical Society*, **113**, 2665. Copyright (1991) American Chemical Society.)

sulfonated polyaniline. In both the redox processes, the steric properties of -SO_3^- groups are reported to be the dominant factors.

The half wave potential, $E_{1/2}$ of the first and second redox processes of 50 % ring sulfonated polyaniline as a function of pH range are shown in Figure 2.24. Both redox processes are found to depend on the solution pH. For the first and second redox processes, slopes of −59 mV/pH and −118 mV/pH, respectively, were observed. These slopes indicate the loss of equal numbers of protons and electrons for the first redox process and the loss of two protons per electron for the second redox process. However, in case of 75 % sulfonated polyaniline, a slope of −59 mV/pH was found in the pH range 1–5 for first redox process and 1–6 for second redox process. At a higher pH range, a slope of −118 mV/pH was attributed to an irreversible oxidation–hydrolysis process resulting in the formation of quinoidal species. In the case of conventional polyaniline, slopes of 0 and −118 mV/pH are reported for the first and second redox processes in the pH range 1–4 [157–160]. The difference in the electrochemical behavior of sulfonated polyaniline is associated with the substitution of ionizable covalently bound sulfonic acid groups.

The electrochemically prepared self-doped polyaniline copolymer of aniline and metanilic acid (monomer ratio of 3:1) is redox active in neutral and basic aqueous solutions [129, 161]. In this case, at pH values greater than 6, the two redox processes of polyaniline overlap and as a result a single redox wave is observed as shown in Figure 2.25. These results suggest a disappearance of emeraldine as a stable oxidation state of polyaniline in neutral pH solution. The peak current of the self-doped polyaniline redox wave decreases in neutral and alkaline pH solution; however, significant redox activity is seen at pH 12. Polyaniline doped with a large external dopant, such as camphorsulfonic acid, is also redox active in neutral and alkaline pH solutions similar to sulfonated polyaniline; however, long term stability is not observed in the case of the large external dopant in contrast to the stable response observed for sulfonated polyaniline [93]. Similarly, self-doped sulfonated polyaniline composites [142] and copolymers [118, 133, 137] are reported to be redox active in neutral pH solution. Sulfonated polyaniline films obtained from direct reaction with concentrated sulfuric acid were reported to be redox active up to pH 10.6 (0.3 M Na_2SO_4) over a wider potential range (−0.2 to 0.6 V vs SCE) [155] than was observed for the copolymers shown in Figure 2.25.

2.5.4 Electronic and Spectroscopic Properties

2.5.4.1 *UV-Vis Spectroscopy*

Introduction of -SO_3H groups on the polyaniline backbone is expected to change the molecular geometry significantly. Sulfonation of the phenyl

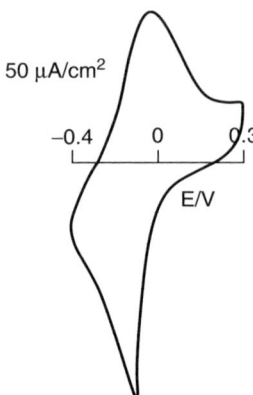

Figure 2.25 Cyclic voltammogram of self-doped polyaniline at pH 9 (0.02 M borate buffer + 0.1 M KCl). Scan rate 25 mV s^{-1}. (Reprinted from *Journal of Electroanalytical Chemistry*, **402**, A. A. Karyakin, I. A. Maltsev, L. V. Lukachova 217. Copyright (1996), with permission from Elsevier.)

rings on the polyaniline backbone can increase the torsional angle between adjacent rings even further to relieve steric strain [162, 163]. This affects the electronic transitions of the polymer. UV-Vis absorption spectra of 50 % ring sulfonated polyaniline and polyaniline doped with HCl films are shown in Figure 2.26 [41]. The absorption peaks due to the $\pi-\pi^*$ transition of the benzenoid ring appears around 320 nm, and the absorptions due to the metallic polaron band transition of the conducting form are observed at 435 and 850 nm. In a comparative study, a hypsochromic shift (blue) of the $\pi-\pi^*$ transition and a bathochromic shift (red) of the polaron band is observed upon transitioning from polyaniline doped with HCl to sulfonated polyaniline. It has been suggested that this change in UV-Vis properties of sulfonated polyaniline is in agreement with the decrease in the extent of conjugation caused by an increase in phenyl ring torsion angle, which results from steric repulsion between the -SO$_3^-$ groups and hydrogens on the adjacent phenyl rings. Chen *et al.* [47] have reported the UV-Vis spectra of self-acid-doped 50 % ring sulfonated polyaniline films cast from water solution in comparison with HCl-doped polyaniline film (cast from N-methylpyrrolidinone). The absorption peak due to the polaron transition is observed at 880 nm for self-acid-doped sulfonated polyaniline similar to results of Epstein and MacDiarmid *et al.* [41]. However, in case of polyaniline, a broad polaron band at 1200 nm is observed. The broader peak in comparison to the sulfonated polyaniline and 320 nm bathochromic shift is reported to be due to either more delocalization

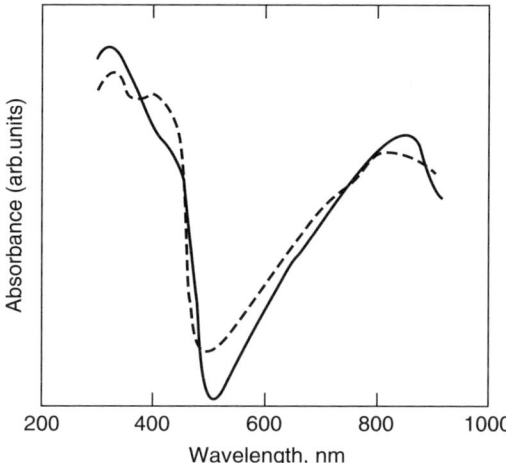

Figure 2.26 UV-Vis absorption spectra of 50 % ring sulfonated polyaniline film (——) cast on a quartz substrate from an aqueous NH$_4$OH solution and then dried in air at room temperature, and polyaniline doped with HCl film (- - -). (Reprinted with permission from *Journal of the American Chemical Society*, 113, 2665. Copyright (1991) American Chemical Society.)

of polarons in HCl-doped polyaniline or more localized polarons in sulfonated polyanilines.

The self-doped ring sulfonated polyaniline can be deprotonated by OH$^-$ in alkaline solution. The UV-Vis spectra of dedoped sulfonated polyaniline are shown in Figure 2.27 [54]. The polaron band observed at 850 nm (Figure 2.26) is blue shifted to ~560 nm due to formation of sulfonated emeraldine base. However, the blue shifts of absorption bands (π–π^* and exciton transition) are higher for 75 % ring sulfonated polyaniline than 50 % ring sulfonated polyaniline. These results suggest that at higher degrees of sulfonation, the band gap increases due to an increase in phenyl ring torsion angle with the decrease in the intrachain interactions, and the strong electron withdrawing nature of the substituent.

Unlike ring sulfonated polyaniline, the fully sulfonated, water-soluble, polyaniline, poly(2-methoxyaniline-5-sulfonic acid), is shown to be remarkably inert to alkaline dedoping and remains in the conducting emeraldine salt form even in 2.0 M NaOH [164]. The UV-Vis near-infrared spectra of poly(2-methoxyaniline-5-sulfonic acid) solution as a function of pH are shown in Figure 2.28. In water, the absence of a localized polaron band in the range of 750–900 nm and presence of a broad absorption in NIR suggests an 'extended coil' conformation

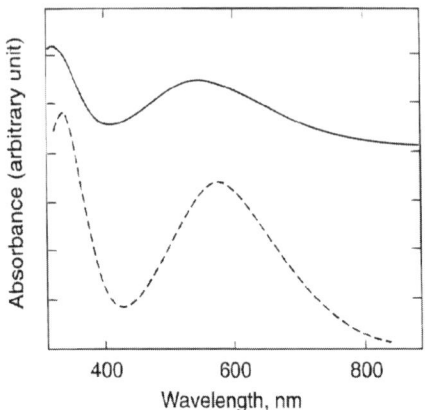

Figure 2.27 UV-Vis spectra of 50 % ring sulfonated (----) and 75 % ring sulfonated (—) polyaniline in 0.1 M NH$_4$OH solution. (Reprinted with permission from *Journal of the American Chemical Society*, 118, 2545. Copyright (1996) American Chemical Society.)

for poly(2-methoxyaniline-5-sulfonic acid) emeraldine salt. In the pH range 3.66 to 8.02, a small decrease in the 473 nm and NIR bands and an increase in the 330 nm band are observed, and are reportedly due to deprotonation of 'free' -SO$_3$H groups on the poly(2-methoxyaniline-5-sulfonic acid) chain. In the pH range 9.15 to 14, the initial NIR and 473 nm bands are replaced by a strong band at 750 nm and a strong band appears at 330 nm. These spectral changes as a function of pH suggest the formation of a 'compact coil' conformation for poly(2-methoxyaniline-5-sulfonic acid) at alkaline pH. Similar results are observed in the presence of alkali and alkaline-earth metal salts.

2.5.4.2 X-Ray Photoelectron Spectroscopy

Polyaniline contains nitrogen in different oxidation and protonation states including amine (–NH–) and imine (–N=) groups connected with benzenoid and quinoid rings of the polymer. X-ray photoelectron spectroscopy has been a valuable technique for determining quantitative proportions of quinoid imine, benzenoid amine and the positively charged nitrogens in polyaniline; and for studying the intrinsic protonation state [165, 166]. This technique has also been useful in studies of both the sulfonation and the doping level of sulfonated polyaniline and investigation of structures. Epstein *et al.* [167, 168] reported the

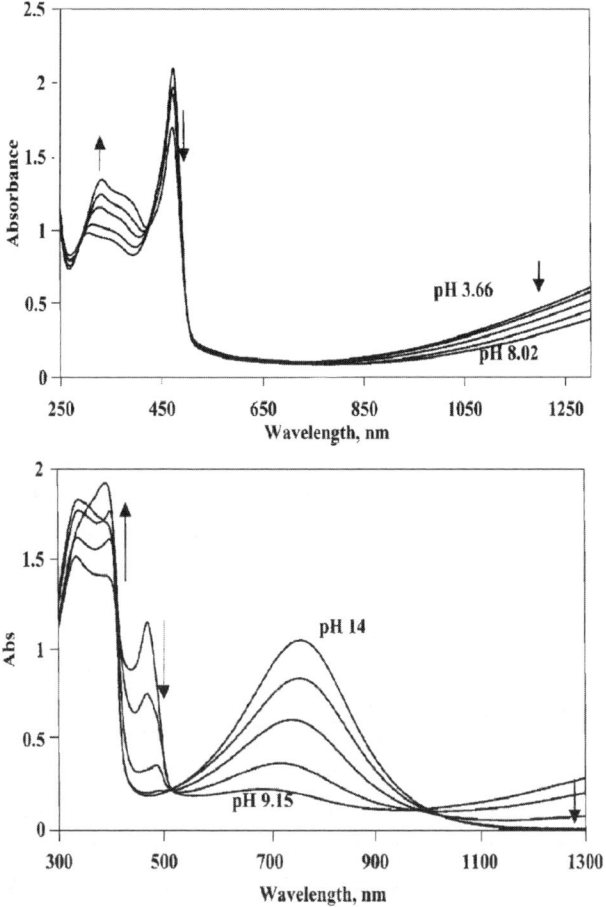

Figure 2.28 UV-Vis–NIR spectra of an aqueous 3×10^{-4} M poly(2-methoxy-aniline-5-sulfonic acid) solution with increasing pH. (Reprinted from *Synthetic Metals*, 135, E. V. Strounina, R. Shepherd, L. A. P. Kane-Maguire, G. G. Wallace, 289. Copyright (2003), with permission from Elsevier.)

X-ray photoelectron spectroscopic measurements of emeraldine base sulfonated polyaniline (50 %) in comparison with chemically synthesized polyaniline base and its chloride salt. For polyaniline base a nearly symmetric N 1s line centered close to 398.6 eV is observed, as expected from the idealized structure of the emeraldine base backbone. The deconvolution of the N 1s core level spectrum shows two peaks with equal intensity, one centered at 398.9 eV due to amine nitrogen, the other located at 397.5 eV due to imine nitrogen atom. Both have an FWHM (full width at half maximum) of 1.6 eV. Similar results were obtained for the sodium salt of emeraldine base sulfonated polyaniline (base form structure is shown in Chapter 1, Figure 1.17, B).

High resolution N 1s X-ray photoelectron spectra of self-doped emeraldine base sulfonated polyaniline and HCl-doped polyaniline are shown in Figure 2.29. In the case of self-doped sulfonated polyaniline (Figure 2.29, A), the nearly symmetric N 1s peak observed in the base form changes to an asymmetric peak. The deconvolution of N 1s spectrum results in two peaks; one at a higher binding energy centered at 400.7 eV with a FWHM of 1.8 eV assigned to the radical cationic nitrogen atoms, and another centered at 399.1 eV with a FWHM of 1.5 eV assigned to uncharged amine sites. The ratio of the integrated areas of the peaks is reportedly 1:1. In the case of HCl-doped polyaniline, the deconvolution of the N 1s spectrum results in three peaks (Figure 2.29, B) located at 402.2 eV (2.0 eV FWHM), 400.4 eV (1.8 eV FWHM) and 399.1 eV (1.6 eV FWHM). The peaks at 402.2, 400.4 and 399.1 eV are attributed to the most positively charged nitrogen sites in the vicinity of chloride ions, nitrogen atoms with delocalized positive charge and uncharged amine sites, respectively. The ratio of the integrated area of these three peaks is reported to be 1:3.3:8.8.

It has been suggested that in all sulfonated samples, the positively charged radical cationic nitrogen atoms have a slightly larger FWHM than that of the amine nitrogen atoms. This wider peak is also asymmetric, which is consistent with the environment of the positively charged nitrogen atoms being less homogeneous than that of the amine nitrogen atoms. This inhomogeneity may be caused by deviations from the average angle between phenyl rings and the planes of the nitrogen sites due to the steric effect of -SO_3^-. The peak observed at 397.5 eV in the base form of polyaniline completely disappears in sulfonated polyaniline (Figure 2.29, A) and a new peak appears at 400.7 eV. These results suggest that the imine nitrogen atoms are completely protonated by the intrinsic sulfonic acid. In addition, the difference in N 1s peaks of HCl-doped polyaniline (Figure 2.29, B) and sulfonated polyaniline (Figure 2.29, A) suggest that the polarons are more localized in sulfonated polyaniline. The localization is reportedly caused by the formation of energetically favourable five- or six-membered rings due to electrostatic attraction between SO_3^- and radical cation nitrogen atoms or amine hydrogen (for structures see Chapter 1, section 1.4.1, Figure 1.25). The S/N ratio is found to be 0.5, which indicates that the sulfonic acid is present with every two phenyl rings of the polyaniline or that the polyaniline is 50 % sulfonated. The amine nitrogen to cationic radical nitrogen atoms ratio (1:1) indicates that the doping level, [SO_3^-]/[N] is 0.5. The

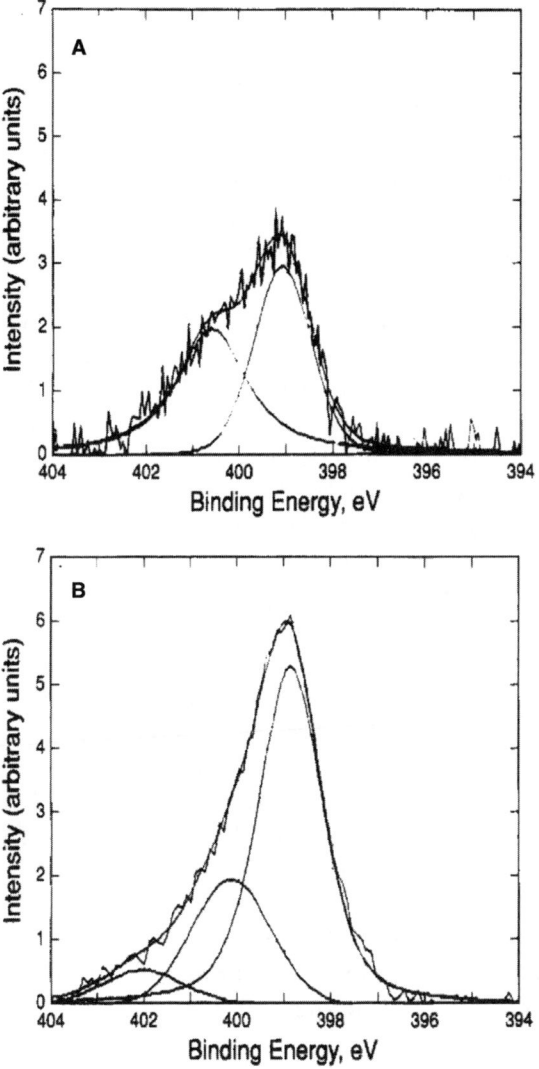

Figure 2.29 N 1s X-ray photoelectron spectra of (A) self-doped emeraldine base sulfonated polyaniline (50%) and (B) polyaniline doped with HCl. (Reprinted with permission from *Macromolecules*, **24**, 4441. Copyright (1991) American Chemical Society.)

doping level of emeraldine-base sulfonated polyaniline determined by Chen *et al.* is lower. They have deconvoluted the N 1s spectrum into three peaks at 401.8 eV (1.9 eV FWHM), 400.2 eV (1.9 eV FWHM) and 399.3 eV (1.6 eV FWHM). The fractional area associated with the radical cation nitrogen (400.2 eV) is about 30 % of the total nitrogen content; hence a doping level of 0.3 is reported.

X-ray photoelectron spectroscopy results indicate a sulfonation level (S/N ratio) of 0.8 ± 0.1 for leucoemeraldine-base sulfonated polyaniline [169]. The N 1s spectrum of the leucoemeraldine-base sulfonated polyaniline can be deconvoluted into four peaks. Two more peaks than observed for the emeraldine-base sulfonated polyaniline are reported at 398.1 eV attributed to dedoped imine units and at 402.6 eV assigned to the protonated amine units which are at a higher binding energy because of the stronger electron localization associated with poorer conjugation at sp^3 bonded sites. The percentage of the total N 1s intensity of the peaks are reported to be 398.1 (5 %), 399.1 (38 %) 400.9 (46 %) and 402.6 eV (11 %). The sum of the fraction of cationic nitrogen and dedoped imine site for leucoemeraldine-base sulfonated polyaniline was found to be the same as those for emeraldine-base sulfonated polyaniline and pernigraniline-base sulfonated polyaniline, which suggests that the leucoemeraldine-base sulfonated polyaniline is still in the emeraldine oxidation state and implies that the emeraldine oxidation state is more stable than the other oxidation states. However, a small portion of imine is not protonated, which might suggest that at the higher sulfonation level the protonation may be more difficult compared with the parent polyaniline in the emeraldine salt form, consistent with sulfonation lowering the electron density at imine nitrogen sites. The S 2p spectrum can be fitted with two peaks at 167.25 eV, attributed to the anionic sulfur resulting from the protonation of the backbone, and 168.55 eV attributed to neutral sulfur in the undissociated sulfonic acid substituents. The percentages of the total S 2p peak intensity are reported to be 167.25 eV (75 %) and 168.55 eV (25 %). These results suggest that approximately 75 % of sulfonic acid protonate the aromatic imine and amine and 25 % are neutral.

2.5.4.3 Electron Spin Resonance Spectroscopy

Magnetic susceptibility and electron spin resonance data of self-doped sulfonated polyaniline shows the presence of 'Curie like' susceptibility and temperature independent 'Pauli like' susceptibility. The product of the spin susceptibility and temperature (χT) versus temperature for the emeraldine base sulfonated polyaniline is shown in Figure 2.30 [41]. Based on the Pauli spin concentration, the density of states at the Fermi level $N(\varepsilon_F)$ (from the slope) is ~ 0.8 state/eV-two rings and an effective Curie spin concentration (from the $T = 0$ intercept) is 0.02 spin/two

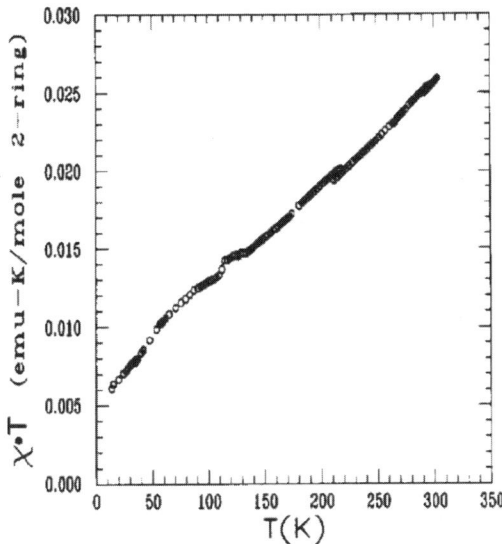

Figure 2.30 Spin susceptibility times temperature as a function of temperature of emeraldine-base sulfonated polyaniline. A core diamagnetism of -1.45×10^{-4} emu/mol two rings was utilized to convert total susceptibility to spin susceptibility. This was determined with the experimental core diamagnetism of the emeraldine backbone plus the Pascal constant value for the sulfonic group. (Reprinted with permission from *Journal of the American Chemical Society*, **113**, 2665. Copyright (1991) American Chemical Society.)

rings. The Pauli susceptibility is consistent with the significant electron delocalization that occurs upon addition of protons from a sulfonic acid group to the imine site. Electron spin resonance data indicate that the sulfonated polyaniline is strongly paramagnetic, which supports the assertion that sulfonated polyaniline is in the polysemiquinone radical cation (polaron energy band) state. The peak-to-peak line width is reported to be 0.3 G at room temperature with intensity similar to that of polyaniline hydrochloride [170, 171].

The temperature dependence of the electron paramagnetic resonance of leucoemeraldine base sulfonated polyaniline has been reported by Epstein *et al.* [169]. The decrease in FWHM height line width, ΔH_{FWHM}, with increasing temperature from 4.9 G at 3 K to 1.0 G at 300 K was observed and is consistent with the observed increase in conductivity with increasing temperature, assuming the electron paramagnetic resonance line width is narrowed by motional effects. For example, as the conductivity increases it is expected that there will be a corresponding increase in the motion of the unpaired electrons or spins, resulting in more effective narrowing of the electron paramagnetic resonance line width. At room temperature, the g-value for leucoemeraldine-base sulfonated polyaniline was found to be $\sim$2.0026, consistent with the unpaired spin delocalized primarily on the phenyl rings [172, 173]. As the temperature decreases, approaching 70 K the g-value increases to $\sim$2.0030, a value typical for heteroatom systems [174, 175], indicating that spins tend to become more localized at the nitrogen sites. These results are consistent with the conductivity data and indicate greater localization of charge at the lower temperatures. In the case of leucoemeraldine-base sulfonated polyaniline, the values for Curie spin concentration and the density of states at the Fermi level are reported to be 0.022 spins/two rings and $\sim$1.0 state/eV two rings, respectively [169].

2.5.5 Molecular Weight

Chemically synthesized polyaniline under ambient conditions typically has molecular weights around 30 000 g/mol [56, 176]. High molecular weight polyaniline approaching 350 000 g/mol has been reported by significantly lowering the reaction temperatures and lengthening the reaction times (−40 °C, 48 h) [177, 178]. However, sulfonated polyanilines synthesized via various methods (section 2.2–2.4) typically exhibit lower molecular weights than does polyaniline. In the postpolymerization sulfonation process using $\sim$30 % fuming sulfuric acid, the prolonged sulfonation process shortens the chain length of the polymer due to the hydrolysis of quinoid structures. The molecular weight of the reduced form of sulfonated polyaniline determined by light scattering was reported to be 11 000 g/mol [179]. The solutions of reduced sulfonated polyaniline were prepared using a 0.1 M aqueous NaOH/hydrazine solution (0.875 hydrazine vol.%). Similar molecular weights (<12 000 g/mol) are obtained using gel permeation chromatography for sulfonated polyanilines prepared chemically and electrochemically by direct polymerization of monomer [70, 86, 118, 180]. Surprisingly, higher molecular weights of 51 000, 96 000 and 160 000 g/mol have been obtained for electropolymerized *para-*, *ortho-* and *meta-*substituted poly(aminobenzene sulfonic acid) homopolymers, respectively [116]. The molecular weight measurements were done using gel permeation chromatography with N,N-dimethylformamide as the mobile phase. High molecular weight copolymers ($\sim$30 000 g/mol) are also prepared by using lower sulfonated monomer content relative to aniline [91, 181]. The molecular weights were determined by gel permeation chromatography in mobile phase N-methyl pyrrolidinone with 0.5 % triethylamine and 0.5 % LiCl.

2.5.6 Thermal Stability

The use of conducting polymers such as polyaniline, polypyrrole, etc. in the thermoplastic industry is limited due to undesirable properties such as loss of dopant and decomposition at higher temperatures. Also, due to loss of conductivity at high temperatures, their use in high temperature conducting coatings, electronic circuits, fuel cells, etc. has been restricted. Self-doped sulfonated polyanilines have shown improved thermal stability relative to the parent polyaniline doped with external dopants, presumably due to the fact that the dopant is covalently attached to the polymer backbone [47, 154, 182]. Self-doped mercaptopropanesulfonic acid-substituted polyanilines [66] are found to be significantly more thermally stable than ring sulfonated polyaniline [39] (Figure 2.31). Three-step decomposition patterns are observed for both of the polymers. This includes an initial loss of water observed below 120 °C. In the case of sulfonated polyaniline, loss of sulfonic acid and major decomposition of the backbone is observed at 185 and 275 °C, respectively. However, in

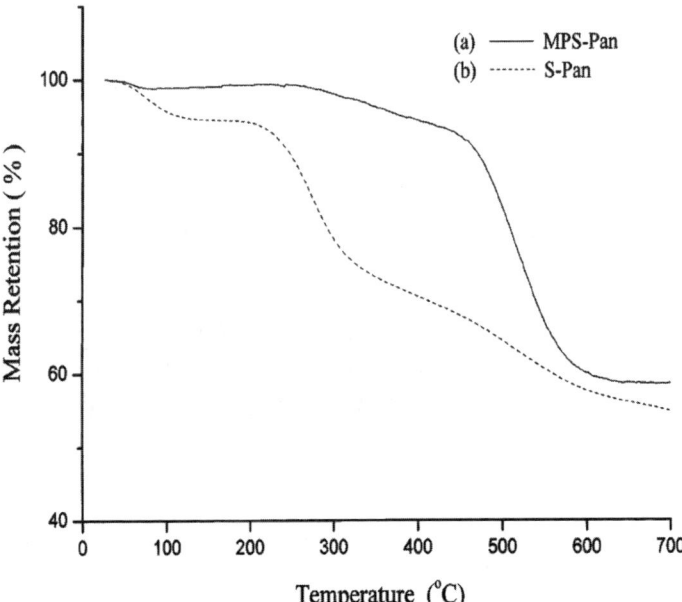

Figure 2.31 Thermogravimetric analysis traces of mercaptopropanesulfonic acid substituted polyaniline (___) and sulfonated polyaniline (- - -). (Reprinted with permission from *Macromolecules*, **36**, 7908. Copyright (2003) American Chemical Society.)

Table 2.2 S/N atomic ratios of MPS-Pan and S-Pan after being heated at different temperatures. (Reprinted with permission from *Macromolecules*, **36**, 7908. Copyright (2003) American Chemical Society.)

Temp[a](° C)	MPS-Pan (S/N)[b]	S-Pan (S/N)[b]
30	0.395	0.558
100	0.393	0.540
210	0.388	0.382
240	0.391	0.224
270	0.298	0.182
340	0.153	0.089
420	0.136	0.038

[a] The samples were heated at the given temperature for 1 h under 1 atm of N2.
[b] The S/N atomic ratio as measured by X-ray photoelectron spectroscopy.

the case of mercaptopropanesulfonic acid substituted polyaniline, minor weight loss is observed at 260–400 °C followed by decomposition at 524 °C. The mechanism behind the thermal stability was determined by Han *et al.* based on the S/N atomic ratio measured at different temperatures using X-ray photoelectron spectroscopy as shown in Table 2.2. In the case of mercaptopropanesulfonic acid substituted polyaniline samples, the S/N ratios remained unchanged below 240 °C, whereas the S/N ratio of sulfonated polyaniline treated at a similar temperature decreased significantly from 0.558 (heated at 30 °C), through 0.382 (at 210 °C), to 0.224 (at 240 °C). These results indicate that the structure of mercaptopropanesulfonic-acid-substituted polyaniline is intact up to at least 240 °C, whereas a significant fraction of the sulfonic acid groups on sulfonated polyaniline is lost at 210 °C. With a further increase in temperature, a steady decrease of the S/N atomic ratio suggests a continuous loss of sulfur-containing moieties from the polyaniline backbone. Further detailed X-ray photoelectron spectroscopy chemical state studies indicate that the reduction in S/N ratio of the mercaptopropanesulfonic-acid-substituted polyaniline initiated at 260 °C is mainly caused by the loss of the terminal -SO_3^- groups, while the sulfide linkage remains intact until around 300 °C. Based on thermal analysis in conjunction with heat treatment and X-ray photoelectron spectroscopy studies, it was concluded that the thermal degradation of sulfonated polyaniline involves the loss of a sulfonic acid group (initiated at 185 °C) followed by backbone decomposition, whereas while mercaptopropanesulfonic-acid-substituted polyaniline degradation also involves the loss of a

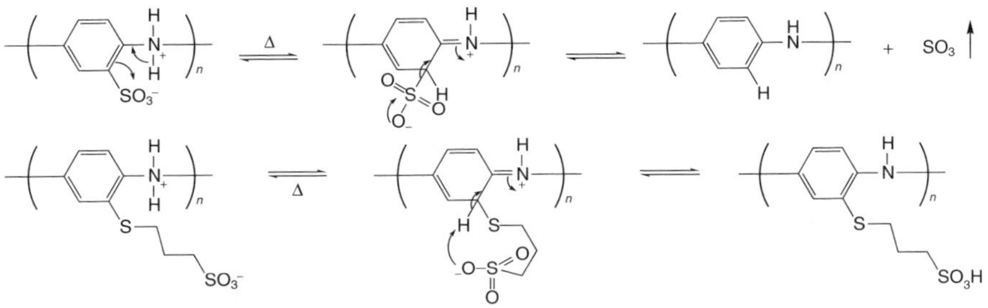

Figure 2.32 Proposed thermal decomposition mechanism of sulfonated polyaniline and mercaptopropanesulfonic acid substituted polyaniline. (Reprinted with permission from *Macromolecules*, 36, 7908. Copyright (2003) American Chemical Society.)

sulfonic acid group, it is initiated at the much higher temperature of 260 °C. The proposed decomposition mechanism of sulfonated polyaniline and mercaptopropanesulfonic acid substituted polyaniline is shown in Figure 2.32. For sulfonated polyaniline, initially a thermally activated, reversible 1,3-hydrogen shift is proposed to occur between the protonated imine site and the sulfonated *ortho*-carbon site, followed by a subsequent irreversible desulfonation due to the evaporation of the SO_3 byproduct at elevated temperatures. For mercaptopropanesulfonic acid substituted polyaniline, in addition to a similar 1,3-hydrogen shift, removal of the *ortho*-proton by the terminal -SO_3^- group (through a relatively more easily accessible eight-member-ring arrangement) was suggested to convert the unstable nonaromatic intermediate to the stable aromatic benzenoid ring, thus neutralizing the decomposition.

2.5.7 Morphology

Similarly to other properties, the position and size of substituents have a profound influence on the morphology of polyaniline [183]. Wu *et al.* [184, 185] suggested that the 'soluble' highly sulfonated polyaniline prepared following Epstein's procedure [54] is not a real solution, but instead is a colloidal dispersion where the solvation, steric and electrostatic effects of -SO_3^- are not sufficient to eliminate the strong backbone $\pi-\pi$ interactions. Colloidal particles with sizes in the range of 50 to 300 nm were observed depending on dispersion environment as shown in Figure 2.33. Atomic force microscopy images confirm that these particles are the aggregates of small particles with dimensions of the order of 10–20 nm. Ruckenstein *et al.* [113, 114] reported that the aggregation of

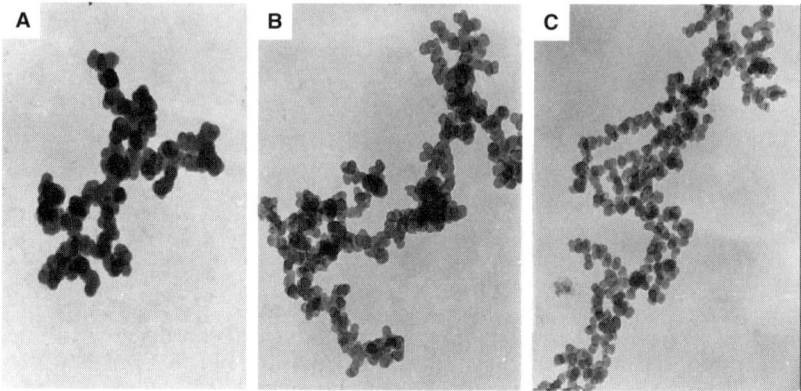

Figure 2.33 Transmission electron microscopy images of sulfonated polyaniline particles from (A) water dispersion, (B) dilute HCl dispersion, pH 3, and (C) dilute NH$_4$OH dispersion, pH 11. (Reprinted from *Synthetic Metals*, 108, Q. Wu, Z. Xue, Z. Qi, F. Wang, 107. Copyright (2000), with permission from Elsevier.)

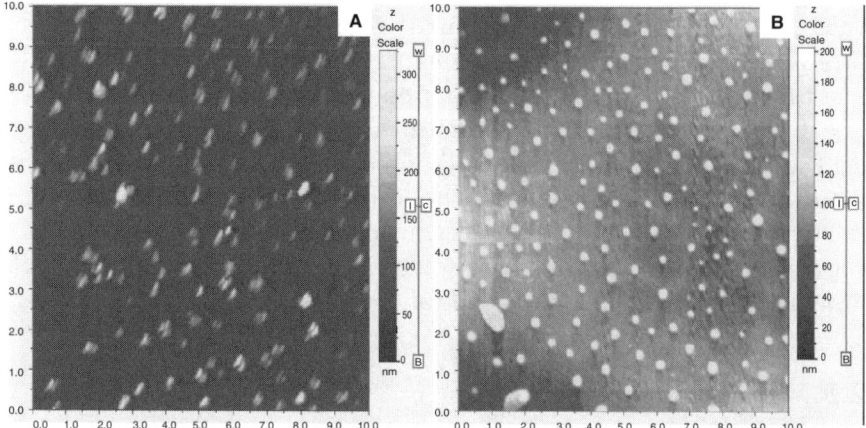

Figure 2.34 Atomic force microscopy images of H-PSDA obtained *via* spin coating on mica plates with a concentration of 0.1 g/L at different pH values: (A) self-doped PSDA at pH 7 and (B) externally doped PSDA at pH 3. (Reprinted with permission from *Macromolecules*, 38, 888. Copyright (2005) American Chemical Society.)

sulfonated polyaniline diblock and graft copolymers in water produces spherical nanoscale particles due to intermolecular and intramolecular self-doping (See section 2.2.2.3). Similarly, pH sensitive nanoscale morphology was observed for hyperbranched sulfonated polydiphenylamine (H-PSDA) as shown in Figure 2.34 [115]. For dedoped H-PSDA in the pH range 7–10, small particles with diameters in the range of 28 to 40 nm were formed. However, after self-doping at pH 7, the particle sizes became nonuniform, attaining an average of about 250 nm (Figure 2.34 A), because of the intermolecular self-doping, which leads to aggregation (See Figure 2.14 in section 2.2.2.3). When the pH is decreased to 3 (Figure 2.34 B), particle size decreases from 250 nm to 120 nm, presumably due to replacement of intermolecular self-doping with external HCl doping. For the electrochemically prepared homo- and copolymers of sulfonated polyanilines, a granular morphology was observed [90, 94, 145, 186, 187]. The formation of a granular morphology indicates that the steric and electrostatic effects of the covalently bound sulfonic acid group lead to a distortion in the polymer chains which, in turn, restricts regular packing during polymer growth and fibrillar morphology.

2.6 SYNTHESIS AND CHARACTERIZATION OF CARBOXYLIC ACID DERIVATIVES

In addition to sulfonic acid groups, carboxylic acid groups as ring substituents results in self-doping of polyaniline and influence properties such as solubility, pH dependent redox activity, conductivity, thermal stability, etc. Sulfonated polyanilines are typically obtained by postpolymerization modifications such as electrophilic and nucleophilic substitution reactions. However, carboxylic-acid-functionalized polyanilines are typically synthesized directly by chemical and electrochemical polymerization of monomer in the form of homopolymer or copolymer with aniline. In contrast to sulfonated polyaniline, very few monomers are available for the synthesis of carboxyl acid functionalized polyaniline. Anthranilic acid (2-aminobenzoic acid) is an important monomer and is often used for the synthesis of carboxyl acid functionalized polyanilines.

2.6.1 Chemical Synthesis

Chan *et al*. [188] first prepared the homopolymer of anthranilic acid and its copolymer with aniline by chemical polymerization in order to improve the solubility of polyaniline, to study the self-doping mechanism, and to evaluate thermal properties. The chemical polymerization of anthranilic acid was carried out using ammonium persulfate as the oxidant in the presence and absence of 1 M HCl. Copolymers of anthranilic acid and aniline were prepared in a similar manner by varying the monomer feed ratios in HCl at a pH of about 0.1. The general structure of copolymers is shown in Figure 2.35. The % yield of pure polyaniline in HCl was reported to be 67 %, whereas increasing anthranilic acid content in the comonomer feed, i.e., aniline to anthranilic acid feed ratio, resulted in a decreased yield of 39 % (1:1) and 32 % (1:2). The yields of the homopolymer poly(anthranilic acid) in the presence and absence of 1 M HCl were reportedly 8.1 and 6.9 %, respectively. The low yield was attributed to the electron withdrawing nature of the carboxylate group which decreases both the reactivity of anthranilic acid and its steric effect. The brownish black powder of the homopolymer of poly(anthranilic acid) prepared in 1 M HCl and the greenish black powder of copolymers were both soluble in N-methylpyrrolidinone and dimethyl sulfoxide. The degree of solubility of the copolymers increased with increasing anthranilic acid content. The conductivities of the homopolymers prepared in 1 M HCl and without HCl were 3×10^{-8} and $<10^{-10}$ S/cm respectively. The conductivity of copolymers (prepared in HCl) decreased from 4 to 2×10^{-5} S/cm with increasing anthranilic acid content.

X-ray photoelectron spectroscopy results show a decrease in the Cl/N ratio with increased anthranilic acid content, suggesting that some of the nitrogen atoms are protonated (self-doped) by intrinsic acid COOH in the copolymers. A similar self-doped mechanism is proposed for sulfonated polyanilines based on X-ray photoelectron spectroscopy [167]. The degree of protonation in the copolymer was around 20–30 %; however, conductivity varied drastically depending on the amount of self-doping vs external doping by HCl. Therefore it can be concluded that self-doping through the carboxylic acid in the copolymer is less effective than doping by HCl. The formation of five- or six-membered chelates between the -COO$^-$ group (Figure 2.36) and the polaronic nitrogen atom and the subsequent delocalization of charge between chelates were suggested to be responsible for reduced self-doping. In addition,

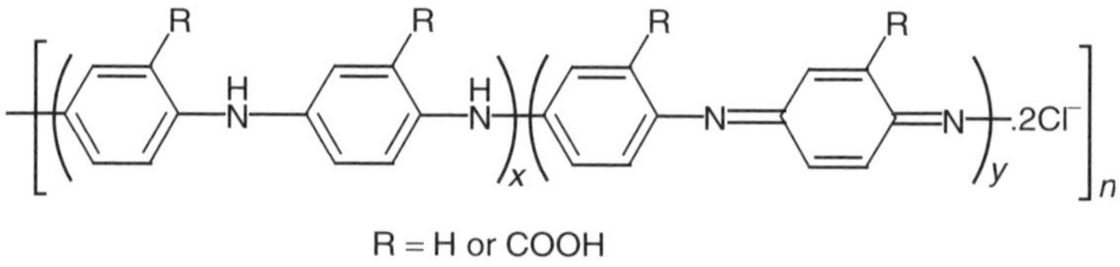

R = H or COOH

Figure 2.35 Structure of aniline/anthranilic acid copolymer. (Reprinted with permission from *Macromolecules*, 25, 6029. Copyright (1992) American Chemical Society.)

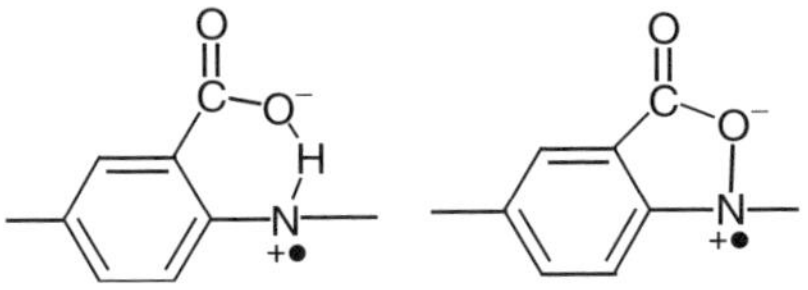

Figure 2.36 Five or six membered chelates formed due to intramolecular interaction between carboxylic acid and polaronic nitrogen atoms. (Reprinted with permission from *Macromolecules*, 25, 6029. Copyright (1992) American Chemical Society.)

other factors such as the disruption of the overlapping of orbitals as a result of steric effects associated with the bulky carboxylate group, and intermolecular interactions between the -COO$^-$ group and hydrogen on the adjacent phenyl ring, reportedly contributed to the decrease of conductivity. UV-Vis absorption spectra of the copolymers revealed a hypsochromic shift in the $\pi-\pi^*$ absorption band with an increase in the content of anthranilic acid, presumably due to steric interactions of the bulky carboxylate group and lowering of the degree of conjugation. The thermal stabilities of the homopolymer and copolymers of anthranilic acid were expected to be higher due to formation of hydrogen bonding between polymer chains. However, the thermograms were similar to polyaniline and no influence of hydrogen bonding on thermal stability was observed.

Diaz *et al.* [70] prepared poly(aniline-co-*o*-anthranilic acid) copolymers following a similar procedure to that of Chan *et al.* [188] with some differences in monomer, oxidant and acid concentrations. The differences were a monomer/oxidant ratio of 1.5 instead of 1.0, 1.2 M HCl instead of 1.0 M, and extensive washing of the product instead of washing with a 'small amount of HCl'. Copolymers prepared with these differences were reported to be soluble in aqueous alkaline solution as well as in organic solvents. The yield obtained by this method was improved over that reported by Chan *et al.*; however, the conductivity and yield of the copolymers decreased with increasing content of anthranilic acid as in the report by Chan *et al.* The molecular weight of the copolymer was in the range 26 000 to 84 000 g/mol depending on the content of anthranilic acid. Barbero *et al.* [96] prepared the copolymers by changing the position of carboxylic acid substituent on the aniline ring. Chemically polymerized poly(aniline-co-2-aminobenzoic acid) and poly(aniline-co-3-aminobenzoic acid) copolymers showed a difference in their properties such as specific charge and fluorescence behavior, due to the different reactivities of 2- and 3-aminobenzoic acid during copolymerization. The reactivities of 2- and 3-aminobenzoic acid were reported to be 12 and 6750 times less than aniline, respectively. Chemical homopolymerization of poly(o-aminobenzoic acid) in a $HCl/H_2O/(NH_4)_2S_2O_8$ solution at 40 °C has been reported [154]. The conductivity of the resulting polymer was reportedly 10^{-6} S/cm and independent of the pH up to 7. There are many reports on the monocarboxyl substituted polyaniline; however, 2,3-dicarboxylic acid substituted polyaniline has not been synthesized, probably due to steric hindrance of the carboxyl group.

Yamamoto et al. [189] chemically synthesized the self-doped oligo (2,3-dicarboxyaniline), which exhibits redox activity up to pH 6. These oligoanilines were prepared in neutral and alkaline media. The molecular weight and yield were in the range 680–1550 and 8–41 %, respectively. Under acidic conditions, 2,3-dicarboxyaniline was not polymerizable reportedly due to steric hindrance by the formation of a six-membered ring through hydrogen bonding with the amino group. The oligo(2,3-dicarboxyaniline) showed redox behaviour similar to polyaniline in aqueous acid solution. Two redox waves were observed up to pH 6 associated with the conversion of leucoemeraldine to the emeraldine salt and its subsequent conversion to pernigraniline. The proposed mechanism is shown in Figure 2.37. Yamamoto et al. suggest that the redox behaviour of oligo(2,3-dicarboxyaniline) is stable up to pH 6 due a different structure of pernigraniline than that of polyaniline. In polyaniline the fully oxidized pernigraniline form results in the degradation of the redox stability at high pH due to a hydrolysis reaction. The redox behaviour was shown to be stable over 10^3 cycles.

The synthesis of poly(4-aminobenzoic acid) catalyzed by horseradish peroxidase in the presence of hydrogen peroxide has been carried out by Tripathy et al. [149], similarly to sulfonated polyaniline (see section 2.4). The authors have suggested that the polymer is water soluble, electroactive, self-doped, and undergoes dedoping in alkaline or ammonia solution. Sathyanarayana et al. [95] chemically synthesized copolymers of 2- and 3-aminobenzoic acid with aniline using an inverse emulsion method in the presence of an organic oxidant, benzoyl peroxide. This approach reportedly overcomes the limitations of the conventional synthesis method such as the use of the strong oxidizing agent ammonium persulfate, difficulty in removing inorganic biproducts, control of reaction temperature and a wide distribution of molecular weights. The advantages of the inverse emulsion process reportedly results from the fact that the reaction takes place in a large number of loci dispersed in a continuous external phase in a heterogeneous system. This allows easier control over the process, less significant thermal and viscosity

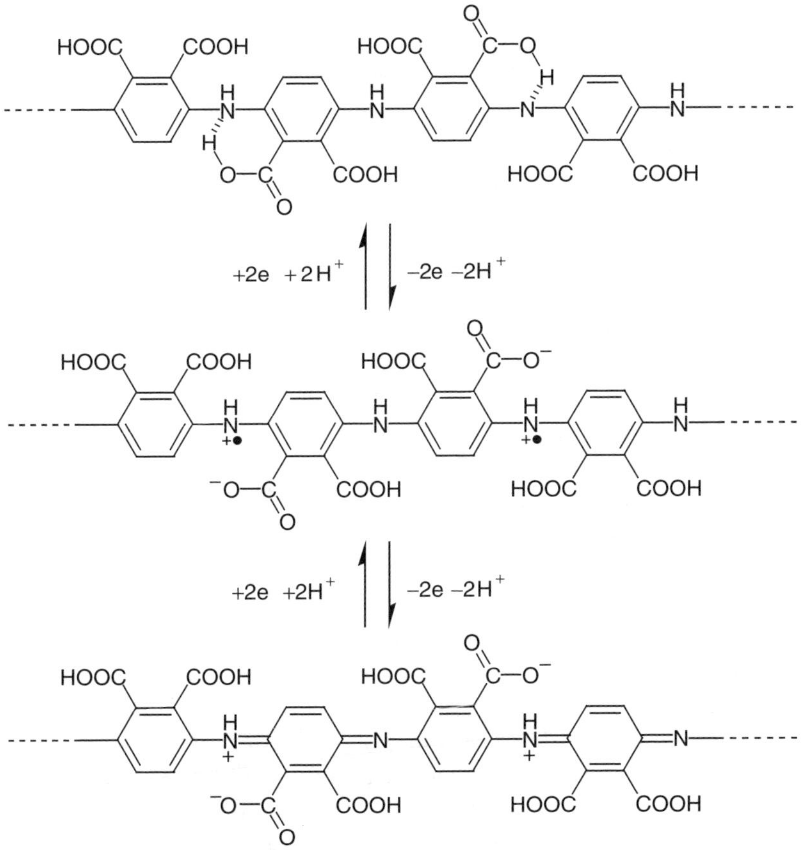

CARBOXYLIC ACID DERIVATIVES 127

Figure 2.37 Reversible conversion of oxidation states in oligo(2,3-dicarboxyaniline). (Reproduced from *Macromolecular Chemistry and Physics*, 2000, **201**, 6. K. Yamamoto, D. Tancichi, with permission from Wiley-VCH.)

problems than in bulk polymerization, and direct use of the product as a solution in the organic solvent without further processing. In the inverse emulsion procedure, copolymers were prepared by adding the oxidant benzoyl peroxide dissolved in chloroform into a mixture of emulsifier (sodium lauryl sulfate) and monomer 2-aminobenzoic acid in water. Aniline was then added in the mixture. Finally, to the milky white emulsion, 1 M HCl was added drop wise with continuous stirring. The polymer in the organic phase was separated and washed with water, and then with anhydrous sodium sulfate to remove traces of water. The viscous organic solution of copolymers was precipitated with acetone and dried in vacuum. The copolymer prepared using the inverse emulsion process resulted in high yield of up to 84 % and a conductivity of 2.5×10^{-1} S/cm, two orders higher than copolymers prepared using conventional ammonium persulfate [81, 188]. In addition, these copolymers are reported to be soluble in organic solvents and the solubility increases with the content of 2-aminobenzoic acid.

2.6.2 Electrochemical Synthesis

In addition to chemical synthesis, the electrochemical homo- and copolymerization of carboxyl acid functionalized polyanilines in aqueous H_2SO_4 and $HClO_4$ solution and their characteristics have been well studied [131, 161, 190–195]. Karyakin et al. [161] reported that the copolymers prepared using 2- and 3-aminobenzoic acid and aniline (1:1) ratio are self-doped and redox active in neutral and basic aqueous solutions up to pH 10. Brett et al. [191] prepared the homopolymers of 2-, 3- and 4-aminobenzoic acid and their copolymers with aniline. In this study, the authors associated the lower conductivity of homo- and copolymers with the formation of short-chain polymers. Scanning electron micrographs of the homopolymers and copolymers are shown in Figure 2.38. The polymer and copolymers exhibit a dendritic morphology that is different from polyaniline, reportedly due to the lower reactivity of carboxyl acid functionalized aniline monomer. A combined diffuse reflectance spectroelectrochemistry setup with quartz crystal microbalance has been shown to be a useful technique for acquiring multidimensional data during the growth and redox switching of the copolymer films by Xie et al. [192]. These studies suggest that preparation of the poly(2-aminobenzoic acid) homopolymer is difficult because of its solubility in the reaction medium (aqueous solution). The crystal frequency and resistance information obtained in this study suggested that the copolymer films were partially self-doped and more rigid than polyaniline. Recently, Knoll et al. [195] studied the polymerization process of aniline, anthranilic acid (2-aminobenzoic acid) and their copolymers using surface plasmon resonance spectroscopy. Surface plasmon resonance spectroscopic studies during polymerization indicate an 'autocatalytic effect' for polyaniline, where polymerization of aniline accelerates after initial formation of oligomers. Oligomers oxidize at lower potentials than does the monomer, which presumably results in an increased rate of polymerization of aniline. However, in the case of poly(anthranilic acid), the polymerization rate decreases after formation of oligomers, which was described as a 'self-inhibiting effect.' However,

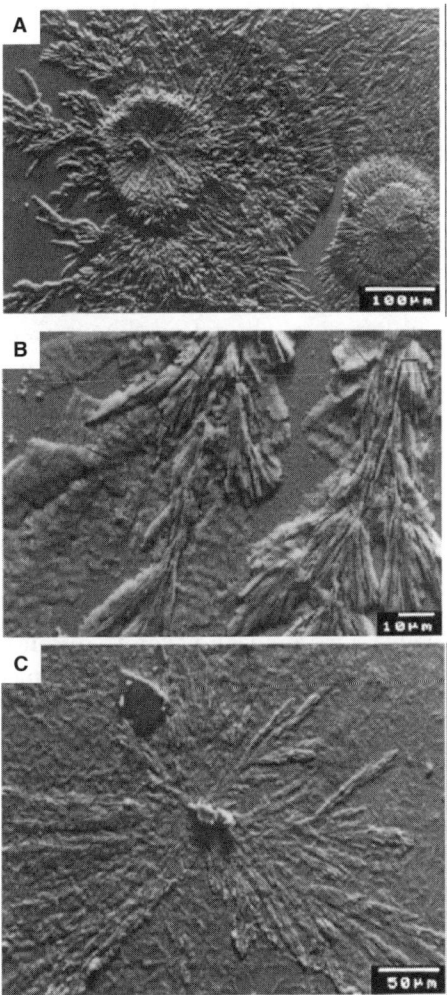

Figure 2.38 Scanning electron micrographs of (A) poly(2-aminobenzoic acid), (B) poly(4-aminobenzoic acid) and (C) poly(2-aminobenzoic acid)/polyaniline, 2:1. (Reprinted from *Synthetic Metals*, 123, C. Thiemann, M. A. Brett, 1. Copyright (2001), with permission from Elsevier.)

in the case of copolymerization, intermediate rates were observed due to the offsetting effect of both monomers. Comparative studies between poly(anthranilic acid) and polyaniline have shown that only the copolymers from aniline and 2-anthranilic acid with appropriate monomer fractions retain a high electroactivity and stability on electrodes in neutral aqueous solution.

2.7 SYNTHESIS AND CHARACTERIZATION OF PHOSPHONIC ACID DERIVATIVES

In comparison with self-doped sulfonic acid and carboxylic acid functionalized polyaniline, phosphonic acid derivatives have not been widely studied. Ng and Chan et al. [196, 197] were the first to prepare self-doped poly(o-aminobenzylphosphonic acid) chemically by direct polymerization of monomer o-aminobenzylphosphonic acid. The polymer was reportedly soluble in its conducting form. The synthetic route to monomer and polymer is shown in Figure 2.39. The monomer (Figure 2.39, 4) was synthesized in three steps from the starting material o-nitrobenzylbromide (Figure 2.39, 1). The self-doped poly(o-aminobenzylphosphonic acid) (Figure, 2.39, 5) was prepared by oxidative coupling of the monomer (Figure 2.39, 4) in an acidic medium. By progressive neutralization of phosphonic acid functional groups with NaOH, hemisodium (Figure 2.40, 1), monosodium (Figure 2.40, 2) and disodium salts of poly(o-aminobenzylphosphonic acid) were obtained as shown in Figure 2.40. Both the hemisodium and monosodium salts were reportedly intrinsically self-doped and water-soluble in the conducting form. However, the disodium salt was an insulating dedoped form of polyaniline. According to X-ray photoelectron spectroscopy results, the polymer as prepared (Figure 2.39, 5), was in the emeraldine oxidation state in which 43 % nitrogen doping occurred solely via internal proton transfer (i.e., self-doping) from the phosphonic acid groups. The hemisodium and monosodium salts of the polymer resulted in 34 and 26 % self-doping, respectively. In the fully neutralized disodium salt polymer, there were no internal protons for protonation of the polymer backbone as the phosphonic acid moiety is doubly ionized.

The conductivity of the as-prepared poly(o-aminobenzylphosphonic acid) (Figure 2.39, 5) was measured in the form of a compact pellet and reported to be approximately 1.5×10^{-3} S/cm and independent of pH up to 6. With progressive neutralization, conductivity decreased to 10^{-7} S/cm for the fully neutralized disodium salt polymer (Figure 2.40, 3). The conductivity of as-prepared (1.5×10^{-3} S/cm) and HCl doped poly(o-aminobenzylphosphonic acid) (10^{-3} S/cm) is lower in comparison with sulfonated polyanilines (0.1 to 1 S/cm) [41, 54] and conventional acid-doped polyaniline. The lower conductivity was attributed to the decrease in conjugation caused by the large steric effect of the bulky -PO_3H_2 and a significant hydrogen bond interaction between -$PO_2(OH)^-$ and $NH^{•+}$ leading to significant charge pinning. The polymer (Figure 2.39, 5) was reported to be electroactive in aqueous media

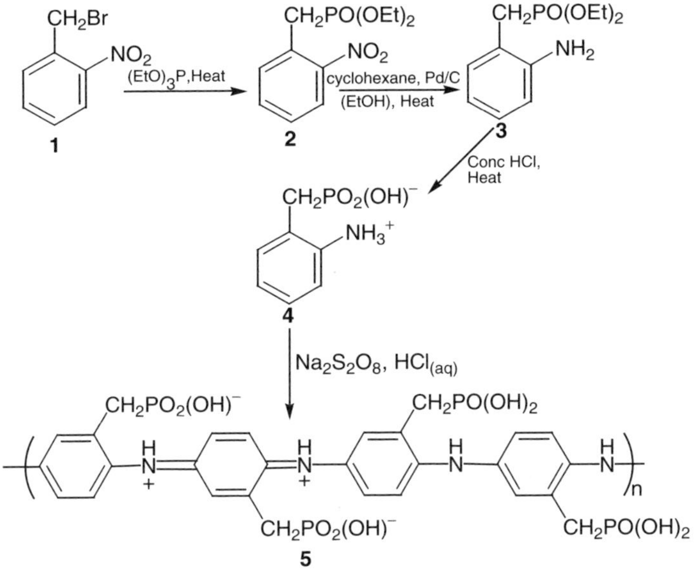

Figure 2.39 Synthesis route of monomer, o-aminobenzylphosphonic acid (4) and self-doped polymer, poly(o-aminobenzylphosphonic acid) (5) (Reprinted with permission from *Journal of the American Chemical Society*, 117, 8517. Copyright (1995) American Chemical Society.)

and highly stable to repeated potential cycling ($>10^2$ cycles) in the range of -0.2 to 1.0 V. During cyclic voltammetry, one redox wave was observed in contrast to conventional polyaniline which shows two sets of redox waves. It was suggested that the emeraldine form of poly(o-aminobenzylphosphonic acid) is exceptionally resistant to oxidation to pernigraniline state due to enhanced hydrogen bond stabilization of the emeraldine salt. FT-IR studies confirmed the hydrogen bonding between $-PO_2(OH)^-$ and $NH^{•+}$ and the progressive neutralization of the phosphonic acid group. The UV-Vis absorption spectra of the polymer shows hypsochromic shifts compared with the parent polyaniline similar to sulfonated polyanilines [41]. However, the shift was reported to be larger for the phosphonic acid derivative due to a decrease in ring conjugation resulting from steric and hydrogen bond interactions. ^{31}P NMR spectra of the D_2O soluble polymer confirmed the presence of unpaired spins (polarons) in the solvated self-doped polymer chains (Figure 2.40, 1 and 2). The protonated hemisodium polymer (Figure 2.40, 1) exhibited broadening of the NMR signal whereas the fully neutralized disodium polymer (Figure 2.40, 3) exhibited a sharp signal. However, the partially

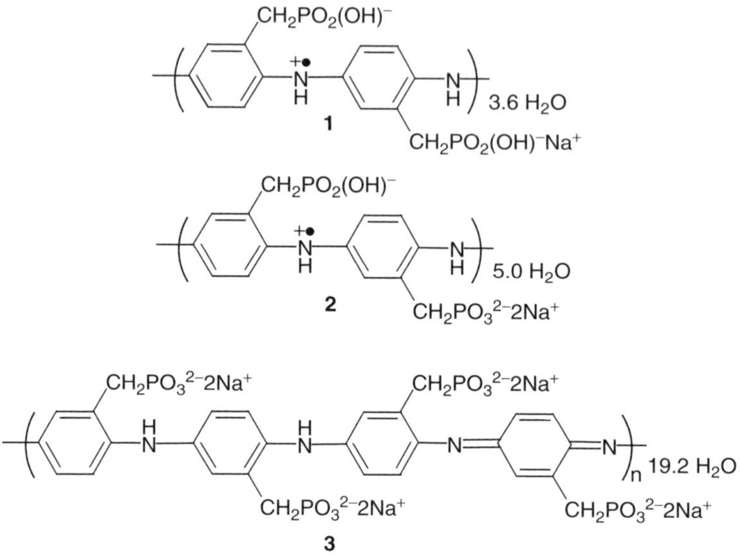

Figure 2.40 Sodium salts of poly(o-aminobenzylphosphonic acid). (Reprinted with permission from *Journal of the American Chemical Society*, 117, 8517. Copyright (1995) American Chemical Society.)

neutralized monosodium polymer (Figure 2.40, **2**) exhibited both broad and sharp signals of protonated and unprotonated segments, respectively. It was suggested that the sharp signals arise from the ^{31}P nuclei in the unprotonated segment of the polymer chains, while broad signals arise from the ^{31}P nuclei interacting with spins in the protonated segment.

2.8 SELF-DOPED POLYANILINE NANOSTRUCTURES

The facile synthesis of materials of nanometer length as well as control over the dimensions of inorganic and organic nanostructures is essential due to their unique size-dependent properties and potential applications in nanodevices, such as field effect transistors [198], sensor/actuator arrays [199, 200], optoelectronic devices [201, 202], and in biotechnology (e.g., delivery agents for pharmaceutical agents), [203, 204] as well as catalytic and analytical systems [205]. Conducting polymers form a unique class of organic materials and are emerging as a promising material for synthesis of nanostructured materials and devices due to their electrical, electronic, magnetic and optical properties. They offer great prospects for practical applications [206–208], which range from chemical and biological sensing and diagnosis to energy conversion and storage, light-emitting display devices, catalysis, drug delivery, separation, microelectronics and optical storage due to their great architectural diversity and flexibility, low cost and ease of synthesis. In recent years, conducting polymer based nanostructured materials in the form of thin films and nanowires have attracted much attention for the construction of fast and inexpensive devices [209–215].

Among all conducting polymers, polyaniline is probably the most widely studied because it has a broad range of electrical and optical tunable properties derived from its structural flexibility. The doping level of polyaniline can be readily controlled through an acid/base dedoping process [157], and it has high conductivity, good environmental stability and ease of preparation. Different morphologies of polyaniline have been obtained through different synthetic approaches or processing routes [216–218]. Low-dimensional nanostructures of polyaniline in various shapes and forms, for example, nanoparticles, nanowires, nanofibers, nanoshells and nanotubes, have been produced by various methods [217, 219–235]. Conducting polyaniline nanostructures are prime candidates for replacing conventional bulk materials in micro- and nanoelectronic devices [236–238] and in chemical [239–242] and biological [219, 243] sensors, because they combine the properties of low-dimensional organic conductors with high surface area materials, resulting in greater sensitivity and faster response times compared with their conventional bulk counterpart. Also, polyaniline nanostructures have metal like and controllable conductivity as well as both thermal and environmental stability.

Self-doped polyanilines are advantageous due to properties such as solubility, pH independence, redox activity and conductivity. These properties make them more promising in various applications such as energy conversion devices, sensors, electrochromic devices, etc. (see Chapter 1, section 1.6). Several studies have focused on the preparation of self-doped polyaniline nanostructures (i.e., nanoparticles, nanofibers, nanofilms, nanocomposites, etc.) and their applications. Buttry and Torresi *et al.* [51, 244, 245] prepared the nanocomposites from self-doped polyaniline, poly(N-propane sulfonic acid, aniline) and V_2O_5 for Li secondary battery cathodes. The self-doped polyaniline was used instead of conventional polyaniline to minimize the anion participation in the charge–discharge process and maximize the transport number of Li^+. In lithium batteries, it is desirable that only lithium cations intercalate into the cathode, because this leads to the use of small amounts of electrolyte that only serves as carrier for the cations to migrate/diffuse from the anode to the cathode, and this permits the fabrication of batteries with higher specific capacity. Furthermore, self-doped polyaniline offers better chemical stability due to the steric protection by the covalently bound negatively charged group on the backbone. This fact reduces the degradation of the quinoid structure during oxidation, increasing the cycleability of the electrode. Two types of nanocomposites have been prepared by the sol–gel method where vanadium tris(isopropoxide) reacted with N-(propane sulfonic acid) aniline, and by adding H_2O_2. A larger specific capacity of 307 Ah/kg, faster reduction kinetics and higher cycleability in comparison to $V_2O_5.16H_2O$ has been reported using nanocomposites.

Wei et al. [246] prepared self-doped copolymer poly(aniline-co-aminonaphthalene sulfonic acid) (polyaniline-ANSA) nanotubes via self-assembly. Copolymers were synthesized chemically at different molar ratios of aniline/ANSA in deionized water using ammonium persulfate as an oxidizing agent. Scanning electron micrograph (SEM) and transmission electron micrograph (TEM) images of self-doped polyaniline-ANSA prepared at different aniline/ANSA molar ratios are shown in Figure 2.41. Nanotubular morphology at a high molar ratio and granular morphology at a low molar ratio of aniline/ANSA were observed. The TEM images of polyaniline-ANSA at a molar ratio of 50 showed hollow nanofibrils with outer diameter 60–100 nm and inner diameter of less than 10 nm. However, with a decrease in aniline/ANSA molar ratio from 50 to 2, a transition from fibrilar to granular morphology was observed. The formation of hollow nanofibrils or nanotubes and transition to granular morphology at lower aniline/ANSA molar ratio were explained based on the aggregation model shown in Figure 2.42. Based on the SEM and TEM studies of initially formed polymer, it was suggested that the nanotubes and granules were composed of smaller nanograins with diameters of several nanometers. For polyaniline-ANSA, each of these nanograins includes many -SO_3H groups on naphthalene units. These -SO_3H groups are probably distributed into two parts based on their function as dopant to nanograins and on their surface. At high molar ratios of aniline/ANSA, the distribution of surface charge is probably asymmetric and nanotubes were formed due to self-assembly. However, at low molar ratios of aniline/ANSA, the symmetric surface-charged nanograins are randomly aggregated and form granular morphology. These nanocomposites are shown to have better thermal stability than HCl-doped polyaniline. The increase in solubility from 36 to 86 (wt%) and self-doped conductivity 5.7×10^{-4} to 4.4×10^{-2} S/cm

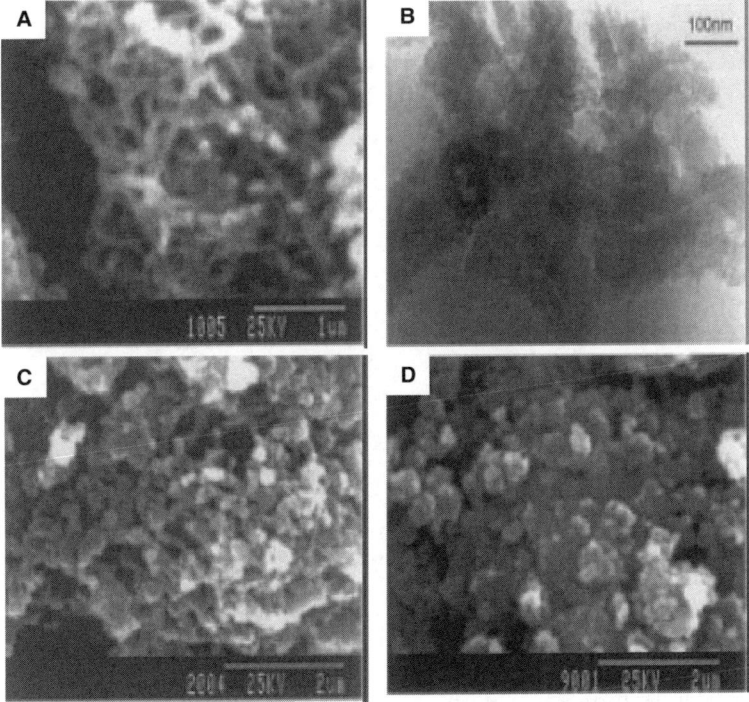

Figure 2.41 SEM and TEM images of self-doped polyaniline-ANSA prepared at different aniline/ANSA molar ratios: (A) 50, SEM, (B) 50, TEM, (C) 10, SEM, and (D) 2, SEM. (Reproduced from *Journal of Applied Polymer Science*, Z. Wei, M. J. Wan, 87, 1297. Copyright (2003), with permission of John Wiley & Sons, Inc.)

were reported with changes in molar ratio of aniline/ANSA from 50 to 2.

In general, the synthesis of polyaniline nanostructures has been carried out both chemically and electrochemically by polymerizing aniline using templates, surfactants, liquid crystals, thiolated cyclodextrins and polyacids, electrospinning, mechanical stretching, coagulating media, interfacial polymerization, seeding and dilute polymerization [217, 219–235]. Surfactants can be used as templates as well as stabilizers for creating nanostructures. Self-doped polymers provide a means for stabilizing nanostructure, for example, Li *et al.* [247] have prepared self-doped copolymer poly(ethylaniline/sulfoanisidine) nanoparticles by chemical polymerization without external stabilizer. The nanoparticle size was controlled by varying the comonomer ratio. Nanoparticles of 120–160 nm are reported to be formed at an ethylaniline/sulfoanisidine

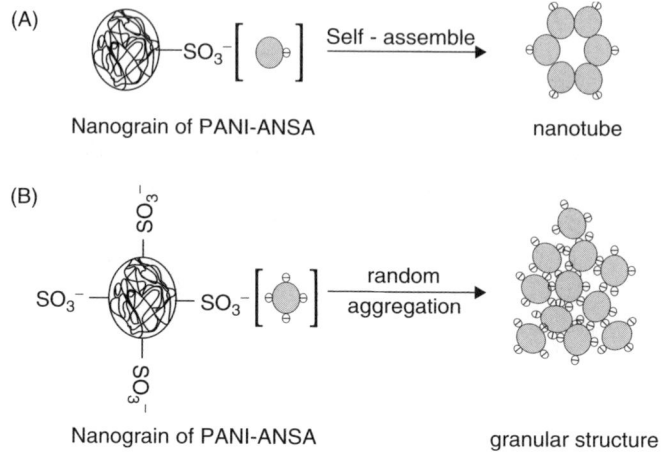

Figure 2.42 Proposed aggregation model of polyaniline-ANSA prepared at different aniline/ANSA molar ratios: (A) self-assembly at a high molar ratio and (B) random aggregation at a lower molar ratio. (Reproduced from *Journal of Applied Polymer Science*, Z. Wei, M. J. Wan, **87**, 1297. Copyright (2003) with permission of John Wiley & Sons, Inc.)

ratio of 70/30. However, at an ethylaniline/sulfoanisidine ratio of 10/90, the particle size of the conducting copolymer increases to 600 nm. In this study, the sulfoanisidine monomers reportedly act as internal stabilizers, which produce electrostatic repulsion between particles and stabilizes small particles. Recently, You and Niu *et al.* [248] prepared self-doped copolymer poly(o-aminobenzenesulfonic acid-co-aniline) microflowers. Electrochemical copolymerization on a Pt coated silicon wafer was carried out using 17 mM aniline and 85 mM o-aminobenzenesulfonic acid without supporting electrolyte. The potential was scanned between −0.1 and 0.9 V vs Ag/AgCl in an unstirred solution at a scan rate of 50 mV/s [248]. The film was washed with pure water and dried under a flow of nitrogen. Then chronoamperometry was carried out in the same monomer solution between 1.1 and −0.2 V with a pulse width of 5 s. The SEM images of self-doped poly(o-aminobenzenesulfonic acid-co-aniline) microflowers are shown in Figure 2.43. The mechanism of microflower formation was not explored in detail; however, it was suggested that they grow in three steps: first, films are generated during the CV scanning; it is then believed that the films became loosened during the application of a high potential (1.1 V); finally the flowers form with the help of nitrogen flow during the drying process. The copolymer films have shown electroactivity at up to pH 13.5.

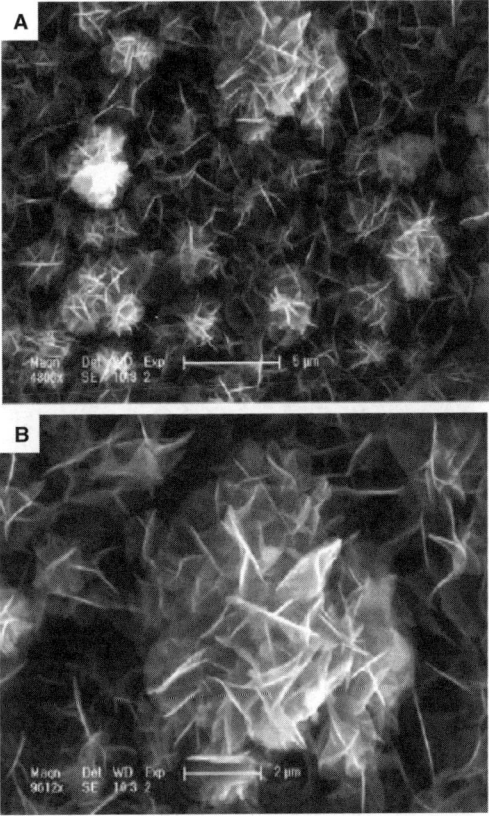

Figure 2.43 SEM images of self-doped poly(o-aminobenzenesulfonic acid-co-aniline) microflowers at (A) low and (B) high magnifications. (Reprinted from *Electrochemistry Communications*, 7, Z. Wang, L. Jiao, T. You, L. Niu, S. Dongs, A. Ivaska, 875. Copyright (2005), with permission from Elsevier.)

Self-doped polyaniline nanofibers have been synthesized recently by Yang *et al.* [249] via a self-assembly process using the self-doped monomer (o-aminobenzenesulfonic acid) and aniline. It was suggested that the sulfonated monomer (o-aminobenzenesulfonic acid) plays a dual role as a monomer and a surfactant in the nanofiber formation. The nanofibers were synthesized chemically using different molar ratios of aniline/o-aminobenzenesulfonic acid without an external dopant. It was proposed that the o-aminobenzenesulfonic acid and anilinium cations form micelles in aqueous solution, which further act as templates to form nanofibers. Self-doped polyaniline nanofibers with an average diameter of 370 nm synthesized using an aniline/o-aminobenzenesulfonic acid

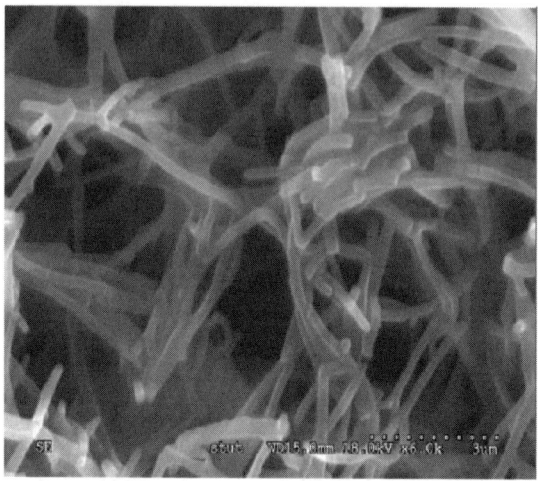

Figure 2.44 SEM image of self-doped polyaniline nanofibers synthesized using an aniline/o-aminobenzenesulfonic acid mole ratio of 1. (Reprinted from *Polymer*, **46**, C. H. Yang, Y. K. Chih, H. E. Cheng, C. H. Chen, 10688. Copyright (2005), with permission from Elsevier.)

mole ratio of 1 is shown in Figure 2.44. A decrease in nanofiber diameter with increasing molar ratio of aniline/o-aminobenzenesulfonic acid was observed. The degree of self-doping was reported to be 21–27 % based on X-ray photoelectron spectroscopy results. However, the conductivity reported was very low, in the range of 10^{-5} to 10^{-6} S/cm. Yang *et al.* [250] have also prepared thin (105 nm) self-doped copolymer films using aniline and o-aminobenzenesulfonic acid. A stable electrochromic device was fabricated using a self-doped copolymer film on indium tin oxide glass with poly(3,4-ethylenedioxythiophene):poly(4-styrenesulfonate). The coulombic efficiency and stability of the optical contrast after the 300th cycle was shown to be 94 and 90 %, respectively. The coulumbic efficiency of 94 % is reportedly much higher than in polythiophene derivative based electrochromic devices (72 %) [251].

Fully sulfonated polyaniline nanoparticles, nanofibers and nanonetworks have been obtained by electrochemical homopolymerization of o-aminobenzenesulfonic acid for the purpose of mediating the electrochemistry of cyctochrome *c* [252]. The nanostructures were synthesized in a mixed solvent of acetonitrile and water (8:1) at 0 °C by following a three-step galvanostatic procedure. First, a relatively high current density is applied to initiate the polymerization. This is followed by two periods during which lower current densities are applied.

A typical procedure involves: 0.06 mA-cm^{-2} for 0.5 h, followed by 0.03 mA-cm^{-2} for 3 h, which was then followed by another 3 h at 0.015 mA-cm^{-2}. Poly(o-aminobenzenesulfonic acid) nanoparticles of 60 nm diameter were obtained at a high galvanostatic current density of 0.06 mA-cm^{-2}. These nanoparticles were used as nucleation sites to grow extended polymer nanofibers by applying a second and third deposition step with a reduced current density. Nanofibers with diameters of 50–100 nm and heights of more than 500 nm standing upright on the electrode surface were obtained. The nanofibers were organized in two- (2D) and three-dimensional (3D) nonperiodic networks with electrical contact. The average distance between contacts was reported to be around 850 and 600 nm for the 2D and 3D systems, respectively.

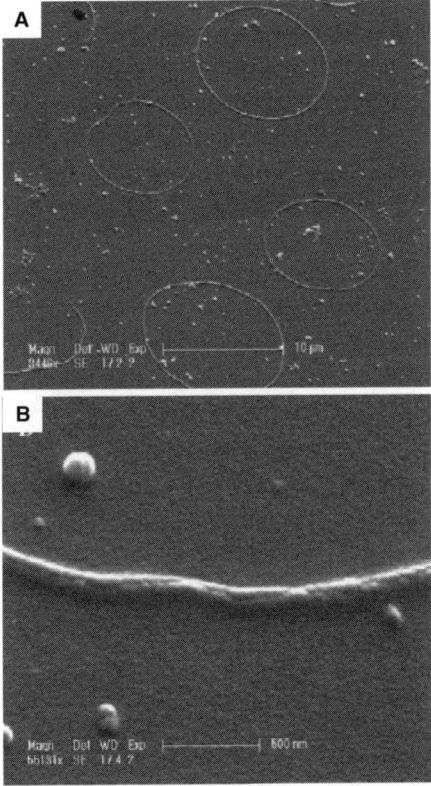

Figure 2.45 SEM images of poly(aniline-co-o-aminobenzesulfonic acid) micro rings prepared using aniline and o-aminobenzesulfonic acid (1:1) at current density of 10 mA cm^2 for 45 min at (A) low and (B) high magnification. (Reproduced from *Nanotechnology*, 2006, 17, 824, J. Song, D. Han, L. Guo, L. Niu, with permission from IOP Publishing Limited.)

The fully sulfonated 3D porous poly(o-aminobenzenesulfonic acid) film coated platinum electrode effectively mediated the redox chemistry of cytochrome c, demonstrating the interaction between cytochrome c and poly(o-aminobenzenesulfonic acid) following the adsorption and immobilization of cytochrome c within the network. Self-doped poly(aniline-co-o-aminobenzesulfonic acid) micro rings have been produced recently by electrochemical synthesis [253]. Copolymer rings with fiber widths of approximately 100 nm and ring diameters that can be varied from several to dozens of micrometers were obtained by applying different current densities. The SEM images of micro rings are shown in Figure 2.45.

REFERENCES

[1] H. Letheby, 'On the production of a blue substance by the electrolysis of sulphate of aniline,' *Journal of the Chemical Society* **1862**, *15*, 161.
[2] A. G. Green, A. E. Woodhead, 'Aniline black and allied compounds. I.,' *Journal of the Chemical Society, Transactions* **1910**, *97*, 2388.
[3] A. G. Green, A. E. Woodhead, 'Aniline black and allied compounds. II.,' *Journal of the Chemical Society, Transactions* **1912**, *101*, 1117.
[4] T. Sotomura, H. Uemachi, K. Takeyama, K. Naoi, N. Oyama, 'New organodisulfide polyaniline composite cathode for secondary lithium battery,' *Electrochimica Acta* **1992**, *37*, 1851.
[5] A. Kaminaga, T. Tatsuma, T. Sotomura, N. Oyama, 'Reactivation and reduction of electrochemically inactivated polyaniline by 2,5-dimercapto-1,3,4-thiadiazole,' *Journal of the Electrochemical Society* **1995**, *142*, L47.
[6] E. M. Genies, S. Picart, 'Is the use of polyaniline associated with sulfurcompounds of interest for battery electrodes,' *Synthetic Metals* **1995**, *69*, 165.
[7] N. Oyama, T. Tatsuma, T. Sato, T. Sotomura, 'Dimercaptan–polyaniline composite electrodes for lithium batteries with high-energy density,' *Nature* **1995**, *373*, 598.
[8] G. Gustafsson, Y. Cao, G. M. Treacy, F. Klavetter, N. Colaneri, A. J. Heeger, 'Flexible light-emitting-diodes made from soluble conducting polymers,' *Nature* **1992**, *357*, 477.
[9] Y. Yang, E. Westerweele, C. Zhang, P. Smith, A. J. Heeger, 'Enhanced performance of polymer light-emitting diodes using high-surface area polyaniline network electrodes,' *Journal of Applied Physics* **1995**, *77*, 694.
[10] R. K. Yuan, S. C. Yang, H. Yuan, R. L. Jiang, H. Z. Qian, D. C. Gui, 'Surface field-effect of polyaniline film,' *Synthetic Metals* **1991**, *41*, 727.
[11] K. S. Lee, G. B. Blanchet, F. Gao, Y. L. Loo, 'Direct patterning of conductive water-soluble polyaniline for thin-film organic electronics,' *Applied Physics Letters* **2005**, *86*, 074102.
[12] P. N. Bartlett, P. R. Birkin, 'The application of conducting polymers in biosensors,' *Synthetic Metals* **1993**, *61*, 15.

[13] J. A. Osaheni, S. A. Jenekhe, H. Vanherzeele, J. S. Meth, Y. Sun, A. G. MacDiarmid, 'Nonlinear optical-properties of polyanilines and derivatives,' *Journal of Physical Chemistry* **1992**, *96*, 2830.

[14] A. J. Epstein, J. A. O. Smallfield, H. Guan, M. Fahlman, 'Corrosion protection of aluminum and aluminum alloys by polyanilines: a potentiodynamic and photoelectron spectroscopy study,' *Synthetic Metals* **1999**, *102*, 1374.

[15] T. Schauer, A. Joos, L. Dulog, C. D. Eisenbach, 'Protection of iron against corrosion with polyaniline primers,' *Progress in Organic Coatings* **1998**, *33*, 20.

[16] B. Wessling, 'Passivation of metals by coating with polyaniline – corrosion potential shift and morphological changes,' *Advanced Materials* **1994**, *6*, 226.

[17] W. K. Lu, R. L. Elsenbaumer, B. Wessling, 'Corrosion protection of mild steel by coatings containing polyaniline,' *Synthetic Metals* **1995**, *71*, 2163.

[18] S. Baek, J. J. Ree, M. Ree, 'Synthesis and characterization of conducting poly(aniline-co-o-aminophenethyl alcohol)s,' *Journal of Polymer Science Part A – Polymer Chemistry* **2002**, *40*, 983.

[19] J. Joo, C. Y. Lee, 'High frequency electromagnetic interference shielding response of mixtures and multilayer films based on conducting polymers,' *Journal of Applied Physics* **2000**, *88*, 513.

[20] K. Gurunathan, A. V. Murugan, R. Marimuthu, U. P. Mulik, D. P. Amalnerkar, 'Electrochemically synthesised conducting polymeric materials for applications towards technology in electronics, optoelectronics and energy storage devices,' *Materials Chemistry and Physics* **1999**, *61*, 173.

[21] T. Kobayashi, H. Yoneyama, H. Tamura, 'Polyaniline film-coated electrodes as electrochromic display devices,' *Journal of Electroanalytical Chemistry* **1984**, *161*, 419.

[22] A. Barnes, A. Despotakis, T. C. P. Wong, A. P. Anderson, B. Chambers, P. V. Wright, 'Towards a 'smart window' for microwave applications,' *Smart Materials and Structures* **1998**, *7*, 752.

[23] A. Watanabe, K. Mori, A. Iwabuchi, Y. Iwasaki, Y. Nakamura, O. Ito, 'Electrochemical polymerization of aniline and N-alkylanilines,' *Macromolecules* **1989**, *22*, 3521.

[24] M. C. Gupta, S. S. Umare, 'Studies on poly(*ortho*-methoxyaniline),' *Macromolecules* **1992**, *25*, 138.

[25] Y. H. Liao, M. Angelopoulos, K. Levon, 'Ring-substituted polyaniline copolymers combining high solubility with high-conductivity,' *Journal of Polymer Science Part A – Polymer Chemistry* **1995**, *33*, 2725.

[26] G. D. Storrier, S. B. Colbran, D. B. Hibbert, 'Chemical and electrochemical syntheses, and characterization of poly(2,5-dimethoxyaniline) (PDMA) – a novel, soluble, conducting polymer,' *Synthetic Metals* **1994**, *62*, 179.

[27] G. Daprano, M. Leclerc, G. Zotti, G. Schiavon, 'Synthesis and characterization of polyaniline derivatives – poly(2-alkoxyanilines) and poly(2,5-dialkoxyanilines),' *Chemistry of Materials* **1995**, *7*, 33.

[28] S. L. Wang, F. S. Wang, X. H. Ge, 'Polymerization of substituted aniline and characterization of the polymers obtained,' *Synthetic Metals* **1986**, *16*, 99.

[29] E. M. Genies, J. F. Penneau, M. Lapkowski, 'Electrochemistry of ortho, meta and para toluidines in NH$_4$F, 2.3HF and conduction properties of products and polymers formed,' *New Journal of Chemistry* **1988**, *12*, 765.

[30] S. Cattarin, L. Doubova, G. Mengoli, G. Zotti, 'Electrosynthesis and properties of ring-substituted polyanilines,' *Electrochimica Acta* **1988**, *33*, 1077.
[31] Y. Wei, W. W. Focke, G. E. Wnek, A. Ray, A. G. MacDiarmid, 'Synthesis and electrochemistry of alkyl ring-substituted polyanilines,' *The Journal of Physical Chemistry* **1989**, *93*, 495
[32] M. Leclerc, J. Guay, L. H. Dao, 'Synthesis and characterization of poly(alkylanilines),' *Macromolecules* **1989**, *22*, 649.
[33] G. Bidan, E. M. Genies, J. F. Penneau, 'Poly(2-propylaniline) – an electroactive polymer, soluble in organic medium in the reduced state,' *Journal of Electroanalytical Chemistry* **1989**, *271*, 59.
[34] E. M. Genies, P. Noel, 'Synthesis and polymerization of ortho-hexylaniline – characterization of the corresponding polyaniline,' *Journal of Electroanalytical Chemistry* **1991**, *310*, 89.
[35] D. Macinnes, B. L. Funt, 'Poly-*ortho*-methoxyaniline – a new soluble conducting polymer,' *Synthetic Metals* **1988**, *25*, 235.
[36] G. Pistoia, G. Montesperelli, P. Nunziante, 'Polyaniline derivatives – an investigation by cyclic voltammetry and impedance spectroscopy,' *Journal of Molecular Electronics* **1990**, *6*, 89.
[37] J. C. Lacroix, P. Garcia, J. P. Audiere, R. Clement, O. Kahn, 'Electropolymerization of methoxyaniline – experimental results and frontier orbital interpretation,' *Synthetic Metals* **1991**, *44*, 117.
[38] N. Ahmad, P. Feng, H. Schursky, S. Shah, D. Antonacci, M. B. Cichowicz, R. K. Kohli, A. G. MacDiarmid, 'Synthesis and characterization of some polyalkyl-anilines and polyalkoxy-anilines,' *Indian Journal of Chemistry Section A – Inorganic Bio-Inorganic Physical Theoretical and Analytical Chemistry* **1993**, *32*, 673.
[39] Y. Jiang, A. J. Epstein, 'Synthesis of self-doped conducting polyaniline,' *Journal of the American Chemical Society* **1990**, *112*, 2800.
[40] J. Y. Bergeron, J. W. Chevalier, L. H. Dao, 'Water-soluble conducting poly (aniline) polymer,' *Chemical Communications* **1990**, 180.
[41] J. Yue, Z. H. Wang, K. R. Cromack, A. J. Epstein, A. G. MacDiarmid, 'Effect of sulfonic acid group on polyaniline backbone,' *Journal of the American Chemical Society* **1991**, *113*, 2665.
[42] C. Barbero, H. J. Salavagione, D. F. Acevedo, D. E. Grumelli, F. Garay, G. A. Planes, G. M. Morales, M. C. Miras, 'Novel synthetic methods to produce functionalized conducting polymers: polyanilines,' *Electrochimica Acta* **2004**, *49*, 3671.
[43] G. Daprano, M. Leclerc, G. Zotti, 'Stabilization and characterization of pernigraniline salt – the acid-doped form of fully oxidized polyanilines,' *Macromolecules* **1992**, *25*, 2145.
[44] P. Snauwaert, R. Lazzaroni, J. Riga, J. J. Verbist, 'Electronic-structure of polyanilines – an XPS study of electrochemically prepared compounds,' *Synthetic Metals* **1986**, *16*, 245.
[45] M. Kaneko, K. Kaneto, 'Electrochemomechanical deformation of polyaniline films doped with self-existent and giant anions,' *Reactive and Functional Polymers* **1998**, *37*, 155.
[46] E. T. Kang, K. G. Neoh, Y. L. Woo, K. L. Tan, 'Self-doped polyaniline and polypyrrole – a comparative-study by X-ray photoelectron-spectroscopy,' *Polymer Communications* **1991**, *32*, 412.

[47] S. A. Chen, G. W. Hwang, 'Structure characterization of self-acid-doped sulfonic acid ring-substituted polyaniline in its aqueous solutions and as solid film,' *Macromolecules* **1996**, *29*, 3950.

[48] C. Barbero, M. C. Miras, B. Schnyder, O. Haas, R. Kotz, 'Sulfonated polyaniline films as cation insertion electrodes for battery applications. 1. Structural and electrochemical characterization,' *Journal of Materials Chemistry* **1994**, *4*, 1775.

[49] C. Barbero, M. C. Miras, R. Kotz, O. Haas, 'Comparative study of the ion-exchange and electrochemical properties of sulfonated polyaniline (SPAN) and polyaniline (PANi),' *Synthetic Metals* **1993**, *55*, 1539.

[50] C. Barbero, M. C. Miras, R. Kotz, O. Haas, 'Sulphonated polyaniline (SPAN) films as cation insertion electrodes for battery applications – Part ii: exchange of mobile species in aqueous and nonaqueous solutions,' *Journal of Electroanalytical Chemistry* **1997**, *437*, 191.

[51] F. Huguenin, R. M. Torresi, D. A. Buttry, J. E. P. da Silva, S. I. C. de Torresi, 'Electrochemical and raman studies on a hybrid organic–inorganic nanocomposite of vanadium oxide and a sulfonated polyaniline,' *Electrochimica Acta* **2001**, *46*, 3555.

[52] H. Varela, S. Maranhao, R. M. Q. Mello, E. A. Ticianelli, R. M. Torresi, 'Comparisons of charge compensation process in aqueous media of polyaniline and self-doped polyanilines,' *Synthetic Metals* **2001**, *122*, 321.

[53] M. C. Bernard, V. T. Bich, S. C. deTorresi, A. HugotLeGoff, 'Spectroelectrochemical characterization (oma and raman) of sulfonic acids doped polyanilines,' *Synthetic Metals* **1997**, *84*, 785.

[54] X. L. Wei, Y. Z. Wang, S. M. Long, C. Bobeczko, A. J. Epstein, 'Synthesis and physical properties of highly sulfonated polyaniline,' *Journal of the American Chemical Society* **1996**, *118*, 2545.

[55] J. Yue, G. Gordon, A. J. Epstein, 'Comparison of different synthetic routes for sulfonation of polyaniline,' *Polymer* **1992**, *33*, 4410.

[56] C. H. Hsu, P. M. Peacock, R. B. Flippen, S. K. Manohar, A. G. MacDiarmid, 'The molecular weight of polyaniline by light-scattering and gel-permeation chromatography,' *Synthetic Metals* **1993**, *60*, 233.

[57] S. Ito, K. Murata, S. Teshima, R. Aizawa, Y. Asako, K. Takahashi, B. M. Hoffman, 'Simple synthesis of water soluble conducting polyaniline,' *Synthetic Metals* **1998**, *96*, 161.

[58] P. Hany, E. M. Genies, C. Santier, 'Polyanilines with covalently bonded alkyl sulfonates as doping agent – synthesis and properties,' *Synthetic Metals* **1989**, *31*, 369.

[59] S. A. Chen, G. W. Hwang, 'Synthesis of water-soluble self-acid-doped polyaniline,' *Journal of the American Chemical Society* **1994**, *116*, 7939.

[60] M. Y. Hua, Y. N. Su, S. A. Chen, 'Water-soluble self-acid-doped conducting polyaniline: poly(aniline-co-n-propylbenzenesulfonic acid-aniline),' *Polymer* **2000**, *41*, 813.

[61] H. K. Lin, S. A. Chen, 'Synthesis of new water-soluble self-doped polyaniline,' *Macromolecules* **2000**, *33*, 8117.

[62] C. C. Han, R. C. Jeng, 'Concurrent reduction and modification of polyaniline emeraldine base with pyrrolidine and other nucleophiles,' *Chemical Communications* **1997**, 553.

[63] C. C. Han, S. P. Hong, K. F. Yang, M. Y. Bai, C. H. Lu, C. S. Huang, 'Highly conductive new aniline copolymers containing butylthio substituent,' *Macromolecules* **2001**, *34*, 587.
[64] C. C. Han, W. D. Hseih, J. Y. Yeh, S. P. Hong, 'Combination of electrochemistry with concurrent reduction and substitution chemistry to provide a facile and versatile tool for preparing highly functionalized polyanilines,' *Chemistry of Materials* **1999**, *11*, 480.
[65] H. Salavagione, G. M. Morales, M. C. Miras, C. Barbero, 'Synthesis of a self-doped polyaniline by nucleophilic addition,' *Acta Polymerica* **1999**, *50*, 40.
[66] C. C. Han, C. H. Lu, S. P. Hong, K. F. Yang, 'Highly conductive and thermally stable self-doping propylthiosulfonated polyanilines,' *Macromolecules* **2003**, *36*, 7908.
[67] H. Zollinger, *Diazo Chemistry I: Aromatic and Heteroaromatic Compounds*, VCH, New York, **1994**.
[68] G. Liu, M. S. Freund, 'Nucleophilic substitution reactions of polyaniline with substituted benzenediazonium ions: a facile method for controlling the surface chemistry of conducting polymers,' *Chemistry of Materials* **1996**, *8*, 1164.
[69] G. A. Planes, G. M. Morales, M. C. Miras, C. Barbero, 'A soluble and electroactive polyaniline obtained by coupling of 4-sulfobenzenediazonium ion and poly(N-methylaniline),' *Synthetic Metals* **1998**, *97*, 223.
[70] S. Shimizu, T. Saitoh, M. Uzawa, M. Yuasa, K. Yano, T. Maruyama, K. Watanabe, 'Synthesis and applications of sulfonated polyaniline,' *Synthetic Metals* **1997**, *85*, 1337.
[71] W. Lee, G. Du, S. M. Long, A. J. Epstein, S. Shimizu, T. Saitoh, M. Uzawa, 'Charge transport properties of fully sulfonated polyaniline,' *Synthetic Metals* **1997**, *84*, 807.
[72] H. S. O. Chan, A. J. Neuendorf, S. C. Ng, P. M. L. Wong, D. J. Young, 'Synthesis of fully sulfonated polyaniline: a novel approach using oxidative polymerisation under high pressure in the liquid phase,' *Chemical Communications* **1998**, 1327.
[73] C. Dearmitt, S. P. Armes, J. Winter, F. A. Uribe, S. Gottesfeld, C. Mombourquette, 'A novel N-substituted polyaniline derivative,' *Polymer* **1993**, *34*, 158.
[74] S. Atkinson, H. S. O. Chan, A. J. Neuendorf, S. C. Ng, T. T. Ong, D. J. Young, 'Synthesis of the water soluble, electrically conducting poly(5-aminonaphthalene-2-sulfonic acid),' *Chemistry Letters* **2000**, 276.
[75] V. George, D. J. Young, 'Synthesis and characterization of novel self-doping water soluble polynaphthylamines,' *Polymer* **2002**, *43*, 4073.
[76] Y. Wei, R. Hariharan, S. A. Patel, 'Chemical and electrochemical copolymerization of aniline with alkyl ring-substituted anilines,' *Macromolecules* **1990**, *23*, 758.
[77] H. Yoon, B. S. Jung, H. Lee, 'Correlation between electrical conductivity, thermal-conductivity, and ESR intensity of polyaniline,' *Synthetic Metals* **1991**, *41*, 699.
[78] S. S. Pandey, S. Annapoorni, B. D. Malhotra, 'Synthesis and characterization of poly(aniline-co-o-anisidine) a processable conducting copolymer,' *Macromolecules* **1993**, *26*, 3190.

REFERENCES

[79] L. H. C. Mattoso, S. K. Manohar, A. G. MacDiarmid, A. J. Epstein, 'Studies on the chemical syntheses and on the characteristics of polyaniline derivatives,' *Journal of Polymer Science Part A – Polymer Chemistry* **1995**, *33*, 1227.

[80] J. A. Conklin, S. C. Huang, S. M. Huang, T. L. Wen, R. B. Kaner, 'Thermal-properties of polyaniline and poly(aniline-co-o-ethylaniline),' *Macromolecules* **1995**, *28*, 6522.

[81] M. T. Nguyen, A. F. Diaz, 'Water-soluble poly(aniline-co-o-anthranilic acid) copolymers,' *Macromolecules* **1995**, *28*, 3411.

[82] P. J. Kinlen, J. Liu, Y. Ding, C. R. Graham, E. E. Remsen, 'Emulsion polymerization process for organically soluble and electrically conducting polyaniline,' *Macromolecules* **1998**, *31*, 1735.

[83] M. R. Huang, X. G. Li, Y. L. Yang, X. S. Wang, D. Y. Yan, 'Oxidative copolymers of aniline with o-toluidine: their structure and thermal properties,' *Journal of Applied Polymer Science* **2001**, *81*, 1838.

[84] P. Savitha, D. N. Sathyanarayana, 'Copolymers of aniline with o- and m-toluidine: synthesis and characterization,' *Polymer International* **2004**, *53*, 106.

[85] H. H. Rehan, S. H. Al-Mazroa, F. F. Al-Fawzan, 'Synthesis and characterization of conducting copolymers from aniline and o-anisidine in HCl solutions,' *Polymer International* **2003**, *52*, 918.

[86] V. Prévost, A. Petit, F. Pla, 'Studies on chemical oxidative copolymerization of aniline and o-alkoxysulfonated anilines: I. Synthesis and characterization of novel self-doped polyanilines,' *Synthetic Metals* **1999**, *104*, 79.

[87] A. Dan, P. K. Sengupta, 'Synthesis and characterization of conducting poly(aniline-co-diaminodiphenylsulfone) copolymers,' *Journal of Applied Polymer Science* **2003**, *90*, 2337.

[88] M. T. Nguyen, P. Kasai, J. L. Miller, A. F. Diaz, 'Synthesis and properties of novel water-soluble conducting polyaniline copolymers,' *Macromolecules* **1994**, *27*, 3625.

[89] H. Mizobuchi, T. Kawai, K. Yoshino, 'Ferromagnetic behavior of self-doping type polyaniline derivatives depending on oxidation state,' *Solid State Communications* **1995**, *96*, 925.

[90] J. H. Fan, M. X. Wan, D. B. Zhu, 'Synthesis and characterization of water-soluble conducting copolymer poly(aniline-co-o-aminobenzenesulfonic acid),' *Journal of Polymer Science Part A – Polymer Chemistry* **1998**, *36*, 3013.

[91] I. Mav, M. Zigon, A. Sebenik, 'Sulfonated polyaniline,' *Synthetic Metals* **1999**, *101*, 717.

[92] B. C. Roy, M. D. Gupta, L. Bhoumik, J. K. Ray, 'Spectroscopic investigation of water-soluble polyaniline copolymers,' *Synthetic Metals* **2002**, *130*, 27.

[93] L. V. Lukachova, E. A. Shkerin, E. A. Puganova, E. E. Karyakina, S. G. Kiseleva, A. V. Orlov, G. P. Karpacheva, A. A. Karyakin, 'Electroactivity of chemically synthesized polyaniline in neutral and alkaline aqueous solutions – role of self-doping and external doping,' *Journal of Electroanalytical Chemistry* **2003**, *544*, 59.

[94] P. S. Rao, D. N. Sathyanarayana, 'Effect of the sulfonic acid group on copolymers of aniline and toluidine with m-aminobenzene sulfonic acid,' *Journal of Polymer Science Part A – Polymer Chemistry* **2002**, *40*, 4065.

[95] P. S. Rao, D. N. Sathyanarayana, 'Synthesis of electrically conducting copolymers of aniline with o/m-amino benzoic acid by an inverse emulsion pathway,' *Polymer* **2002**, *43*, 5051.

[96] H. J. Salavagione, D. F. Acevedo, M. C. Miras, A. J. Motheo, C. A. Barbero, 'Comparative study of 2-amino and 3-aminobenzoic acid copolymerization with aniline synthesis and copolymer properties,' *Journal of Polymer Science Part A – Polymer Chemistry* **2004**, *42*, 5587.

[97] K. G. Neoh, E. T. Kang, K. L. Tan, 'Coexistence of external protonation and self-doping in polyaniline,' *Synthetic Metals* **1993**, *60*, 13.

[98] Y. H. Park, H. C. Shin, Y. Lee, Y. Son, D. H. Baik, 'Electrochemical preparation of polypyrrole copolymer films from PSPMS precursor,' *Macromolecules* **1999**, *32*, 4615.

[99] A. I. Nazzal, G. B. Street, 'Pyrrole styrene graft-copolymers,' *Chemical Communications* **1985**, 375.

[100] G. B. Street, S. E. Lindsey, A. I. Nazzal, K. J. Wynne, 'The structure and mechanical properties of polypyrrole,' *Molecular Crystals and Liquid Crystals* **1985**, *118*, 137.

[101] D. Stanke, M. L. Hallensleben, L. Toppare, 'Electrically conductive poly (methyl methacrylate-g-pyrrole) via chemical oxidative polymerization,' *Synthetic Metals* **1993**, *55*, 1108.

[102] U. Geissler, M. L. Hallensleben, L. Toppare, 'Conductive polymer composites and copolymers of pyrrole and N-vinylcarbazole,' *Synthetic Metals* **1991**, *40*, 239.

[103] U. Geissler, M. L. Hallensleben, L. Toppare, 'Electrochemical studies on carbazole/pyrrole-copolymers,' *Synthetic Metals* **1993**, *55*, 1483.

[104] N. Kizilyar, L. Toppare, A. Onen, Y. Yagci, 'Synthesis of conducting PPy/PTHF copolymers,' *Journal of Applied Polymer Science* **1999**, *71*, 713.

[105] C. Ozdilek, J. Hacaloglu, L. Toppare, Y. Yagci, 'Structural and thermal characterization of PTSA doped polypyrrole-polytetrahydrofuran graft copolymer,' *Synthetic Metals* **2004**, *140*, 69.

[106] G. Abbati, E. Carone, L. D'Ilario, A. Martinelli, 'Polyurethane-polyaniline conducting graft copolymer with improved mechanical properties,' *Journal of Applied Polymer Science* **2003**, *89*, 2516.

[107] E. Carone, L. D'Ilario, A. Martinelli, 'New conducting thermoplastic elastomers. I. Synthesis and chemical characterization,' *Journal of Applied Polymer Science* **2002**, *83*, 857.

[108] E. Carone, L. D'Ilario, A. Martinelli, 'New conductive thermoplastic elastomers. Part II. Physical and chemical-physical characterization,' *Journal of Applied Polymer Science* **2002**, *86*, 1259.

[109] F. R. De Risi, L. D'Ilario, A. Martinelli, 'Synthesis and characterization of epoxidized polybutadiene/polyaniline graft conducting copolymer,' *Journal of Polymer Science Part A – Polymer Chemistry* **2004**, *42*, 3082.

[110] E. T. Kang, K. G. Neoh, K. L. Tan, Y. Uyama, N. Morikawa, Y. Ikada, 'Surface modifications of polyaniline films by graft copolymerization,' *Macromolecules* **1992**, *25*, 1959.

[111] W. S. Yin, E. Ruckenstein, 'Water-soluble self-doped conducting polyaniline copolymer,' *Macromolecules* **2000**, *33*, 1129.

[112] W. J. Bae, K. H. Kim, Y. H. Park, W. H. Jo, 'A novel water soluble and self-doped conducting polyaniline graft copolymer,' *Chemical Communications* **2003**, 2768.
[113] F. J. Hua, E. Ruckenstein, 'Synthesis of a water soluble diblock copolymer of polysulfonic diphenyl aniline and poly(ethylene oxide),' *Journal of Polymer Science Part A – Polymer Chemistry* **2004**, *42*, 2179.
[114] F. J. Hua, E. Ruckenstein, 'Preparation of densely grafted poly(aniline-2-sulfonic acid-co-aniline)s as novel water-soluble conducting,' *Journal of Polymer Science Part A – Polymer Chemistry* **2005**, *43*, 1090.
[115] F. J. Hua, E. Ruckenstein, 'Hyperbranched sulfonated polydiphenylamine as a novel self-doped conducting polymer and its pH response,' *Macromolecules* **2005**, *38*, 888.
[116] A. Kitani, K. Satoguchi, H. Q. Tang, S. Ito, K. Sasaki, 'Electrosynthesis and properties of self-doped polyaniline,' *Synthetic Metals* **1995**, *69*, 129.
[117] H. Q. Tang, A. Kitani, S. Ito, 'Electrochemical copolymerization of aniline and aniline-2,5-disulfonic acid,' *Electrochimica Acta* **1997**, *42*, 3421.
[118] H. Q. Tang, T. Yamashita, A. Kitani, S. Ito, 'Electrosynthesis of water-soluble self-doped poly(aniline-2,5-disulfonic acid),' *Electrochimica Acta* **1998**, *43*, 2237.
[119] D. Z. Zhou, P. C. Innis, G. G. Wallace, S. Shimizu, S. I. Maeda, 'Electrosynthesis and characterisation of poly(2-methoxyaniline-5-sulfonic acid) – effect of pH control,' *Synthetic Metals* **2000**, *114*, 287.
[120] R. Guo, J. N. Barisci, P. C. Innis, C. O. Too, G. G. Wallace, D. Zhou, 'Electrohydrodynamic polymerization of 2-methoxyaniline-5-sulfonic acid,' *Synthetic Metals* **2000**, *114*, 267.
[121] A. Malinauskas, R. Holze, 'Deposition and characterisation of self-doped sulphoalkylated polyanilines,' *Electrochimica Acta* **1998**, *43*, 521.
[122] E. V. Strounina, L. A. P. Kane-Maguire, G. G. Wallace, 'Optically active sulfonated polyanilines,' *Synthetic Metals* **1999**, *106*, 129.
[123] J. Y. Lee, C. Q. Cui, X. H. Su, M. S. Zhou, 'Modified polyaniline through simultaneous electrochemical polymerization of aniline and metanilic acid,' *Journal of Electroanalytical Chemistry* **1993**, *360*, 177.
[124] J. Y. Lee, X. H. Su, C. Q. Cui, 'Characterization of electrodeposited copolymers of aniline and metanilic acid,' *Journal of Electroanalytical Chemistry* **1994**, *367*, 71.
[125] J. Y. Lee, C. Q. Cui, 'Electrochemical copolymerization of aniline and metanilic acid,' *Journal of Electroanalytical Chemistry* **1996**, *403*, 109.
[126] C. H. Yang, T. C. Wen, 'Polyaniline derivative with external and internal doping via electrochemical copolymerization of aniline and 2,5-diaminobenzenesulfonic acid on IrO_2-coated titanium electrode,' *Journal of the Electrochemical Society* **1994**, *141*, 2624.
[127] Y. Wei, G. W. Jang, K. F. Hsueh, R. Hariharan, S. A. Patel, C. C. Chan, C. Whitecar, 'Effects of *p*-phenylenediamine and other additives on polymerization of aniline and its derivatives and on the properties of resultant polymers,' *Polymeric Materials Science and Engineering* **1989**, *61*, 905.
[128] H. Q. Tang, A. Kitani, S. Maitani, H. Munemura, M. Shiotani, 'Electropolymerization of aniline modified by *para*-phenylenediamine,' *Electrochimica Acta* **1995**, *40*, 849.

[129] A. A. Karyakin, I. A. Maltsev, L. V. Lukachova, 'The influence of defects in polyaniline structure on its electroactivity: optimization of "self-doped" polyaniline synthesis,' *Journal of Electroanalytical Chemistry* **1996**, *402*, 217.
[130] R. M. Q. Mello, R. M. Torresi, S. I. C. de Torresi, E. A. Ticianelli, 'Ellipsometric, electrogravimetric, and spectroelectrochemical studies of the redox process of sulfonated polyaniline,' *Langmuir* **2000**, *16*, 7835.
[131] C. M. A. Brett, C. Thiemann, 'Conducting polymers from aminobenzoic acids and aminobenzenesulphonic acids: influence of pH on electrochemical behaviour,' *Journal of Electroanalytical Chemistry* **2002**, *538*, 215.
[132] C. Thiemann, C. M. A. Brett, 'Electropolymerisation and properties of conducting polymers derived from aminobenzenesulphonic acids and from mixtures with aniline,' *Synthetic Metals* **2001**, *125*, 445.
[133] R. Mazeikiene, G. Niaura, A. Malinauskas, 'Voltammetric study of the redox processes of self-doped sulfonated polyaniline,' *Synthetic Metals* **2003**, *139*, 89.
[134] G. Niaura, R. Mazeikiene, A. Malinauskas, 'Structural changes in conducting form of polyaniline upon ring sulfonation as deduced by near infrared resonance raman spectroscopy,' *Synthetic Metals* **2004**, *145*, 105.
[135] T. Kawai, H. Mizobuchi, N. Yamasaki, H. Araki, K. Yoshino, 'Optical and magnetic-properties of polyaniline derivatives having ionic groups,' *Japanese Journal of Applied Physics Part 2 – Letters* **1994**, *33*, L357.
[136] R. Mazeikiene, A. Malinauskas, 'Voltammetric study of some redox processes at electrodes, modified with new self-doped polyanilines,' *Bulletin of Electrochemistry* **2003**, *19*, 547.
[137] R. Mazeikiene, A. Malinauskas, 'Electrochemical preparation and study of novel self-doped polyanilines,' *Materials Chemistry and Physics* **2004**, *83*, 184.
[138] R. Mazeiklene, G. Niaura, A. Malinauskas, 'Surface enhanced resonance raman spectroelectrochemical study of electrochemically generated copolymer films of aniline and aminonaphthalenedisulfonates,' *Journal of Electroanalytical Chemistry* **2005**, *580*, 87.
[139] R. Mazeikiene, G. Niaura, A. Malinauskas, 'Raman spectroelectrochemical study of self-doped copolymers of aniline and selected aminonaphthalenesulfonates,' *Electrochimica Acta* **2006**, *51*, 1917.
[140] C. Barbero, R. Kotz, 'Electrochemical formation of a self-doped conductive polymer in the absence of a supporting electrolyte – the copolymerization of o-aminobenzenesulfonic acid and aniline,' *Advanced Materials* **1994**, *6*, 577.
[141] D. E. Grumelli, E. S. Forzani, G. M. Morales, M. C. Miras, C. A. Barbero, E. J. Calvo, 'Microgravimetric study of electrochemically controlled nucleophilic addition of sulfite to polyaniline,' *Langmuir* **2004**, *20*, 2349.
[142] A. Kitani, K. Satoguchi, K. Iwai, S. Ito, 'Electrochemical behaviors of polyaniline/polyaniline-sulfonic acid composites,' *Synthetic Metals* **1999**, *102*, 1171.
[143] A. Kitani, K. Iwai, S. Ito, 'Electrochemical behaviors of polyaniline/poly(aniline-2,5-disulfonic acid) composites,' *Electrochemistry* **1999**, *67*, 1262.
[144] E. Kim, M. Lee, M. H. Lee, S. B. Rhee, 'Liquid-crystalline assemblies from self-doped polyanilines,' *Synthetic Metals* **1995**, *69*, 101.

[145] Y. Sahin, K. Pekmez, A. Yildiz, 'Electrochemical preparation of soluble sulfonated polymers and aniline copolymers of aniline sulfonic acids in dimethylsulfoxide,' *Journal of Applied Polymer Science* **2003**, *90*, 2163.

[146] Y. Sahin, K. Pekmez, A. Yildiz, 'Electrochemical copolymerization of aniline and anilinesulfonic acids in FSO_3H/acetonitrile solution,' *Journal of Applied Polymer Science* **2002**, *85*, 1227.

[147] Y. Sahin, K. Pekmez, A. Yildiz, 'Electrochemical synthesis of self-doped polyaniline in fluorosulfonic acid/acetonitrile solution,' *Synthetic Metals* **2002**, *129*, 107.

[148] Y. Sahin, K. Pekmez, A. Yildiz, 'Electropolymerization and *in situ* sulfonation of aniline in water–acetonitrile mixture containing FSO_3H,' *Synthetic Metals* **2002**, *131*, 7.

[149] K. S. Alva, K. A. Marx, J. Kumar, S. K. Tripathy, 'Biochemical synthesis of water soluble polyanilines: poly(*p*-aminobenzoic acid),' *Macromolecular Rapid Communications* **1996**, *17*, 859.

[150] K. S. Alva, J. Kumar, K. A. Marx, S. K. Tripathy, 'Enzymatic synthesis and characterization of a novel water-soluble polyaniline: poly(2,5-diaminobenzenesulfonate),' *Macromolecules* **1997**, *30*, 4024.

[151] W. Liu, J. Kumar, S. Tripathy, K. J. Senecal, L. Samuelson, 'Enzymically synthesized conducting polyaniline,' *Journal of the American Chemical Society* **1999**, *121*, 71.

[152] W. Liu, A. L. Cholli, R. Nagarajan, J. Kumar, S. Tripathy, F. F. Bruno, L. Samuelson, 'The role of template in the enzymatic synthesis of conducting polyaniline,' *Journal of the American Chemical Society* **1999**, *121*, 11345.

[153] A. G. MacDiarmid, J. C. Chiang, A. F. Richter, A. J. Epstein, 'Polyaniline – a new concept in conducting polymers,' *Synthetic Metals* **1987**, *18*, 285.

[154] X. H. Wang, J. Li, L. X. Wang, X. B. Jing, F. S. Wang, 'Structure and properties of self-doped polyaniline,' *Synthetic Metals* **1995**, *69*, 147.

[155] C. M. Li, S. L. Mu, 'The electrochemical activity of sulfonic acid ring-substituted polyaniline in the wide pH range,' *Synthetic Metals* **2005**, *149*, 143.

[156] Z. H. Wang, H. H. S. Javadi, A. Ray, A. G. MacDiarmid, A. J. Epstein, 'Electron localization in polyaniline derivatives,' *Physical Review B* **1990**, *42*, 5411.

[157] W. S. Huang, B. D. Humphrey, A. G. MacDiarmid, 'Polyaniline, a novel conducting polymer – morphology and chemistry of its oxidation and reduction in aqueous electrolytes,' *Journal of the Chemical Society-Faraday Transactions I* **1986**, *82*, 2385.

[158] J. F. Wolf, C. E. Forbes, S. Gould, L. W. Shacklette, 'Proton-dependent electrochemical-behavior of oligomeric polyaniline compounds,' *Journal of the Electrochemical Society* **1989**, *136*, 2887.

[159] S. H. Glarum, J. H. Marshall, 'Electron delocalization in poly(aniline),' *Journal of Physical Chemistry* **1988**, *92*, 4210.

[160] W. W. Focke, G. E. Wnek, 'Conduction mechanisms in polyaniline (emeraldine salt),' *Journal of Electroanalytical Chemistry* **1988**, *256*, 343.

[161] A. A. Karyakin, A. K. Strakhova, A. K. Yatsimirsky, 'Self-doped polyanilines electrochemically active in neutral and basic aqueous solutions – electropolymerization of substituted anilines,' *Journal of Electroanalytical Chemistry* **1994**, *371*, 259.

[162] J. M. Ginder, A. J. Epstein, A. G. MacDiarmid, 'Ring-rotational defects in polyaniline,' *Solid State Communications* **1989**, *72*, 987.
[163] J. M. Ginder, A. J. Epstein, 'Role of ring torsion angle in polyaniline – electronic structure and defect states,' *Physical Review B* **1990**, *41*, 10674.
[164] E. V. Strounina, R. Shepherd, L. A. P. Kane-Maguire, G. G. Wallace, 'Conformational changes in sulfonated polyaniline caused by metal salts and OH,' *Synthetic Metals* **2003**, *135*, 289.
[165] A. G. MacDiarmid, J. C. Chiang, M. Halpern, W. S. Huang, S. L. Mu, N. L. D. Somasiri, W. Q. Wu, S. I. Yaniger, 'Polyaniline – interconversion of metallic and insulating forms,' *Molecular Crystals and Liquid Crystals* **1985**, *121*, 173.
[166] A. G. MacDiarmid, A. J. Epstein, 'Polyanilines – a novel class of conducting polymers,' *Faraday Discussions* **1989**, 317.
[167] J. Yue, A. J. Epstein, 'XPS study of self-doped conducting polyaniline and parent systems,' *Macromolecules* **1991**, *24*, 4441.
[168] X. L. Wei, A. J. Epstein, 'Synthesis of highly sulfonated polyaniline,' *Synthetic Metals* **1995**, *74*, 123.
[169] X. L. Wei, M. Fahlman, K. J. Epstein, 'XPS study of highly sulfonated polyaniline,' *Macromolecules* **1999**, *32*, 3114.
[170] A. J. Epstein, J. M. Ginder, F. Zuo, R. W. Bigelow, H. S. Woo, D. B. Tanner, A. F. Richter, W. S. Huang, A. G. MacDiarmid, 'Insulator-to-metal transition in polyaniline,' *Synthetic Metals* **1987**, *18*, 303.
[171] J. M. Ginder, A. F. Richter, A. G. MacDiarmid, A. J. Epstein, 'Insulator-to-metal transition in polyaniline,' *Solid State Communications* **1987**, *63*, 97.
[172] I. B. Goldberg, H. R. Crowe, P. R. Newman, A. J. Heeger, A. G. MacDiarmid, 'Electron spin resonance of polyacetylene and arsenic pentafluoride-doped polyacetylene,' *Journal of Chemical Physics* **1979**, *70*, 1132.
[173] J. C. Scott, P. Pfluger, M. T. Krounbi, G. B. Street, 'Electron spin resonance studies of pyrrole polymers – evidence for bipolarons,' *Physical Review B* **1983**, *28*, 2140.
[174] H. H. S. Javadi, R. Laversanne, A. J. Epstein, R. K. Kohli, E. M. Scherr, A. G. MacDiarmid, 'ESR of protonated emeraldine – insulator to metal transition,' *Synthetic Metals* **1989**, *29*, E439.
[175] A. Carrington, A. D. McLachlan, *Introduction to Magnetic Resonance; With Applications to Chemistry and Chemical Physics*. Chapman and Hall, London, **1967**.
[176] A. G. MacDiarmid, C. K. Chiang, A. F. Richter, N. L. D. Somasiri, A. J. Epstein, *Conducting Polymers: Special Applications*, 1st edn, D. Reidel Publishing Co., Dordrecht, **1987**.
[177] L. H. C. Mattoso, A. G. MacDiarmid, A. J. Epstein, 'Controlled synthesis of high-molecular-weight polyaniline and poly(o-methoxyaniline),' *Synthetic Metals* **1994**, *68*, 1.
[178] P. N. Adams, A. P. Monkman, 'Characterization of high molecular weight polyaniline synthesized at $-40\,°C$ using a 0.25:1 mole ratio of persulfate oxidant to aniline,' *Synthetic Metals* **1997**, *87*, 165.
[179] C. H. Hsu, P. M. Peacock, R. B. Flippen, J. Yue, A. J. Epstein, 'The molecular weight of sulfonic acid ring substituted polyaniline by laser light scattering,' *Synthetic Metals* **1993**, *60*, 223.

[180] F. Masdarolomoor, P. C. Innis, S. Ashraf, G. G. Wallace, 'Purification and characterisation of poly(2-methoxyaniline-5-sulfonic acid),' *Synthetic Metals* 2005, *153*, 181.
[181] I. Mav, M. Zigon, A. Sebenik, J. Vohlidal, 'Sulfonated polyanilines prepared by copolymerization of 3-aminobenzenesulfonic acid and aniline: the effect of reaction conditions on polymer properties,' *Journal of Polymer Science Part A – Polymer Chemistry* 2000, *38*, 3390.
[182] J. Yue, A. J. Epstein, Z. Zhong, P. K. Gallagher, A. G. MacDiarmid, 'Thermal stabilities of polyanilines,' *Synthetic Metals* 1991, *41*, 765.
[183] S. M. Yang, J. H. Chiang, 'Morphological-study of alkylsubstituted polyaniline,' *Synthetic Metals* 1991, *41*, 761.
[184] Q. J. Wu, L. X. Wu, Z. N. Qi, F. S. Wang, 'The fractal structures in highly sulfonated polyaniline,' *Synthetic Metals* 1999, *105*, 13.
[185] Q. J. Wu, Z. J. Xue, Z. N. Qi, F. S. Wang, 'The microscopic morphology of highly sulfonated polyaniline,' *Synthetic Metals* 2000, *108*, 107.
[186] B. C. Roy, M. D. Gupta, L. Bhowmik, J. K. Ray, 'Studies on water soluble conducting polymer – aniline initiated polymerization of *m*-aminobenzene sulfonic acid,' *Synthetic Metals* 1999, *100*, 233.
[187] P. S. Rao, D. N. Sathyanarayana, 'Synthesis of electrically conducting copolymers of *o-/m*-toluidines and *o-/m*-amino benzoic acid in an organic peroxide system and their characterization,' *Synthetic Metals* 2003, *138*, 519.
[188] H. S. O. Chan, S. C. Ng, W. S. Sim, K. L. Tan, B. T. G. Tan, 'Preparation and characterization of electrically conducting copolymers of aniline and anthranilic acid – evidence for self-doping by X-ray photoelectronspectroscopy,' *Macromolecules* 1992, *25*, 6029.
[189] K. Yamamoto, D. Taneichi, 'A self-doped oligoaniline with two stable redox couples in a wide pH range,' *Macromolecular Chemistry and Physics* 2000, *201*, 6.
[190] D. M. Zhou, J. J. Xu, H. Y. Chen, H. Q. Fang, 'Ascorbate sensor based on "self-doped" polyaniline,' *Electroanalysis* 1997, *9*, 1185.
[191] C. Thiemann, C. M. A. Brett, 'Electrosynthesis and properties of conducting polymers derived from aminobenzoic acids and from aminobenzoic acids and aniline,' *Synthetic Metals* 2001, *123*, 1.
[192] L. Jiang, Q. J. Xie, L. Yang, X. Y. Yang, S. Z. Yao, 'Simultaneous EQCM and diffuse reflectance UV-Visible spectroelectrochemical measurements: poly(aniline-co-*o*-anthranilic acid) growth and property characterization,' *Journal of Colloid and Interface Science* 2004, *274*, 150.
[193] A. Benyoucef, F. Huerta, J. L. Vazquez, E. Morallon, 'Synthesis and *in situ* MRS characterization of conducting polymers obtained from aminobenzoic acid isomers at platinum electrodes,' *European Polymer Journal* 2005, *41*, 843.
[194] S. Patra, N. Munichandraiah, 'Insoluble poly(anthranilic acid) confined in nafion membrane by chemical and electrochemical polymerization of anthranilic acid,' *Synthetic Metals* 2005, *150*, 285.
[195] Y. J. Wang, W. Knoll, '*In situ* electrochemical and surface plasmon resonance (SPR) studies of aniline carboxylated aniline copolymers,' *Analytica Chimica Acta* 2006, *558*, 150.

[196] S. C. Ng, H. S. O. Chan, H. H. Huang, P. K. H. Ho, 'Poly(o-aminobenzylphosphonic acid) – a novel water-soluble, self-doped functionalized polyaniline,' *Chemical Communications* **1995**, 1327.
[197] H. S. O. Chan, P. K. H. Ho, S. C. Ng, B. T. G. Tan, K. L. Tan, 'A new water-soluble, self-doping conducting polyaniline from poly(o-aminobenzylphosphonic acid) and its sodium salts – synthesis and characterization,' *Journal of the American Chemical Society* **1995**, *117*, 8517.
[198] M. C. McAlpine, R. S. Friedman, S. Jin, K. H. Lin, W. U. Wang, C. M. Lieber, 'High-performance nanowire electronics and photonics on glass and plastic substrates,' *Nano Letters* **2003**, *3*, 1531.
[199] E. Lindner, V. V. Cosofret, S. Ufer, R. P. Buck, R. P. Kusy, R. B. Ash, H. T. Nagle, 'Flexible (kapton-based) microsensor arrays of high-stability for cardiovascular applications,' *Journal of the Chemical Society – Faraday Transactions* **1993**, *89*, 361.
[200] H. Sakai, R. Baba, K. Hashimoto, A. Fujishima, A. Heller, 'Local detection of photoelectrochemically produced H_2O_2 with a wired horseradish peroxidase microsensor,' *Journal of Physical Chemistry* **1995**, *99*, 11896.
[201] M. Trau, N. Yao, E. Kim, Y. Xia, G. M. Whitesides, I. A. Aksay, 'Microscopic patterning of orientated mesoscopic silica through guided growth,' *Nature* **1997**, *390*, 674.
[202] J. H. Fendler, 'Self-assembled nanostructured materials,' *Chemistry of Materials* **1996**, *8*, 1616.
[203] D. D. Lasie, *Liposomes: From Physics to Applications*. Plenum Press, New York, **1993**.
[204] J. M. Schnur, 'Lipid tubules – a paradigm for molecularly engineered structures,' *Science* **1993**, *262*, 1669.
[205] D. I. Gittins, D. Bethell, D. J. Schiffrin, R. J. Nichols, 'A nanometre-scale electronic switch consisting of a metal cluster and redox addressable groups,' *Nature* **2000**, *408*, 67.
[206] A. G. MacDiarmid, 'Nobel lecture: "Synthetic metals": a novel role for organic polymers,' *Reviews of Modern Physics* **2001**, *73*, 701.
[207] K. Doblhofer, K. Rajeshwar, *Handbook of Conducting Polymers*, 2nd edn, Marcel Dekker, New York, **1998**.
[208] D. Kumar, R. C. Sharma, 'Advances in conductive polymers,' *European Polymer Journal* **1998**, *34*, 1053.
[209] V. Saxena, B. D. Malhotra, 'Prospects of conducting polymers in molecular electronics,' *Current Applied Physics* **2003**, *3*, 293.
[210] H. X. He, J. S. Zhu, N. J. Tao, L. A. Nagahara, I. Amlani, R. Tsui, 'A conducting polymer nanojunction switch,' *Journal of the American Chemical Society* **2001**, *123*, 7730.
[211] R. F. Service, 'Assembling nanocircuits from the bottom up,' *Science* **2001**, *293*, 782.
[212] J. Janata, M. Josowicz, 'Conducting polymers in electronic chemical sensors,' *Nature Materials* **2003**, *2*, 19.
[213] D. T. McQuade, A. E. Pullen, T. M. Swager, 'Conjugated polymer based chemical sensors,' *Chemical Reviews* **2000**, *100*, 2537.
[214] H. Q. Zhang, S. Boussaad, N. Ly, N. J. J. Tao, 'Magnetic field assisted assembly of metal/polymer/metal junction sensors,' *Applied Physics Letters* **2004**, *84*, 133.

[215] A. Lodha, R. Singh, 'Prospects of manufacturing organic semiconductor based integrated circuits,' *IEEE Transactions on Semiconductor Manufacturing* **2001**, *14*, 281.
[216] B. Vincent, J. Waterson, 'Colloidal dispersions of electrically conducting, spherical polyaniline particles,' *Journal of the Chemical Society-Chemical Communications* **1990**, 683.
[217] J. Stejskal, P. Kratochvil, S. P. Armes, S. F. Lascelles, A. Riede, M. Helmstedt, J. Prokes, I. Krivka, 'Polyaniline dispersions. 6. Stabilization by colloidal silica particles,' *Macromolecules* **1996**, *29*, 6814.
[218] C. Y. Yang, P. Smith, A. J. Heeger, Y. Cao, J. E. Osterholm, 'Electron diffraction studies of the structure of polyaniline dodecylbenzenesulfonate,' *Polymer* **1994**, *35*, 1142.
[219] C. G. Wu, T. Bein, 'Conducting polyaniline filaments in a mesoporous channel host,' *Science* **1994**, *264*, 1757.
[220] J. C. Michaelson, A. J. McEvoy, 'Interfacial polymerization of aniline,' *Chemical Communications* **1994**, 79.
[221] Z. Wei, Z. Zhang, M. Wan, 'Formation mechanism of self-assembled polyaniline micro/nanotubes,' *Langmuir* **2002**, *18*, 917.
[222] L. M. Huang, Z. B. Wang, H. T. Wang, X. L. Cheng, A. Mitra, Y. X. Yan, 'Polyaniline nanowires by electropolymerization from liquid crystalline phases,' *Journal of Materials Chemistry* **2002**, *12*, 388.
[223] S. J. Choi, S. M. Park, 'Electrochemical growth of nanosized conducting polymer wires on gold using molecular templates,' *Advanced Materials* **2000**, *12*, 1547.
[224] J. M. Liu, S. C. Yang, 'Novel colloidal polyaniline fibrils made by template guided chemical polymerization,' *Chemical Communications* **1991**, 1529.
[225] A. G. MacDiarmid, W. E. Jones, I. D. Norris, J. Gao, A. T. Johnson, N. J. Pinto, J. Hone, B. Han, F. K. Ko, H. Okuzaki, M. Llaguno, 'Electrostatically-generated nanofibers of electronic polymers,' *Synthetic Metals* **2001**, *119*, 27.
[226] H. X. He, C. Z. Li, N. J. Tao, 'Conductance of polymer nanowires fabricated by a combined electrodeposition and mechanical break junction method,' *Applied Physics Letters* **2001**, *78*, 811.
[227] E. M. Scherr, A. G. MacDiarmid, S. K. Manohar, J. G. Masters, Y. Sun, X. Tang, M. A. Druy, P. J. Glatkowski, V. B. Cajipe, et al., 'Polyaniline: oriented films and fibers,' *Synthetic Metals* **1991**, *41*, 735.
[228] X. Y. Zhang, W. J. Goux, S. K. Manohar, 'Synthesis of polyaniline nanofibers by 'nanofiber seeding',' *Journal of the American Chemical Society* **2004**, *126*, 4502.
[229] V. Gupta, N. Miura, 'Large area network of polyaniline nanowires prepared by potentiostatic deposition process,' *Electrochemistry Communications* **2005**, *7*, 995.
[230] N. R. Chiou, A. J. Epstein, 'Polyaniline nanofibers prepared by dilute polymerization,' *Advanced Materials* **2005**, *17*, 1679.
[231] J. X. Huang, R. B. Kaner, 'A general chemical route to polyaniline nanofibers,' *Journal of the American Chemical Society* **2004**, *126*, 851.
[232] J. Kameoka, R. Orth, Y. N. Yang, D. Czaplewski, R. Mathers, G. W. Coates, H. G. Craighead, 'A scanning tip electrospinning source for deposition of oriented nanofibres,' *Nanotechnology* **2003**, *14*, 1124.

[233] Z. X. Wei, M. X. Wan, T. Lin, L. M. Dai, 'Polyaniline nanotubes doped with sulfonated carbon nanotubes made via a self-assembly process,' *Advanced Materials* **2003**, *15*, 136.
[234] Y. Long, Z. J. Chen, N. L. Wang, Y. J. Ma, Z. Zhang, L. J. Zhang, M. X. Wan, 'Electrical conductivity of a single conducting polyaniline nanotube,' *Applied Physics Letters* **2003**, *83*, 1863.
[235] A. L. Briseno, S. B. Han, I. E. Rauda, F. M. Zhou, C. S. Toh, E. J. Nemanick, N. S. Lewis, 'Electrochemical polymerization of aniline monomers infiltrated into well ordered truncated eggshell structures of polyelectrolyte multilayers,' *Langmuir* **2004**, *20*, 219.
[236] C. J. Martinez, B. Hockey, C. B. Montgomery, S. Semancik, 'Porous tin oxide nanostructured microspheres for sensor applications,' *Langmuir* **2005**, *21*, 7937.
[237] E. S. Forzani, H. Q. Zhang, L. A. Nagahara, I. Amlani, R. Tsui, N. J. Tao, 'A conducting polymer nanojunction sensor for glucose detection,' *Nano Letters* **2004**, *4*, 1785.
[238] Y. F. Ma, J. M. Zhang, G. J. Zhang, H. X. He, 'Polyaniline nanowires on Si surfaces fabricated with DNA templates,' *Journal of the American Chemical Society* **2004**, *126*, 7097.
[239] S. Virji, J. X. Huang, R. B. Kaner, B. H. Weiller, 'Polyaniline nanofiber gas sensors: examination of response mechanisms,' *Nano Letters* **2004**, *4*, 491.
[240] J. Huang, S. Virji, B. H. Weiller, R. B. Kaner, 'Nanostructured polyaniline sensors,' *Chemistry – A European Journal* **2004**, *10*, 1315.
[241] J. X. Huang, S. Virji, B. H. Weiller, R. B. Kaner, 'Polyaniline nanofibers: facile synthesis and chemical sensors,' *Journal of the American Chemical Society* **2003**, *125*, 314.
[242] H. Q. Liu, J. Kameoka, D. A. Czaplewski, H. G. Craighead, 'Polymeric nanowire chemical sensor,' *Nano Letters* **2004**, *4*, 671.
[243] A. Morrin, O. Ngamna, A. J. Killard, S. E. Moulton, M. R. Smyth, G. G. Wallace, 'An amperometric enzyme biosensor fabricated from polyaniline nanoparticles,' *Electroanalysis* **2005**, *17*, 423.
[244] F. Huguenin, M. T. D. Gambardella, R. M. Torresi, S. I. de Torresi, D. A. Buttry, 'Chemical and electrochemical characterization of a novel nanocomposite formed from V_2O_5 and poly(N-propane sulfonic acid aniline), a self-doped polyaniline,' *Journal of the Electrochemical Society* **2000**, *147*, 2437.
[245] F. Huguenin, M. J. Giz, E. A. Ticianelli, R. M. Torresi, 'Structure and properties of a nanocomposite formed by vanadium pentoxide containing poly(n-propane sulfonic acid aniline),' *Journal of Power Sources* **2001**, *103*, 113.
[246] Z. X. Wei, M. X. Wan, 'Synthesis and characterization of self-doped poly(aniline-co-aminonaphthalene sulfonic acid) nanotubes,' *Journal of Applied Polymer Science* **2003**, *87*, 1297.
[247] X. G. Li, M. R. Huang, W. Feng, M. F. Zhu, Y. M. Chen, 'Facile synthesis of highly soluble copolymers and sub-micrometer particles from ethylaniline with anisidine and sulfoanisidine,' *Polymer* **2004**, *45*, 101.
[248] Z. J. Wang, L. S. Hao, T. Y. You, L. Niu, S. J. Dong, A. Ivaska, 'Electrochemical preparation of self-doped poly(o-aminobenzenesulfonic acid-co-aniline) microflowers,' *Electrochemistry Communications* **2005**, *7*, 875.

[249] C. H. Yang, Y. K. Chih, H. E. Cheng, C. H. Chen, 'Nanofibers of self-doped polyaniline,' *Polymer* **2005**, *46*, 10688.
[250] C. H. Yang, Y. K. Chih, W. C. Wu, C. H. Chen, 'Molecular assembly engineering of self-doped polyaniline film for application in electrochromic devices,' *Electrochemical and Solid State Letters* **2006**, *9*, C5.
[251] A. S. Ribeiro, D. A. Machado, P. F. D. Filho, M. A. De Paoli, 'Solid-state electrochromic device based on two poly(thiophene) derivatives,' *Journal of Electroanalytical Chemistry* **2004**, *567*, 243.
[252] L. Zhang, X. Jiang, L. Niu, S. J. Dong, 'Syntheses of fully sulfonated polyaniline nano-networks and its application to the direct electrochemistry of cytochrome c,' *Biosensors & Bioelectronics* **2006**, *21*, 1107.
[253] J. X. Song, D. X. Han, L. P. Guo, L. Niu, 'Direct electrochemical generation of conducting polymer micro-rings,' *Nanotechnology* **2006**, *17*, 824.

3
Boronic Acid Substituted Self-Doped Polyaniline

3.1 INTRODUCTION

Complexation of boric and boronic acids groups with diol containing compounds such as carbohydrates has been known since the 1950s [1–4]. Boronic acids form covalent bonds with 1,2- or 1,3-diols to generate five- or six-membered cyclic esters in nonaqueous or basic aqueous media [5]. Boronic ester formation from a boronic acid and the vicinal *cis*-diol of a carbohydrate is reported to be one of the strongest single pair, reversible functional group, interactions in an aqueous environment among organic compounds and, as a result, has been widely used for the development of carbohydrate sensors [6–14]. Similarly, the strong complexation of boronic acid with diols has been used for the construction of carbohydrate transporters [15] and as affinity ligands for the separation of carbohydrates and glycoproteins [16–24]. The reversible boronate ester formation is shown in Figure 3.1. The complexation of saccharides (as well as alkyl and aromatic diols) with aromatic boronic acids produces a stable ester, where the association constant as well as the relative concentrations of the neutral trigonal ester and the tetrahedral boronate ester are dependent on the pH [25], electrolyte concentration, and pK_a of aromatic boronic acid among other factors [6]. Addition of a Lewis base such as fluoride [23, 26, 27] or amines [28, 29] can be used to enhance the complexation of saccharide under less basic conditions and facilitates the formation of anionic esters.

Self-Doped Conducting Polymers M.S. Freund and B.A. Deore
© 2007 John Wiley & Sons, Ltd

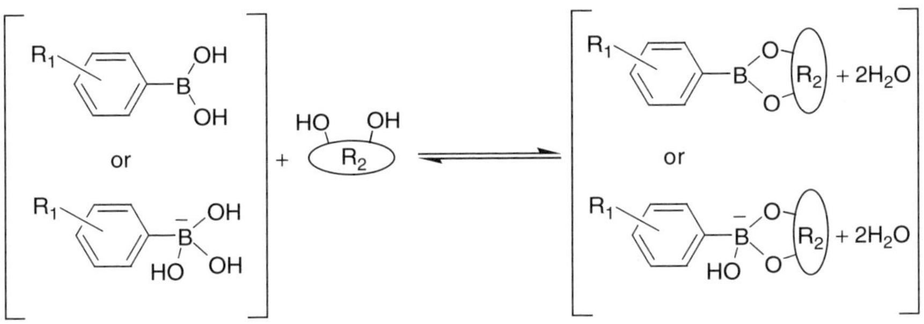

INTRODUCTION

Figure 3.1 Boronate ester formation. (Reprinted from *Tetrahedron*, **58**, G. Springsteen, B. Wang, 5291. Copyright (2002), with permission from Elsevier.)

Boronic acid substituted polyaniline, i.e., poly(anilineboronic acid) was first synthesized by Wolfbeis *et al.* as a copolymer with aniline [30] and Fabre *et al.* as a homopolymer [31]. These reports on boronic-acid-substituted polyaniline opened up the possibility of coupling this boronic acid chemistry to a conducting polymer and thereby developing new sensing strategies for fluoride [31], saccharides [7, 8], biogenic amines [32–34] and nucleotides [35]. Wolfbeis *et al.* [30] demonstrated that aminophenylboronic acid can be chemically polymerized as a copolymer with aniline under standard conditions. Fabre *et al.* [31] prepared poly(anilineboronic acid) electrochemically in the presence of fluoride and suggested that the polymer should be self-doped. Based on this study, self-doped poly(anilineboronic acid) was prepared by Freund *et al.* chemically and electrochemically exploiting interactions between saccharide and the boronic acid substituent on polyaniline in the presence of fluoride [36, 37].

The self-doping mechanism in the case of boronic-acid-substituted polyaniline is different than the conventional mechanism reported for sulfonic acid, carboxylic acid and phosphonic acid substituted polyaniline. The self-doped state of boronic acid substituted polyaniline is controlled via complexation between boronic acid along the backbone with Lewis bases including *cis*-diols, fluoride, alcohols, amines, etc., resulting in specific orbital changes from sp^2 to the more stable sp^3 hybridized boron. The formation of anionic tetrahedral boronate ester acts as an inner dopant and balances the positive charge on the nitrogen of polyaniline backbone. The redox and self-doping mechanism of poly(anilineboronic acid) (PABA) in the presence of saccharides (A) and fluoride (B) is shown in Figure 3.2. The self-doping process is reversible and able to switch the polymer between self-doped

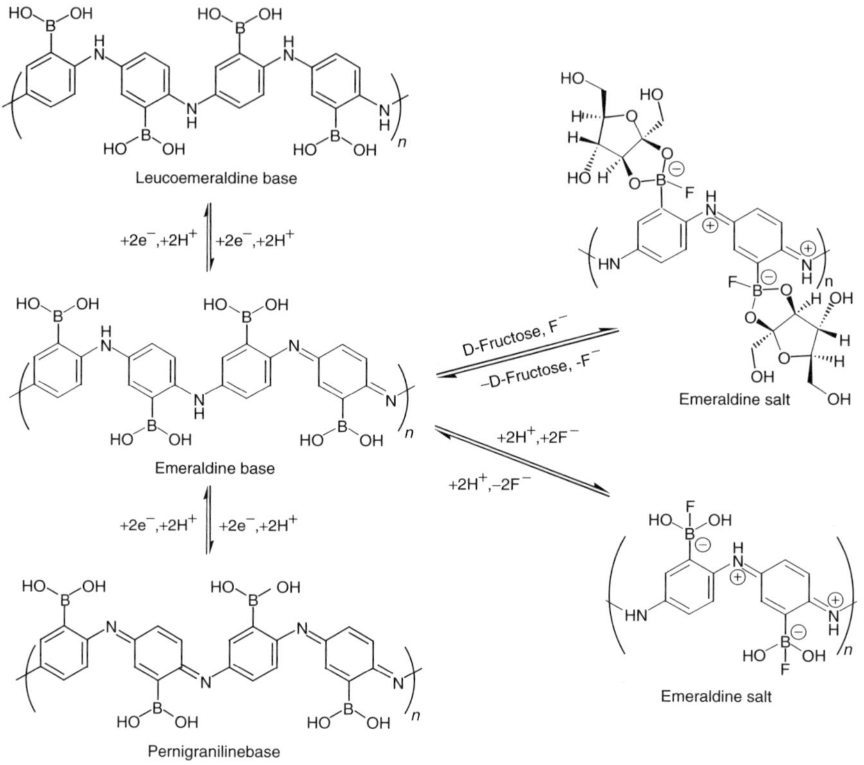

Figure 3.2 Proposed redox and self-doping mechanism of PABA.

and nonself-doped states. The self-doped PABA (B) is soluble under polymerization conditions at neutral pH. The polymer has good conductivity, electroactivity up to pH 12, and high molecular weight and thermal and mechanical stability [37] as discussed in more detail below.

3.2 SYNTHESIS

3.2.1 Electrochemical Synthesis

Fabre *et al.* electrochemically polymerized PABA using a fluoride catalyzed reaction [31]. The electropolymerization of 3-aminophenylboronic acid in 0.5 M H_2SO_4 aqueous solution was found to be inefficient (Figure 3.3, A). However, addition of fluoride strongly enhanced the polymerization rate (Figure 3.3, B). This effect was not observed with Cl^- or with unsubstituted polyaniline in the presence of fluoride.

SYNTHESIS

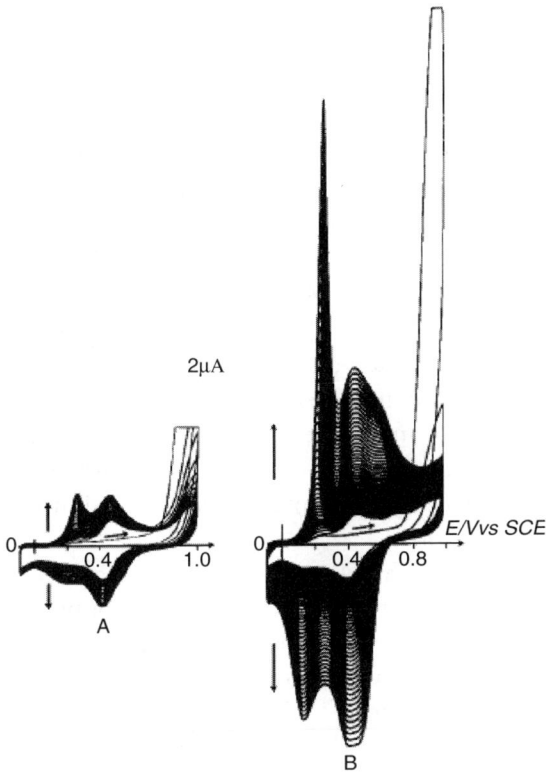

Figure 3.3 Cyclic voltammograms of electropolymerization of 0.1 M 3-aminophenylboronic acid (A) in the absence and (B) in the presence of 0.1 M KF; electrolyte: 0.5 M H_2SO_4 aqueous solution; potential scan rate: 100 mV/s (Reproduced from *European Journal of Organic Chemistry*, **2000**, 1703, M. Nicolas, B. Fabre, G. Marchand, J. Simone, with permission from Wiley-VCH.)

The enhanced reactivity of 3-aminophenylboronic acid in the presence of fluoride was reportedly due to interaction between fluoride and boron. The complexation of fluoride with the boronic acid moiety substantially reduced the oxidation potential required for polymerization, thereby avoiding deleterious side reactions that occur at more positive potentials. The binding of fluoride was confirmed by Fabre *et al.* [31] with ^{11}B NMR spectroscopy of monomer solution before and after addition of fluoride, since the chemical shift is sensitive to the hybridization of the boron. In monomer solution without fluoride, a ^{11}B NMR signal was observed at 27.7 ppm with reference to borontrifluoride etherate. However, after addition of fluoride, a new signal was observed at 3.4 ppm. The ^{11}B chemical shift of a tetrahedral boronate formed in the presence of fluoride was approximately 24 ppm up field from the trigonal boronic acid signal similar to that reported by Smith *et al.* [23]. The magnitude of signal increased with increase in concentration of fluoride. These results confirm the transformation of sp^2 to sp^3 hybridized boron upon binding of fluoride. This is consistent with the literature, where it is reported that the boronic acid and boronate groups are able strongly to bind hard Lewis base anions such as fluoride, resulting in specific orbital changes from sp^2 to the more stable sp^3 hybridized boron [5, 26, 27, 38–40]. As a result, PABA prepared in the presence of fluoride under acidic conditions can be considered a self-doped polymer. The concentration of fluoride used during the polymerization can impact the properties of the resulting polymer. For example, Freund *et al.* prepared PABA electrochemically in the presence of sodium fluoride in 0.5 M HCl solution for saccharide sensing. The sensor sensitivity was improved significantly by increasing the concentration of sodium fluoride during the polymerization [7, 8].

Self-doped polyaniline has the unique property of maintaining high conductivity and redox activity at high pH (See Chapter 2, Section 2.5). However, it has never been demonstrated that these properties should allow electrochemical polymerization at elevated pH. The electrochemical polymerization of self-doped PABA at weakly acidic pH (pH 5.0) in the presence of 300 mM fluoride and at neutral pH (pH 7.4) in the presence of 10 M saccharide and one equivalent of fluoride to monomer has been achieved, based on the formation of an anionic boronate complex [36]. Anionic boronate esters form the basis of a self-doped polyaniline with stable redox behavior under these conditions. Figure 3.4 (a and b) shows the cyclic voltammogram of 3-aminophenylboronic acid in the presence of different concentrations of fluoride in 10 X phosphate buffered saline solution (0.12 M potassium phosphate monobasic, 1.4 M sodium chloride and 0.03 M potassium chloride, PBS) at pH 5.0. At lower concentrations of fluoride (40 mM) (Figure 3.4, a), the voltammogram looks similar to the oxidation of 3-aminophenylboronic acid under acidic condition without fluoride. As seen in the voltammogram, although repetitive cycling results in the deposition of a redox active film with surface waves around 0.1 V, continual cycling results in a positive shift and a gradual decrease in the cathodic peak near 0.0 V. However, at higher fluoride concentrations up to 300 mM (near saturation point under these conditions, Figure 3.4, b), significant and continuous polymer growth is observed. In contrast, polymerization is not observed at a pH value of 7.4, even at elevated fluoride concentration (Figure 3.4, c).

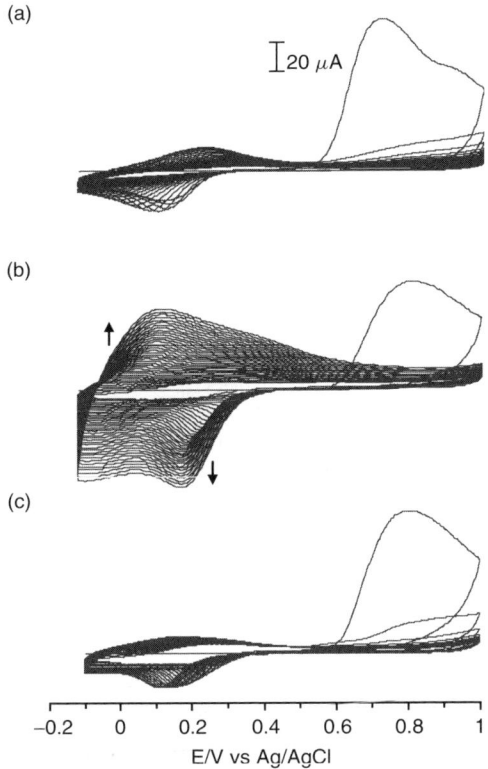

Figure 3.4 Cyclic voltammogram of 40 mM 3-aminophenylboronic acid plus (a) 40 mM NaF and (b) 300 mM NaF in pH 5.0 10 X PBS (pH was adjusted with 0.5 M HCl) and (c) 300 mM NaF in pH 7.4 10 X PBS at GC electrode. Scan rate: 100 mV/s (Analyst, 2003, **128**, 803, B. A. Deore, M. S. Freund. Reproduced with permission of The Royal Society of Chemistry.)

In order to determine the structure of the monomer, ^{11}B NMR study was carried out under conditions similar to those used to obtain the voltammograms shown in Figure 3.4. A ^{11}B NMR chemical shift for 3-aminophenylboronic acid in the presence of 40 mM (Figure 3.4, a) and 300 mM (Figure 3.4, b) fluoride at pH 5.0 is observed at 28.7 and 3.7 ppm, respectively (Table 3.1). At higher concentrations of fluoride, a 22.8 ppm upfield shift is observed relative to that observed for lower fluoride concentrations due to formation of tetrahedral boronate species. These results suggest that at sufficiently high fluoride concentration, a self-doped polymer is produced. However, upon increasing the pH to 7.4 in the presence of 300 mM of fluoride (Figure 3.4, c), the boronic acid reverts back to a neutral trigonal boronic acid structure by a 22 ppm

Table 3.1 NMR chemical shifts of 3-aminophenylboronic acid (3-APBA) adducts. (*Analyst*, 2003, 128, 803, B. A. Deore, M. S. Freund. Reproduced with permission of the Royal Society of Chemistry.)

Solution conditions	pH	Boron chemical shift (ppm)
40 mM 3-APBA	7.4	29.2
40 mM 3-APBA + 40 mM NaF	7.4	28.7
40 mM 3-APBA + 300 mM NaF	7.4	26.8
40 mM 3-APBA + 300 mM NaF	5.0	3.7
40 mM 3-APBA + 10 M D-fructose	7.4	6.4
40 mM 3-APBA + 40 mM NaF + 10 M D-fructose	7.4	6.4
40 mM 3-APBA + 40 mM NaF + 10 M D-fructose	5.0	6.4

Shifts measured relative to reference borontrifluoride etherate.

downfield shift back to a value close to that observed in the absence of fluoride. These results indicate that a 1:1 molar ratio of fluoride to monomer is not sufficient to allow sustained polymerization, while increasing the fluoride concentration eventually results in polymerization at pH 5.0.

Similarly, self-doped PABA can be prepared using excess of saccharide and one equivalent of fluoride to monomer. Complexation between saccharides and aromatic boronic acids is highly pH dependent, presumably due to the tetrahedral intermediate involved in complexation [25]. Because the pK_a of 3-aminophenylboronic acid is 8.75, complexation requires pH values above 8.6. This pH range is not compatible with the electrochemical synthesis of polyaniline, which is typically carried out near a pH value of 0. However, Smith *et al.* have shown that the addition of fluoride can stabilize the complexation of molecules containing *vicinal* diols with aromatic boronic acids [23]. Based on this work, it was postulated that the electrochemical polymerization of a saccharide complex with 3-aminophenylboronic acid in the presence of one molar equivalent of fluoride at pH values lower than 8 is possible if a self-doped polymer is produced in the process.

Cyclic voltammograms of 3-aminophenylboronic acid in 10 M D-fructose in 10 X PBS without and with fluoride is shown in Figure 3.5. No oxidation of monomer is observed without fluoride in pH 5.0 10 X PBS even in the presence of relatively high D-fructose concentrations (Figure 3.5, a). This is in contrast to the clear oxidation of 3-aminophenylboronic acid under identical conditions in the absence of fluoride and D-fructose (Figure 3.6, a). Similar behavior was observed at pH 7.4, in the absence of fluoride (Figure 3.6, b). However, after

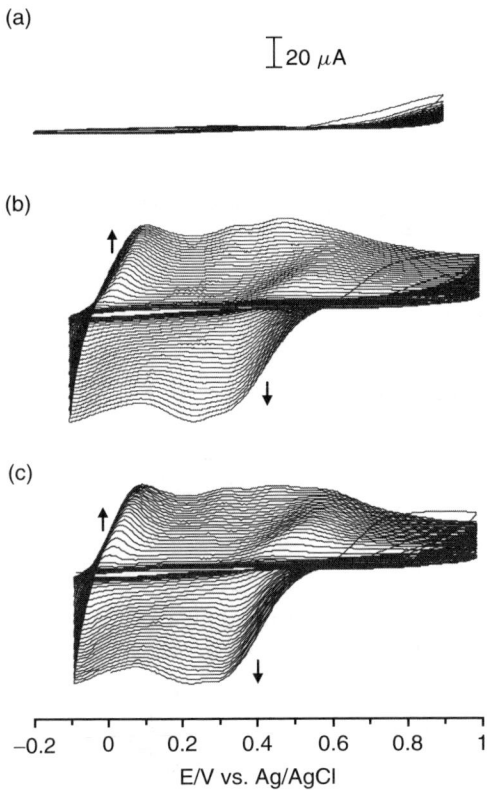

Figure 3.5 Cyclic voltammogram of 40 mM 3-aminophenylboronic acid plus (a) 10 M fructose and (b) 40 mM NaF plus 10 M fructose in pH 5.0 10 X PBS at GC electrode. (pH adjusted with 0.5 M HCl) and (c) 40 mM NaF plus 10 M fructose in pH 7.4 10 X PBS. Scan rate: 100 mV/s. (Analyst, 2003, **128**, 803, B. A. Deore, M. S. Freund. Reproduced with permission of The Royal Society of Chemistry.)

addition of one equivalent of fluoride with respect to monomer in solution with excess D-fructose at pH 5.0 and 7.4, efficient and sustained electropolymerization is obtained (Figure 3.5, b and c). ^{11}B NMR of 3-aminophenylboronic acid with 10 M D-fructose in 10 X PBS suggests that the tetrahedral boronate ester is formed even without fluoride (Table 3.1). After addition of fluoride, boron remains in a tetrahedral structure. However, polymerization is not observed in the absence of fluoride. The observed behavior might be explained by intermolecular amine–boronic acid interactions that would be disrupted by the addition of fluoride. Under these conditions (in the absence of fluoride), although

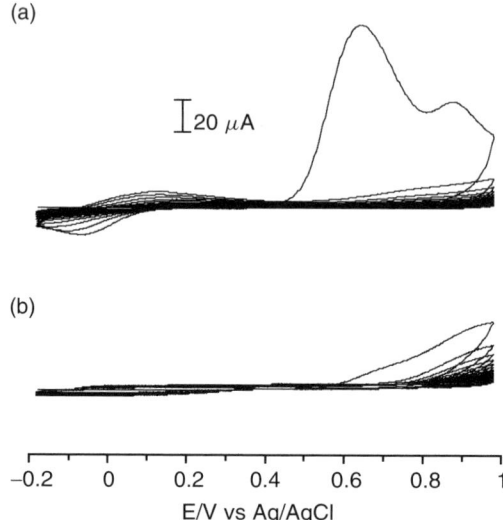

Figure 3.6 Cyclic voltammogram of 40 mM 3-aminophenylboronic acid (a) without fluoride and D-fructose in pH 5.0 and (b) with 10 M D-fructose in pH 7.4 10 X PBS at GC electrode. Scan rate: 100 mV-s^{-1}.

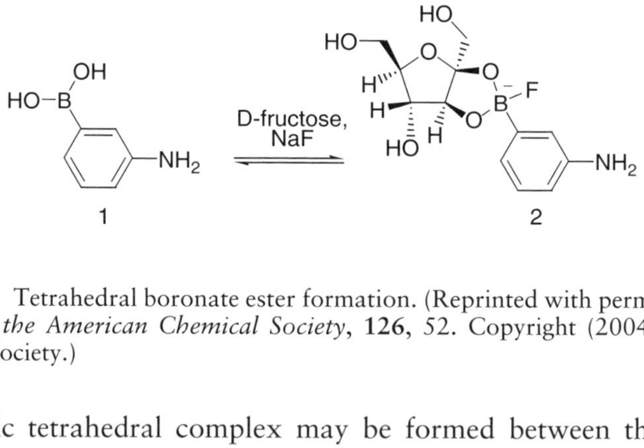

Figure 3.7 Tetrahedral boronate ester formation. (Reprinted with permission from *Journal of the American Chemical Society*, **126**, 52. Copyright (2004) American Chemical Society.)

an anionic tetrahedral complex may be formed between the boronic acid group and the saccharide, polymerization would not occur since the amine would be tied up in the complex and, perhaps, because of steric complications. This behavior suggests that a complex involving both fluoride and D-fructose (Figure 3.7, 2) is involved in the polymerization since the presence of neither fluoride nor D-fructose at these concentrations alone is sufficient to allow electropolymerization (see Figure 3.4a and 3.5a, respectively). Similarly, polymerization was observed at pH

7.4, as shown in Figure 3.5c, suggesting that complexation is sufficient throughout this pH range to enable polymerization of a self-doped polyaniline. Interestingly, this is reported to be the first example of the electropolymerization of a self-doped polyaniline under neutral pH conditions.

3.2.2 Chemical Synthesis

Wolfbeis *et al.* chemically synthesized PABA/polyaniline copolymer thin films on the surface of polystyrene cuvettes for optical sensing of saccharides [30]. The polymerization was carried out using ammonium persulfate and ferric chloride as oxidizing agents. Complete and faster polymerization was observed using ammonium persulfate. Recently, Freund *et al.* carried out oxidative homopolymerization of tetrahedral boronate complex (Figure 3.7, 2) under ambient conditions with the addition of ammonium persulfate as oxidizing agent [37]. In the presence of D-fructose and fluoride, PABA was synthesized by reaction of an aqueous solution of 40 mM 3-aminophenylboronic acid and 40 mM sodium fluoride in 10 M D-fructose (19.5 mL) with 40 mM ammonium persulfate (0.5 mL) added over 10 minutes. The pH of the reaction mixture was 7 before adding an oxidizing agent. The mixture was stirred overnight at room temperature. In a similar manner PABA was also prepared in PBS (pH 7.4). The water-soluble PABA formed using both methods was soluble under the polymerization conditions and easily passed through a 0.02 μm Anotop filter. The water soluble PABA produced under these conditions was readily precipitated by diluting the solution (~3 times) and in turn reducing the fluoride and D-fructose concentration. Following centrifugation and rinsing in water, the precipitate was washed with 0.5 M HCl to remove D-fructose. The PABA precipitate was soluble in the conducting form in water in the presence of D-fructose and fluoride. In addition, PABA was highly soluble in alkaline aqueous solution and partially soluble and highly dispersible in alcohols. The conductivity of PABA measured using four-probe measurement on a dry pellet was approximately 0.2 S/cm. In the absence of D-fructose under acidic conditions PABA has been prepared in a similar fashion using 40 mM 3-aminophenylboronic acid and 200 mM fluoride in 0.5 M HCl. Under these conditions, the polymer precipitated during polymerization. However, PABA was soluble in the conducting form in water in the presence of D-fructose and fluoride and alkaline pH solution.

3.3 PROPERTIES OF SELF-DOPED PABA

3.3.1 pH Dependent Redox Behavior

The redox behavior of chemically and electrochemically prepared PABA in acidic solution (0.5 M HCl) is reportedly similar to that observed for unsubstituted polyaniline [41]. Two sets of redox waves are observed, at 0.18 and 0.74 V, suggesting facile conversion of leucoemeraldine to emeraldine and subsequent conversion to pernigraniline oxidation states. These results suggest that the boronic acid substituent and polymerization conditions have no detrimental influence on the electronic properties of the polymer. This is in contrast to sulfonated polyaniline where the redox couples are more closely spaced than for polyaniline due to the electronic and steric effects of the -SO_3^- groups on the backbone of the polymer (for details see, Chapter 2, section 2.5.3).

Self-doping in sulfonated polyaniline occurs when the conjugated backbone is sufficiently basic and the acid functionality is sufficiently strong such that protonation of the polymer backbone occurs. This intramolecular doping greatly enhances the redox activity over a wider range of pH values. In the case of PABA, self-doping occurs through the formation of an anionic tetrahedral boronate ester. Therefore, the redox activity of electrochemically prepared, self-doped, PABA in the presence of D-fructose and fluoride as a function of pH is significantly different from polyaniline. Similarly to sulfonated polyaniline, PABA is redox active over the wider pH range of 1–13 as shown in Figure 3.8. However, the electroactivity exhibits more complex pH dependent behavior, suggesting a transition between species involved in the self-doping process. At low pH values in the range of 1–4 (Figure 3.8, a–d), cyclic voltammograms exhibit two sets of redox waves due to conversion of oxidation states. The cyclic voltammograms are identical to polyaniline and PABA in acidic (aqueous HCl) solution with the exception that the two sets of redox waves are more closely spaced with increasing pH, similarly to sulfonated [42] and alkyl substituted [43] polyaniline. At pH values greater than 4, only one set of peaks is observed until a neutral pH is reached. This result suggests that the emeraldine form is not stable in the pH range 4–7 (Figure 3.8, d–g) and PABA is converted directly from the fully reduced leucoemeraldine to the fully oxidized pernigraniline form. In contrast to sulfonated polyaniline, above neutral pH the single set of waves split again and a well behaved cyclic voltammogram similar to that obtained in acidic pH is observed in the pH range 8–11 (Figure 3.8, h–k). The magnitude of current observed changes significantly and reversibly as a function of pH. The maximum peak current

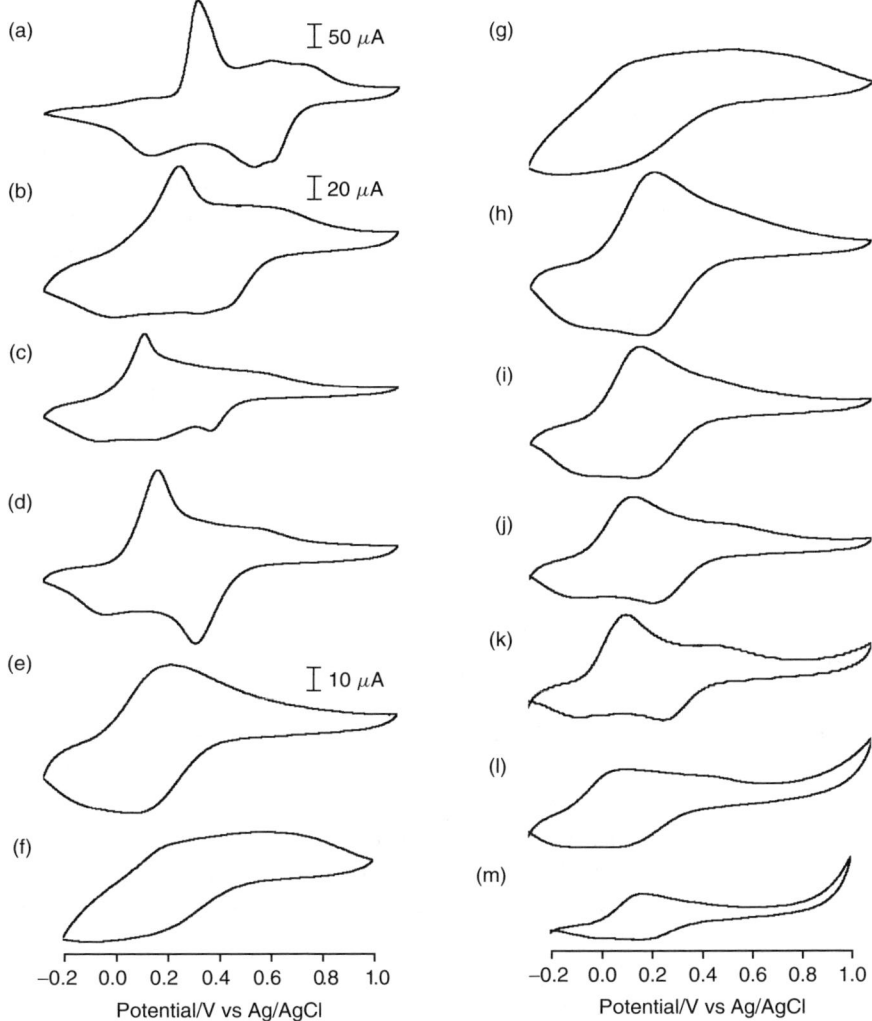

Figure 3.8 Cyclic voltammograms of PABA film in PBS with 10 M D-fructose and 40 mM NaF as a function of pH: (a) 1.0, (b) 2.5, (c) 3.5, (d) 4.2, (e) 5.5, (f) 6.4, (g) 7.0, (h) 8.0, (i) 9.0, (j) 10.0, (k) 11.0, (l) 12.0, and (m) 13.0. (Reprinted with permission from *Chemistry of Materials*, **16**, 1427. Copyright (2004) American Chemical Society.)

is observed at low pH 1; however, the peak current gradually decreases with increasing pH up to 5.

The redox behavior of PABA in the presence of D-fructose and fluoride up to neutral pH values is typical of the emeraldine salt form of polyaniline [41]. Under weakly acidic and neutral pH conditions, the individual oxidation and reduction peaks are merged into a single redox wave; this is indicative of the emeraldine base form of polyaniline where little deprotonation and protonation occurs during the redox process. In the case of polyaniline, complete loss of electroactivity is observed at basic pH. In contrast, self-doped forms of polyaniline are redox active up to a pH of 12; however, they show only a single redox wave above pH 3 [44, 45]. The redox chemistry of self-doped PABA is clearly quite different from that of previously reported forms of self-doped polyaniline. This is probably due to the pH dependent nature of the boronic acid group of PABA and the presence of various boronic acid species.

The role of the various boronic acid species present in the polymer in the observed redox behavior is demonstrated in the cyclic voltammograms of PABA films under various electrolyte solution conditions. Cyclic voltammograms of PABA exhibit two sets of waves similar to polyaniline in the absence and in the presence of fluoride in PBS at pH 2 (Figure 3.9, A; (i) and (ii), respectively). In contrast, the PABA films are electrochemically inactive in PBS at pH 10.5 with or without fluoride (Figure 3.9, B; (i) and (ii), respectively). Further, these films are no longer redox active when returned to acidic solutions, suggesting irreversible degradation (overoxidation) of PABA film in an alkaline pH medium. Only in the first anodic sweep, one broad anodic peak around 0 V is observed. The disappearance of the peak in subsequent scans is typical voltammetric behavior of a polymer upon overoxidation in an aqueous alkaline solution. However, PABA films remain redox active at both pH values, with the addition of D-fructose both in the absence of fluoride (Figures 3.9, A and 3.9B; iii) and in the presence of fluoride (Figures 3.9, A and 3.9B; iv). The redox behavior of PABA is found to be reversible and stable regardless of exposure and cycling at different pHs in the presence of D-fructose and fluoride. These results indicate that the changes in redox mechanism of self-doped PABA as a function of pH and in different electrolyte solutions are a result of structural changes in the complexation of boronic acid with D-fructose.

The role of D-fructose, fluoride, and pH in the structure formed with boronic acid has been investigated with NMR spectroscopy of monomer solutions at different pH values in the presence of D-fructose, with and without fluoride. For example, ^{11}B and ^{19}F NMR can be used to determine the structure of the boron (neutral trigonal vs anionic tetrahedral) and the involvement of fluoride in the complex (free vs complexed fluoride). The ^{11}B chemical shift of a tetrahedral boronate appears approximately 20 ppm upfield from the trigonal boronic acid

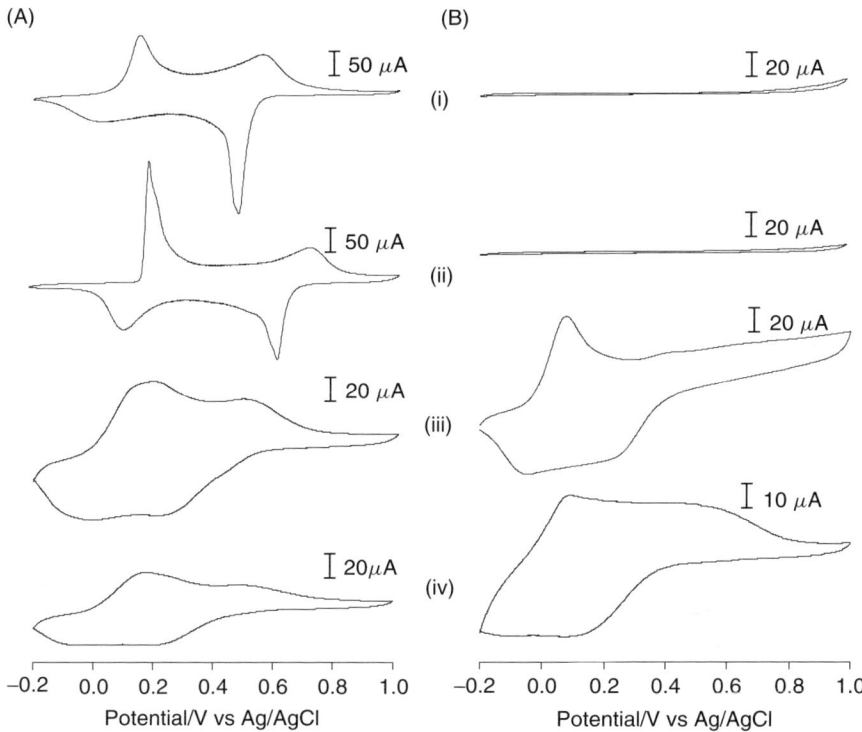

Figure 3.9 Cyclic voltammograms of PABA film in (i) PBS, (ii) PBS with 40 mM NaF, (iii) PBS with 10 M D-fructose, and (iv) PBS with 10 M D-fructose and 40 mM NaF, at pH (A) 2.0 and (B) 10.5. (Reprinted with permission from *Chemistry of Materials*, **16**, 1427. Copyright (2004) American Chemical Society.)

signal [46–48]. Similarly, the ^{19}F chemical shift of tetrahedral phenylboronate (RBF_3^-) is approximately 20 ppm upfield from the free fluoride (KF) as reported by James *et al.* [39]. The chemical shifts observed for ^{11}B and ^{19}F under solution conditions similar to those in Figures 3.8 and 3.9 and similar to those used during the electrochemical polymerization of the monomer are shown in Table 3.2. The ^{11}B NMR chemical shift for tetrahedral boron is observed below 7 ppm, which is around 22 ppm upfield from trigonal boronic acid (29.2 ppm). The characteristic ^{19}F NMR signals of free and complexed fluoride are observed at −40.1 and −22.3 ppm, respectively. At a pH value of 1.4 in the presence of fluoride, the chemical shift of 3.7 ppm indicates the presence of tetrahedral boronate. ^{19}F NMR under the same conditions indicates the involvement of fluoride in a tetrahedral complex. In the absence of fluoride, a

Table 3.2 ^{11}B and ^{19}F NMR chemical shifts of 3-aminophenylboronic acid (3-APBA) adducts. (Reprinted with permission from *Chemistry of Materials*, 16, 1427. Copyright (2004) American Chemical Society.)

Solution conditions	Chemical shift (ppm)		
	pH	^{11}B	^{19}F
40 mM 3-APBA + 40 mm NaF + 10 M Fructose	1.4	3.7	−22.3
40 mM 3-APBA + 10 M Fructose	1.4	29.2	–
40 mM 3-APBA + 40 mM NaF + 10 M Fructose	6.4	6.4	−20.3 −38.6
40 mM 3-APBA + 10 M Fructose	6.4	6.4	–
40 mM 3-APBA + 40 mM NaF + 10 M Fructose	10	4.1	−40.1
40 mM 3-APBA + 10 M Fructose	10	4.1	–

Shifts measured relative to reference borontrifluoride etherate and hexafluorobenzene

25.5 ppm downfield shift in the boron resonance indicates a transition to the trigonal form of boronic acid, consistent with the involvement of fluoride in the complex. At near neutral and alkaline pH conditions both with and without fluoride, the boron chemical shifts indicate that the monomer is converted to a tetrahedral form. ^{19}F NMR under the same conditions indicates the presence of both free and complexed fluoride suggesting partial involvement of fluoride in complexation at pH 6.4. However, at alkaline pH, the chemical shifts indicate that fluoride is not involved in complexation. According to studies under aqueous neutral to alkaline solutions (pH values between 7.4 and 12), the D-fructose complex with boronic acid involves three hydroxyl groups of the β-D-fructofuranose moiety at the C-2, C-3, and C-6 positions [49, 50]. All of these results are consistent with the formation of a self-doped form of the polymer throughout the pH range of 1–10 and the following: (i) under acidic conditions, the polymer is self-doped with the boronate anion (Figure 3.10, **1**), (ii) under neutral conditions there is a mixture of the anions (Figure 3.10, **1** and **2**) and (iii) under basic conditions, the polymer contains the anion (Figure 3.10, **2**).

^{11}B and ^{19}F NMR results also reveal the structure of the monomer involved in the electropolymerization process under various pH conditions. According to the ^{11}B and ^{19}F NMR results, under the electrochemical [36] and chemical [37] polymerization conditions (third entry in Table 3.2) the formation of tetrahedral boronate ester involves fluoride (Figure 3.10, **1**) as well as hydroxide (Figure 3.10, **2**). Polymerization does not occur at elevated pH where tetrahedral boronate involves hydroxide, nor does it occur at neutral pH in the absence of

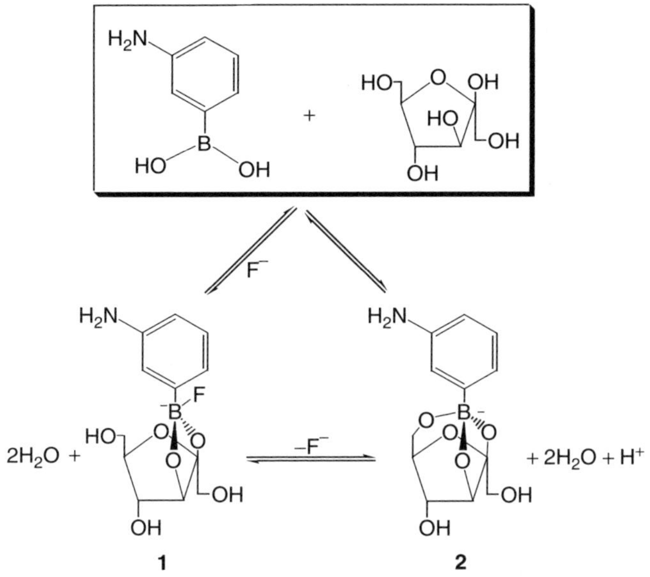

Figure 3.10 Tetrahedral boronate ester formation. (Reprinted with permission from *Chemistry of Materials*, **16**, 1427. Copyright (2004) American Chemical Society.)

fluoride, since the form consisting of tetrahedral boronate with fluoride is the species involved in polymerization under neutral conditions. Although the synthesis of the self-doped polymer occurs only through the formation of poly(**1**), once formed, the conversion of poly(**1**) to poly(**2**) extends the electroactivity of the polymer to neutral and alkaline pH conditions. The transition in the redox behavior of the polymer near neutral pH values (Figure 3.8, f and g) is in turn probably due to transition from poly(**1**) to poly(**2**) and the mixture of anionic species that results. The roles of the mixed species present clearly impact the charge compensation process that occurs during redox switching.

Regardless of the nature of the anionic species involved in the doping process, the redox process remains coupled to proton transfer as is the case with the parent polyaniline. The pH sensitivity of PABA was demonstrated in open-circuit potential measurements on glassy carbon electrodes. The pH dependence was reported to be linear throughout the entire pH range with a near Nernstian slope of 58 mV/pH (Figure 3.11). This is similar to the behavior reported for polyaniline [51, 52] where one proton per electron is transferred during the conversion of the leucoemeraldine to the pernigraniline form of the polymer.

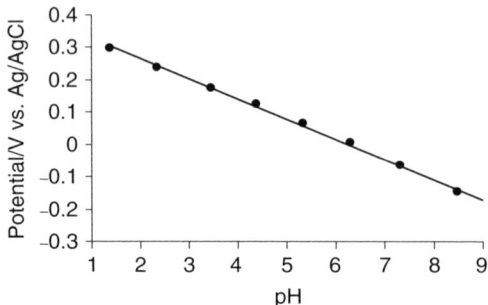

Figure 3.11 Potential of PABA film in PBS with 40 mM NaF and 10 M D-fructose as a function of pH. Slope = −58 mV/pH. (Reprinted with permission from *Chemistry of Materials*, 16, 1427. Copyright (2004) American Chemical Society.)

3.3.2 Spectroscopy

Self-doped PABA has several advantages, including good solubility as well as redox activity and similar conductivity to that of sulfonated polyaniline. In addition, self-doped PABA chemically polymerized in the presence of D-fructose and fluoride (poly (**2**), Figure 3.7) has the unique ability to switch between a self-doped and a nonself-doped state reversibly. In its self-doped state, the polymer is soluble under the polymerization conditions and can be easily and reversibly converted into the insoluble nonself-doped form. The UV-Vis spectra of the soluble form of PABA, as well as thin films of the precipitated form, are reportedly similar to those of the emeraldine salt form of unsubstituted polyaniline, exhibiting absorption bands near 350 and 820 nm due to the $\pi-\pi^*$ and bipolaron band transitions, respectively [53, 54]. The existence of these bands in pure water indicates that PABA exists in a self-doped state. PABA (poly(**2**), Figure 3.7) was converted to the base form of PABA (poly(**1**), Figure 3.7) by the removal of D-fructose from the polymer in PBS. The appearance of an absorption band at 600 nm and the disappearance of the peak at 820 nm, as shown in Figure 3.12, indicates the conversion to the nonself-doped state and subsequently to the emeraldine base form of the polymer. The switching behavior is reversible by exposing the base form of PABA (poly(**1**), Figure 3.7) to the original concentrations of D-fructose and fluoride either in PBS or pure water.

The changes in the redox behavior of electrochemically prepared self-doped PABA as a function of pH of solution in the presence of D-fructose and fluoride are a result of structural changes in the complexation

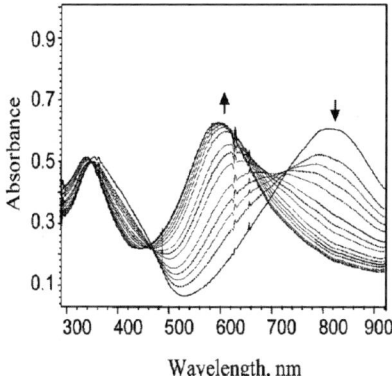

Figure 3.12 UV-Vis spectral changes of a chemically prepared PABA thin film as a function of time upon exposure to pH 7.4 PBS over the course of an hour. (Reprinted with permission from *Journal of the American Chemical Society*, **126**, 52. Copyright (2004) American Chemical Society.)

of boronic acid with D-fructose. As expected, a corresponding electrochromic change of PABA was observed with *in situ* UV-Vis measurements as the polymer was stepped to potentials between −0.2 and 1.0 V. The time required to reach equilibrium (i.e., current to approach zero) using the step technique was different in acidic solution (60–120 s) compared with neutral solutions (400 s), probably due to an increase in the electronic resistance of PABA in the latter case. This was consistent with bulk resistance measurements and with the apparent ohmic induced broadening of the peaks in Figures 3.8f and 3.8g.

The spectral changes of PABA films as a function of potential are shown in Figure 3.13 and are consistent with previous studies of polyaniline in acidic solution [55–57]. In the potential window of −0.2 to 1.0 V, spectral changes in the range of 400–450 nm and 650–800 nm were observed. The absorption peaks observed at 400–450 and 750–800 nm have been attributed to polaron and bipolaron transitions, respectively [58]. The presence of these localized and delocalized cation radicals confirms the self-doped state of PABA throughout the entire pH range (Figure 3.13a–c). In the potential range of −0.2 to 0.3 V, the oxidation increases the absorbance at 750–800 nm. At higher potentials and further oxidation, the band at 750–800 nm is displaced to 600–650 nm where the quinoneimine form of polyaniline absorbs. These spectral changes indicate the transition of the oxidation state of PABA from leucoemeraldine to emeraldine, and subsequent conversion to pernigraniline. In the case of polyaniline, the polaron band at

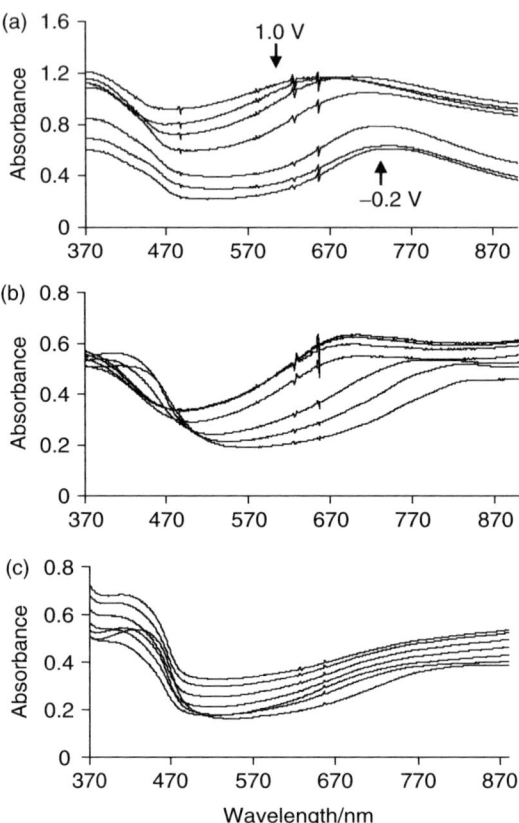

Figure 3.13 UV-Vis spectra of PABA film in PBS with 10 M D-fructose and 40 mM NaF at potentials from −0.2 to 1.0 V with a difference of 0.2 V as a function of pH: (a) 1.3, (b) 6.4, and (c) 9.0. (Reprinted with permission from *Chemistry of Materials*, **16**, 1427. Copyright (2004) American Chemical Society.)

∼400–450 nm is distinct only at low pH values. The bipolaron band at 780–800 nm shows a strong blue shift with increasing potentials at pH greater than 1. However, with increasing pH, the extent of blue shift of the band at 780 nm decreases gradually and the band disappears even at high oxidation potentials. A new band at ∼600 nm begins to appear at pH greater than 4 [59]. At higher pH values, the band near 600 nm shifts to shorter wavelength, and an increase in absorption near 330 nm is observed [55–57]. The disappearance of the polaron and bipolaron bands, and the evolution of a new band at ∼600 nm with increasing pH, are indicative of the existence of a quinoid structure. In contrast, the spectra in Figure 3.13 for all pH values are similar to those observed for polyaniline at low pH values, with the exception of a decrease in the intensity observed for PABA. Interestingly, the polaron and bipolaron bands (at ~400 and 780 nm, respectively), which are characteristic of the emeraldine salt form, are observed throughout the pH range at lower potentials. These results suggest that the emeraldine salt form of self-doped PABA is not very susceptible to pH changes. The decrease in intensity of absorption with increasing pH is probably due to a decrease in the number of redox centers, and is consistent with the change in the magnitude of current observed in the cyclic voltammograms of PABA shown in Figure 3.8. This change in the magnitude of absorption corresponds to the transitions in the oxidation states, and the magnitude of the redox current returns to the original magnitude in acidic solution.

3.3.3 Molecular Weight

The physical, chemical and mechanical properties of a polymeric materials are known to depend strongly on molecular weight and molecular weight distribution. Unfortunately, chemically synthesized polyanilines (parent, substituted and self-doped) typically suffer from low molecular weight, probably due to precipitation of low molecular weight species during polymerization. Postpolymerization modification in the case of sulfonated polyaniline can compound this problem, since the prolonged sulfonation process using ~30 % fuming sulfuric acid shortens the chain length of the polymer due to the hydrolysis of quinoid structures [60, 61]. Furthermore, there is lack of control over the degree of sulfonation as well as the distribution of functional groups along the polymer backbone. These issues are avoided in the case of self-doped PABA, since it is soluble under polymerization conditions, thereby resulting in a high molecular weight homopolymer [37].

The molecular weight of PABA (0.033 mg/mL) in N-methylpyrrolidone was determined using gel permeation chromatography. In general, molecular weight determination studies of polyaniline using gel permeation chromatography have used the reduced leucoemeraldine base form of polymer dissolved in 0.5 % LiCl-N-methylpyrrolidone mobile phase [62–66]. The reduced leucoemeraldine form is typically prepared by reduction of the emeraldine base form of polyaniline, either by dissolving in N-methylpyrrolidone containing various amounts of LiCl [62, 63] or by reduction with aqueous hydrazine [64–66]. In these studies, the reduced form was investigated at relatively high concentrations (above 0.5 mg/mL), and in turn the addition of lithium salt was reportedly required to prevent aggregation [63, 67, 68]. The concentration of the

PABA solution used for molecular weight determination was significantly lower (0.033 mg/mL) than the value where polyaniline is reported to aggregate (above 1 mg/mL); therefore, the lithium salt was not deemed necessary. Gel permeation chromatography indicated that a number average molecular weight of 1 676 000 g/mol, a weight average molecular weight of 1 760 000 g/mol and a polydispersity of approximately 1.05 could be achieved with chemically synthesized PABA.

Postpolymerization modification of PABA was used to confirm that the high molecular weight observed is not a result of intermolecular anhydride formation, resulting in crosslinking under the conditions used for gel permeation chromatography analysis (column temperature, 70 °C) [69]. The boronic acid groups on the polymer were removed via an *ipso*-substitution reaction involving hydrogen peroxide and iodine as shown in Figure 3.14 [70–72]. The molecular weight of the iodo- and hydroxy-substituted polyaniline was determined under identical conditions. Figure 3.15 shows the gel permeation chromatograph of PABA, iodo- and hydroxyl-substituted polyaniline. A single peak corresponding to a weight average molecular weight of 1 851 000 and 1 702 000 g/mol for iodo- and hydroxyl-substituted polymers, respectively, was observed. The majority of the polymer reportedly retained its high molecular weight even after the removal of the boronic acid group. These results suggest that the high molecular weight observed for PABA is not due to significant boronic acid anhydride crosslinking. For polyaniline, typically a bimodal chromatogram is observed with a high molecular weight fraction in the leucoemeraldine [62–66] and emeraldine base [68] forms in N-methylpyrrolidone elution. The addition of 0.5 % LiCl to the N-methylpyrrolidone solution of the polymer results in a unimodal

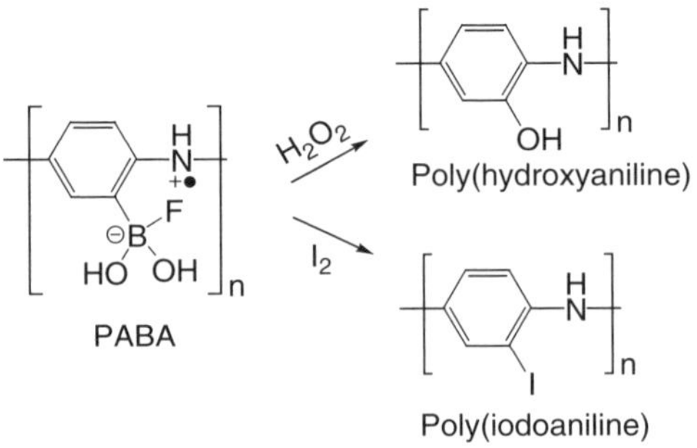

Figure 3.14 *ipso*-substitution reaction of PABA with hydrogen peroxide and iodine. (Reprinted with permission from *Langmuir*, **17**, 7183. Copyright (2001) American Chemical Society.)

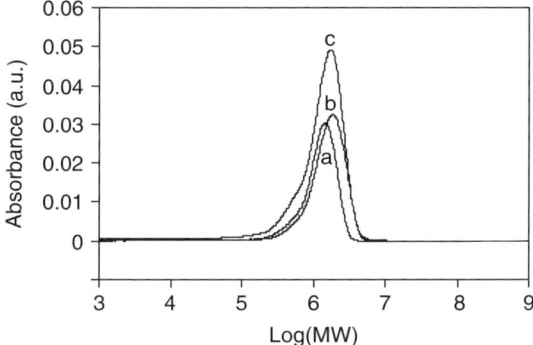

Figure 3.15 GPC chromatograms of PABA dissolved in N-methylpyrrolidone before and after (b) exposure to 30 % H_2O_2 for 10 min and (c) iodination reaction. (Reprinted with permission from *Macromolecules*, 38, 10022. Copyright (2005) American Chemical Society.)

molecular weight distribution, which indicates that the high molecular weight fraction is caused by polymer aggregation [67, 68]. However, in the case of PABA, it is asserted that the high molecular weight results from the fact that the polymer is soluble under polymerization conditions where termination of polymer via precipitation [73–78] is prevented.

3.4 SELF-CROSSLINKED SELF-DOPED POLYANILINE

3.4.1 Introduction

To date, attempts to crosslink electronically conducting polymers have not been very successful in yielding enhanced mechanical properties while maintaining good conductivity. For example, crosslinking with formaldehyde [79], elastomers [80], condensation at high temperature [81] and thermal annealing in N-methylpyrrolidone [82] results in significant losses in conductivity. This is probably due to the loss of conjugation and/or the incompatibility of the volume shrinkage and the presence of counterions. For example, the volume change required to accommodate the presence of anions and associated water in polyaniline films has been shown to be in the range of 10–30 % [83]. Therefore any significant volume change would probably be at the expense of dopant and, in turn, conductivity.

Self-doped polyaniline offers several unique properties due to the fact that the counterion that acts as the dopant is covalently attached to the backbone [42, 84]. PABA exhibits unique self-doping properties through the formation of tetrahedral boronate species in the presence of fluoride [31, 37, 85]. The presence of boronic acid groups along the conjugated backbone of polyaniline offers many possibilities for chemical modification and reactivity. For example, aryl boronic acid chemistry is rich with examples of the formation of boron chelates involving alcohols [86], amino acids [87], hydroxyoximes [88], salycylaldehyde and azomethines [89]. The pertinent examples are the fluorine-containing salicaldimine–boron complex (Figure 3.16, B) [90] (Figure 3.16, A) and the dimer of 2-aminophenylboronic acid formed in aprotic solvents and in the solid state [91]. These compounds consist of a six-member heterocyclic complex containing a boron-imine dative bond, and are air stable.

Based on this chemistry, a new strategy for the generation of a crosslinked, self-doped conducting polymer has been demonstrated [92]. The intermolecular reaction between boronic acid groups and imines in PABA containing fluoride (Figure 3.17, A) results in crosslinks with an analogous structure involving tetravalent boron. This results in a self-doped, self-crosslinked polyaniline (Figure 3.17, B). The key advantage of such a structure is that the crosslink site can also act as a dopant site. In turn, this structure can accommodate the volume shrinkage associated with crosslinking while maintaining the anionic dopant required for

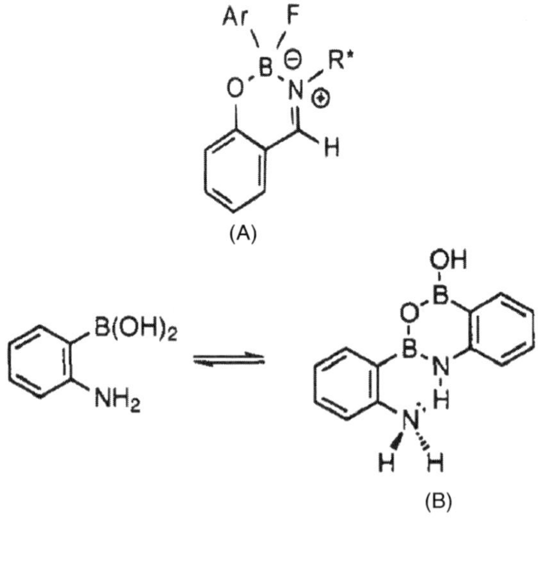

Figure 3.16 Dimer formation of 2-aminophenylboronic acid in aprotic solvents in the solid state. (Reprinted with permission from (A) *Journal of the American Chemical Society*, **122**, 3047. Copyright (2000) American Chemical Society; (B) *Journal of the American Chemical Society*, **116**, 7597. Copyright (1994) American Chemical Society.)

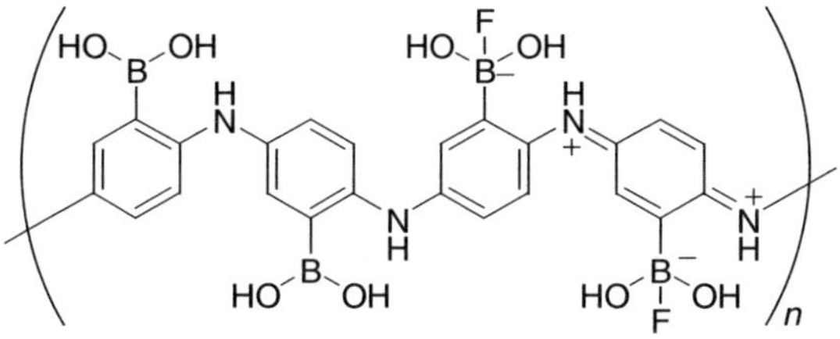

Figure 3.17 Emeraldine salt form of PABA (A), self-doped in the presence of fluoride. Proposed cross-link (B), resulting from an interchain dehydration reaction between a boronic acid-imine and a boronic acid moiety, hence maintaining a self-doped state. (Reprinted with permission from *Chemistry of Materials*, **17**, 3803. Copyright (2005) American Chemical Society.)

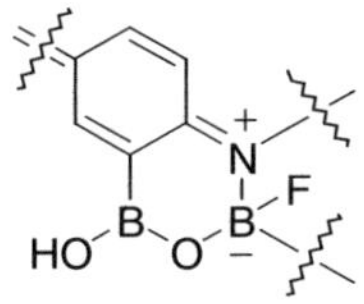

conductivity. Further, this approach does not require the use of fillers or other nonconducting crosslinking agents that will ultimately reduce conductivity. The approach has been demonstrated via self-crosslinking of PABA, which results in the creation of tetravalent boron through the formation of a dative boron-imine bond. The self-doped, self-crosslinked, PABA reportedly exhibits a significant degree of conductivity, unprecedented hardness, surpassing more common 'hard' polymers, and thermal stability, which is discussed in the following sections.

3.4.2 Synthesis and Characterization

Self-doped PABA has been prepared in water in the presence of excess fructose, and one equivalent of fluoride to monomer, under ambient conditions (for details see Chapter 3, Section 3.2.2) [37]. The resulting water-soluble polymer was precipitated by dilution in pure water. Following filtration and rinsing with water, the precipitate was washed with 0.5 M HCl to remove D-fructose, and dried in air. Pellets of air-dried PABA were produced at 10 000 psi for 5 min and crosslinked at 100 °C under vacuum for 24 h. The atomic percent of boron and fluorine in a heat treated pellet as determined by X-ray photoelectron spectroscopy

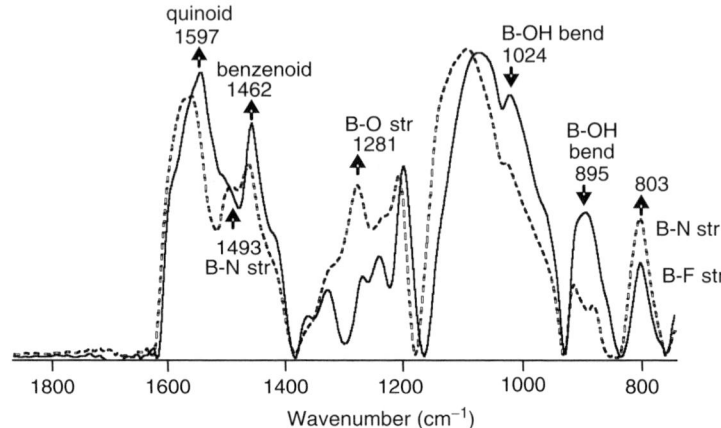

Figure 3.18 Fourier transform infrared attenuated total reflectance absorption spectra of PABA pellet (solid line) air-dried and (dash line) heat treated. (Reprinted with permission from *Chemistry of Materials*, 17, 3803. Copyright (2005) American Chemical Society.)

of a freshly polished surface revealed a B:F ratio of approximately 2:1, indicating that a significant amount of fluoride remained complexed in the polymer, even after washing [92].

The crosslinked polymer was characterized using infrared attenuated total reflectance spectroscopy (Figure 3.18). The spectra of air dried and heat treated PABA pellets revealed changes consistent with increased crosslinking while maintaining the basic polyaniline structure. Vibration characteristics of polyaniline were observed at 1597, 1462 and 1130 cm^{-1}, corresponding to quinoid, benzenoid and C–N stretching ring modes, respectively [93]. The vibrations at 895, 1024 and 803 cm^{-1} have been attributed to B–OH and B–F stretching modes, respectively [94]. The spectrum of the heat treated pellet showed an additional peak at 1493 cm^{-1} assigned to a B–N stretching mode [95, 96] and an increase in the intensity of 1281 and 803 cm^{-1} vibrations assigned to B–O and the B–N stretching mode of a dative bond [95, 96]. While B–F and B–N stretch bands overlap at 803 cm^{-1}, the increase in intensity can only be associated with an increase in the formation of B–N dative bonds. All of the spectral observations are consistent with increased crosslinking involving the formation of boronic acid anhydride and boron–nitrogen dative bonds (Figure 3.17, B). Furthermore, the relatively unchanged ratio of vibrations at 1597 and 1462 cm^{-1} indicate that the ratio of quinoid to benzenoid structures in the film remains the same after heat treatment and, in turn, the polymer remains in an oxidized state.

3.4.3 Mechanical Properties

3.4.3.1 Hardness Measurement

Microhardness measurement (e.g., Vickers hardness) is a standard method for measuring the mechanical properties of materials ranging from 'hard' polymers to 'superhard' materials such as diamond and cubic boron nitride (Figure 3.19). This kind of measurement is especially useful for characterizing polymers since it is related to yield stress, modulus of elasticity and some secondary relaxation transitions [97]. Common 'hard' plastics typically exhibit hardness values ≤0.3 GPa. The hardness of PVC, polymethyl methacrylate, polycarbonate, polystyrene and acetal are 0.16, 0.2, 0.3, 0.15 and 0.16 GPa, respectively [98, 99]. In order to achieve increased hardness, more complex materials such as functionally graded polymer composites, containing significant amounts of much harder fillers, have been explored. Hardness values of 0.6 GPa can be achieved with composites containing 45 % SiC filler [100]. Microhardness measurements of bulk polyaniline have not been investigated in detail; however, in one set of studies, surface microhardness

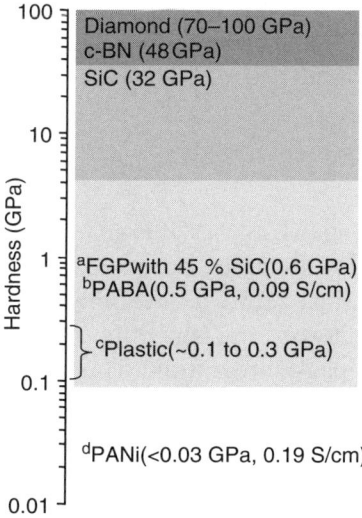

Figure 3.19 Vickers hardness of crosslinked PABA in comparison with other materials. a: functionally graded polymer composite with 45 % SiC; b: heat treated crosslinked PABA; c: common hard plastics including PVC, PMMA, polycarbonate, polystyrene and acetal; d: heat treated polyaniline (polyaniline). (Reprinted with permission from *Chemistry of Materials*, 17, 3803. Copyright (2005) American Chemical Society.)

measurements of pristine and modified crosslinked polyaniline thin films were carried out by Kang *et al.* [101, 102]. The surface microhardness of as cast emeraldine base and crosslinked emeraldine base films from N-methylpyrrolidone where reported to be around 1–2 and 6 GPa, respectively. Surface modification of emeraldine base films via graft copolymerization with acrylic acid and styrenesulfonic acid enhanced the microhardness to 20 GPa. These values are interfacial values and are not obtained with bulk samples. Hardness values obtained using the standard Vickers hardness technique for air dried and heat treated PABA pellets can reach 0.3 and 0.5 GPa, respectively [92]. These values are dramatically higher than polyaniline and those reported for other bulk polymers and approach values achieved with composites (Figure 3.19). The increase in hardness of the heat treated PABA was attributed to increased crosslinking facilitated by heating and removal of water (a product of the crosslinking reaction) under vacuum. The hardness measurements represent a lower limit since the material probably yields due to the breakdown of intergranular crosslinks prior to the plastic flow of polymer chains. The granular nature of the pressed pellets made from the powder may be different from that observed for polymer films prepared from a solvent such as N-methylpyrrolidone. It is expected that crosslinking between grains will ultimately limit the hardness in the pellet and therefore the indentation measurement is expected to have a lower limit.

3.4.4 ^{11}B NMR

^{11}B magic angle spinning (MAS) NMR is an ideal method for quantifying the coordination environment of boron [103, 104]. At a sufficiently strong magnetic field (i.e., ≥ 11.7 T) the signals corresponding to three and four coordinate boron are generally resolved and yield relative populations by direct peak integration. The application of MAS NMR to conducting samples is rare due to the potential for sample heating during rapid spinning, and probe damage from the requisite high radio frequency fields. However, no anomalous effects were reported in the case of PABA.

The ^{11}B NMR spectrum of a coarse ground pellet (14 mg) was obtained at 192.4 MHz (14.1 T) on a Varian Inova 600 spectrometer. For the one dimensional MAS experiment spinning at 12 kHz, a pulse of 1.5 μs with a radiofrequency (rf) field of 25 kHz (<15° tip angle) was used to ensure homogeneous excitation of all boron sites [105]. The spectrum was the result of 800 transients separated by a relaxation delay of 10 s. For the two dimensional multiple quantum (MQ) MAS spectrum, multiple quantum excitation and conversion were achieved using 7 μs and 3 μs pulses with an 83 kHz rf field, respectively, and a 15 μs z-filter at 8 kHz rf. A total of 128 t_1 increments of 24 transients each were collected on a sample spinning at 18.2 kHz. No ^{1}H or ^{19}F decoupling was required to achieve optimal peak narrowing at these spinning speeds. The sample temperature was maintained at 22 °C. The chemical shift axis is presented relative to BF_3–OEt_2, as measured by secondary reference 0.1 M boric acid, which appeared at +19.6 ppm. Relative populations of four and three coordinate borons were evaluated by direct integration of the two major peaks, and corrected for small effects arising from multiple transitions present in the quadrupolar ^{11}B spins [106].

Figure 3.20 shows the one-dimensional (A) and two-dimensional MQMAS (B) ^{11}B NMR spectra obtained for heat treated PABA. The broad signal centered at 16.5 ppm and a sharper peak at 1.5 ppm are observed corresponding to three-coordinate and four-coordinate boron, respectively. A two-dimensional MQMAS [107] experiment indicated that two four-coordinate boron sites are present in the sample (Figure 3.20, B), one of which (6.6 ppm) is partly obscured by the quadrupole-broadened, three-coordinate, boron signal in the MAS spectrum. While the precise identities of the four-coordinate boron species were not determined, the chemical shifts were reportedly consistent with

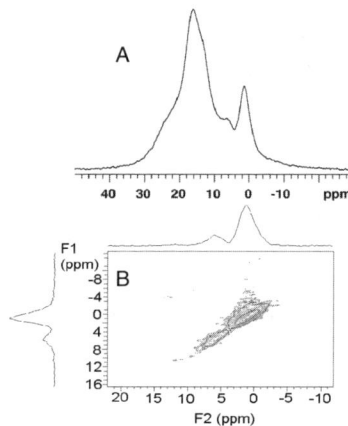

Figure 3.20 ^{11}B NMR spectra of heat treated PABA pellet obtained at 192.4 MHz (14.1 T) A: One-dimensional MAS experiment B: Two-dimensional sheared MQMAS experiment optimized to excite preferentially four-coordinate boron sites (displayed region) (Reprinted with permission from *Chemistry of Materials*, **17**, 3803. Copyright (2005) American Chemical Society.)

the local atomic connectivities indicated in Figure 3.20. Fitting the one dimensional MAS spectral intensities using the site information from the two dimensional MQMAS experiment revealed that 21 ± 2 % of the boron in the sample is four coordinate. Since such sites can act as a dopant (Figure 3.17, B), the NMR results suggest that the heat treated PABA is doped to that level.

3.4.5 Thermal Properties

Enhanced thermal stability of PABA was expected due to the high molecular weight and crosslinked self-doped nature of the polymer produced upon heating or drying under vacuum. Specifically, the anionic tetrahedral boronic acid, which acts as the dopant, is covalently linked to the polyaniline backbone, is electron donating and serves to crosslink the polymer. In general, the thermal properties of polyaniline are largely limited by the volatility of the dopants (for details see Chapter 1, Section 1.5.6). Self-doped sulfonated polyanilines are reported to be much more thermally stable than unsubstituted polyaniline doped with HCl [42, 84, 108, 109]. Han *et al.* [110] have reported that the mercaptopropanesulfonic-acid-substituted self-doped polyaniline has better thermal stability than that of HCl doped polyaniline and sulfonated polyaniline (For details see Chapter 2, Section 2.4.6). The covalently bonded, electron-donating, mercaptopropanesulfonic acid constituent group at the polymer backbone reportedly stabilizes the aromatic benzenoid ring, thereby reducing the decomposition of the backbone. The polymer retains its sulfonic acid group until ~260 °C, whereas sulfonated polyaniline starts to lose its sulfonic acid groups at a much lower temperature of about 185 °C [110].

Thermal stability of HCl-doped polyaniline with a molecular weight of 30 000 g/mol produced by a standard method [111] and a molecular weight of 100 000 g/mol (commercially obtained from Aldrich) in comparison with PABA was studied as shown in Figure 3.21. HCl-doped polyaniline with a molecular weight of 30 000 g/mol (Figure 3.21, a) and 100 000 g/mol (Figure 3.21, b) both exhibit three major weight loss steps. The first weight loss below 100 °C is attributed to the evaporation of water. The second weight loss near 200 °C is due to the removal of HCl, resulting in the concomitant loss of conductivity. The third weight loss above 400 °C is associated with the decomposition of the polyaniline backbone. These transitions are similar to those reported for HCl doped polyaniline under nitrogen [108]. While PABA

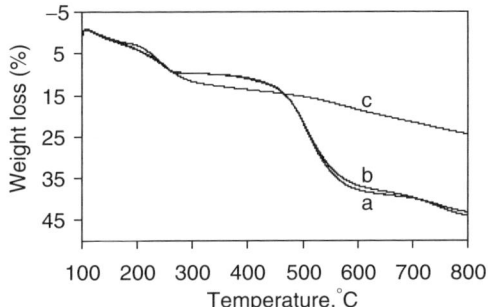

Figure 3.21 Thermograms of (a) polyaniline (M_w 30 000 g/mol) produced by the conventional method, (b) polyaniline (M_w 100 000 g/mol) purchased from Aldrich and (c) PABA (Mw 1 760 000 g/mol) dried in air. (Reprinted with permission from *Macromolecules*, 38, 10022. Copyright (2005) American Chemical Society.)

(Figure 3.21, c) exhibits a step near 200 °C, the weight loss above 500 °C is small relative to polyaniline. Interestingly, a slow rate of weight loss is observed throughout the thermogram, attributed to the slow crosslinking process and the associated loss of water. Above 500 °C, the decomposition of the polymer occurs slowly, resulting in only 8 % weight loss. The total weight loss for PABA was about 24 % of the initial weight, which is significantly less than the nearly 50 % weight loss exhibited by HCl-doped polyaniline. These results suggest that the self-doped PABA has greater thermal stability than HCl-doped polyaniline as well as self-doped sulfonated and mercaptopropanesulfonic acid substituted polyaniline [108, 110, 112]. It is suggested that the enhanced thermal stability is a result of anhydride formation as a result of crosslinking (Figure 3.17, B).

3.4.6 Temperature Dependent Conductivity

According to a two-dimensional MQMAS ^{11}B NMR experiment, 21 ± 2 % of the boron in heat treated PABA is four coordinate. Since such negatively charged crosslink sites (Figure 3.17, B) can act as a dopant, the NMR results suggest that the PABA is self-doped to that level [92]. Thermal studies of PABA suggest that the greater thermal stability of the polymer is due to crosslinking. Based on thermogravimetric and ^{11}B NMR results, PABA should be self-doped and in turn conducting at high temperature. Therefore, the stability of conductivity of PABA (M_w 1 760 000 g/mol) in comparison with the various forms of polyaniline, (HCl doped polyaniline with a molecular weight of 30 000 and

Table 3.3 Conductivities of PABA and polyaniline. (Reprinted with permission from *Macromolecules*, 38, 10022. Copyright (2005) American Chemical Society.)

Dry condition	Conductivity (S/cm)		
	Polyaniline (M_w 30 000)	Polyaniline (M_w 100 000)	PABA (M_w 1 760 000)
Air	7.95	4.68	0.96
Vacuum 100 °C	0.19	0.13	0.094
Vacuum 500 °C	0.00039	0.00033	0.009

100 000 g/mol) was determined by heating polymer at 100 and 500 °C under vacuum for 24 h. The conductivities of the heat-treated polymers in the form of a compressed pellet measured using four probe measurement are shown in Table 3.3. The conductivity of HCl doped polyaniline with M_w 30 000 and 100 000 g/mol pellets treated at 100 °C decreased from 7.95 to 0.19 and from 4.68 to 0.13 S/cm, respectively. The decrease in conductivity was supported by the thermogravimetric studies, corresponding to the loss of the volatile Cl$^-$ dopant at 100 °C. These results are consistent with previous reports that the HCl-doped polyaniline loses its conductivity above 100 °C due to evaporation of dopant, and reports that the polymer is insulating above 400 °C where the backbone is completely decomposed [113–119]. However, similarly to the thermogravimetric studies, no significant effect of the molecular weight on thermal stability of the conductivity of polyaniline was observed.

The conductivity of self-doped PABA without heat treatment was observed to be around 0.96 S/cm. This conductivity value is consistent with the 21 % doping suggested by ^{11}B NMR based on the conductivities of other forms of self-doped polyanilines [110, 113]. However, the conductivity was lower than HCl doped polyaniline, probably due to distortion of the polymer backbone by the presence of the boronic acid substituent [116–118]. After heat treatment at 100 and 500 °C, a decrease in conductivity from 0.96 S/cm (without heat treatment) to 0.094 and 0.009 S/cm, respectively was observed. However, in the case of polyaniline, the conductivities of the heat treated polyaniline declined significantly compared with that of the self-doped PABA. The relative decrease in conductivity of heat treated PABA was less than that of HCl-doped polyaniline, probably due to the formation of a thermally stable boronic acid anhydride crosslink. In the case of polyaniline, the dramatic decrease in conductivity was a result of the decomposition of the backbone above 420 °C, as seen in the thermograms. In contrast, the process of crosslinking in the PABA polymer above 100 °C may make a more thermally stable backbone when heated to high temperatures. These results suggest that the conductivity of PABA is not as high as that of native polyaniline under ambient conditions; however, it is sufficiently high for many applications and has the added benefit of functioning at extreme temperatures.

3.5 APPLICATIONS

The electrochemically and chemically synthesized self-doped PABA has been used for sensing applications based on the reversible complexation of the boronic acid substituent on the polymer with fluoride-[31] and diol-containing compounds such as saccharides[7, 8], catecholamines [32, 33] and glycoproteins [120]. Also, biogenic amine vapor detection has been carried out based on boronate-alkylammonium electrostatic interactions [34]. PABA-RNA multilayer films have also been prepared using a layer-by-layer deposition technique involving soluble self-doped PABA and are used for controlled release of RNA. These approaches involve the formation of self-doped PABA and the resulting impact on the electrochemical potential is described in the following sections.

3.5.1 Saccharide Sensor

Saccharides play significant roles in biological processes. The development of saccharide sensors is highly desirable in order to understand cellular activity and to diagnose disease. The detection of saccharides, including D-glucose and D-fructose, has been the focus of considerable research due to the importance of saccharide monitoring in such diverse areas as medical diagnostics [121, 122] and bioprocessing [123]. One of the most successful electrochemical approaches for detection of saccharides is the use of enzyme-based electrodes. These have been under development since the 1960s [124]. The most advanced glucose sensors are based on glucose oxidase coupled to electrochemical systems [122, 125–128]. However, the enzyme-based sensors have inherent problems such as requiring mediators, mass transport dependent responses, and the fact that they consume the analyte of interest.

Sensors based on the complexation of boron compounds with saccharides are an attractive alternative to enzymatic approaches. This covalent molecular recognition has proven important for saccharide detection. A number of sensing strategies including direct pH measurements [129], fluorescence [130, 131], UV-Vis [40, 132], circular dichroism [133, 134], near-IR [30], surface plasmon resonance spectroscopy [135], potentiometry [7, 8, 136, 137], conductance measurements [1, 135, 138], and quartz crystal microbalance measurements [135, 139] have been developed. These approaches overcome many of the issues that plague enzymatic systems as described above. Most importantly, these approaches are 'reagentless' and their sensitivity is not dependent on the mass transport of analyte.

Recently, Fabre *et al.* [31] and Freund *et al.* [7, 8] used electrochemically deposited, self-doped, boronic-acid-substituted, conducting polymers for saccharide and fluoride detection. Freund *et al.* prepared a potentiometric sensor for saccharides using self-doped PABA [7, 8]. The transduction mechanism in that system is reportedly the change in pK_a of polyaniline that accompanies complexation, and the resulting change in the electrochemical potential. Sensors produced with this approach exhibit reversible responses with selectivity to various saccharides and 1,2-diols (Figure 3.22) that reflect their binding constants with phenylboronic acid observed in bulk solutions. The sensitivity

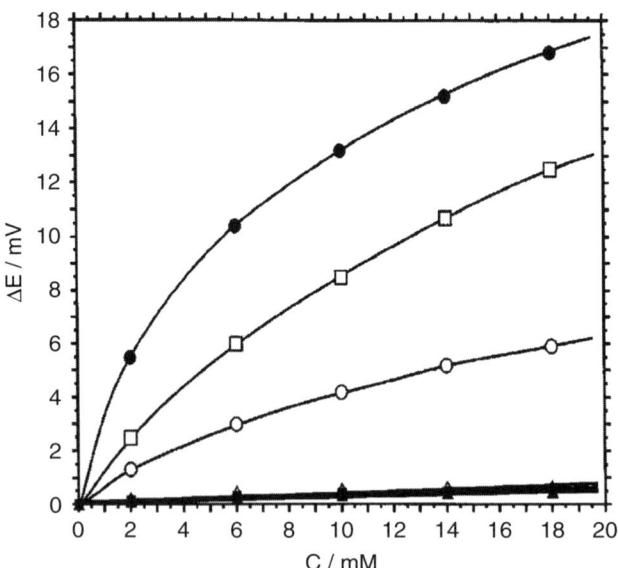

Figure 3.22 Sensitivity of PABA to D-fructose (●), D-glucose (○), *cis*-1,2-cyclopentanediol (□), *trans*-1,2-cyclopentanediol (△), *cis*-1,2-cyclohexandiol (■), and *trans*-1,2-cyclohexanediol (▲) and in PBS. (Reprinted with permission from *Journal of the American Chemical Society*, **124**, 12486. Copyright (2002) American Chemical Society.)

and mechanical stability of PABA films are reportedly enhanced significantly by increasing the fluoride concentration during electrochemical polymerization. These results suggest that the formation of self-doped polymer under polymerization conditions is important for increasing the sensitivity for diols.

3.5.2 Nucleotide Sensors

Nucleotides, such as β-nicotinamide adenine dinucleotide (NADH) and its oxidized form (NAD$^+$), are ubiquitous in all living systems and are required for the reactions of more than 400 oxidoreductases [140, 141]. Thus, a vast number of studies have been directed at understanding the factors that may control the redox activity of these biological cofactors. In particular, the complex redox behavior exhibited by the NAD$^+$/NADH couple (Figure 3.23 [142]) under physiological conditions has prompted extensive electrochemical studies of the reaction. The electrochemical oxidation of NADH at low overpotentials is of special interest in biosensor development and, as a result, it has been investigated for decades as a means for creating enzyme-based electrocatalytic systems for quantification of various substrates. A major complication in this field is the fact that direct oxidation of NADH at unmodified electrode surfaces proceeds at high overpotentials (>0.5 V) and generally leads to fouling of the electrode surface, as well as the formation of a mixture of oxidation products, including products of radical coupling. Attempts to develop catalytic electrode surfaces for the oxidation of NADH to enzymatically active NAD$^+$ have concentrated on the use of redox mediator species [143] and conducting polymers [144–146].

The interaction of borate and boronic acid with a number of ribose-containing nucleotides and cofactors such as NAD$^+$, NADH, adenosine triphosphate (ATP) and adenosine monophosphate (AMP) have been

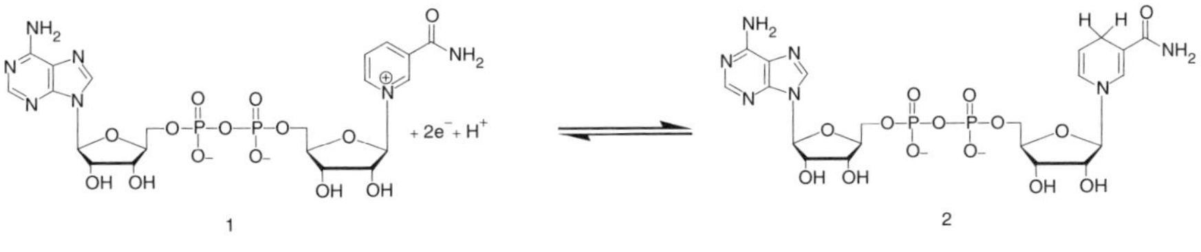

Figure 3.23 Interconversion of NAD$^+$ (1) and NADH (2). (Reprinted with permission from *Journal of the American Chemical Society*, **103**, 2379. Copyright (1981) American Chemical Society.)

studied using capillary electrophoresis [147], electrospray ionization spectrometry and ^{11}B NMR spectroscopy [148, 149]. Recently, Freund *et al.* studied the reactivity of PABA with NAD^+ and NADH [35]. In that study, PABA was in a self-doped state after complexation with the nucleotide at neutral pH through ester formation; however, the nature of species responsible for self-doping in the cases of NAD^+ and NADH were different. The structural changes associated with boronic acid/boronate binding were investigated using various techniques including cyclic voltammetry, open-circuit potential, quartz crystal microbalance, ^{11}B NMR, and polarization modulated infrared reflection absorption spectroscopic measurements. Details are given in the following sections.

3.5.2.1 Cyclic Voltammetry

Figure 3.24 shows the redox behavior of PABA thin films observed at neutral pH in the presence of NADH and NAD^+. The PABA film was redox inactive at neutral pH (Figure 3.24,a) due to deprotonation and loss of dopant as with polyaniline [150, 151]. However, in the presence of NADH (Figure 3.24, b) and NAD^+ (Figure 3.24, c), PABA films became redox active due to complexation of boronic acid with *cis*-2,3-ribose diols and subsequent formation of self-doped polymer. In the presence of NADH, the cyclic voltammogram of PABA thin film exhibited a single redox couple at Ep_a 0.05 and Ep_c −0.10 V. In contrast, a second redox couple was observed in the presence of NAD^+ at Ep_a 0.34 and Ep_c

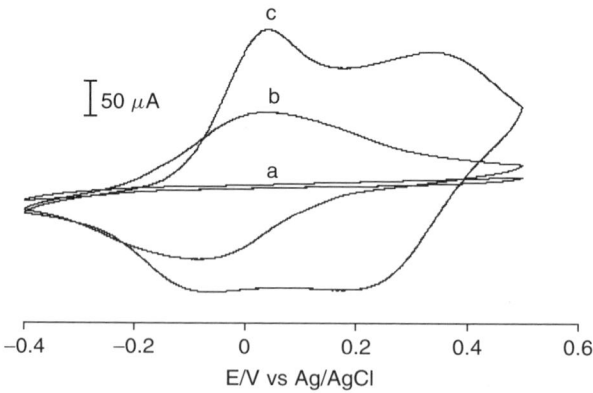

Figure 3.24 Cyclic voltammograms of PABA modified gold electrode in PBS at pH 7.4 (a) in presence of 10 mM NADH (b) and NAD^+ (c). Scan rate 100 mV/s. (Reprinted with permission from *Chemistry of Materials*, **17**, 2918. Copyright (2005) American Chemical Society.)

0.18 V. The redox behavior of a PABA thin film observed in the presence of NAD^+ was very similar to that of polyaniline [41, 152] and PABA [8] in acidic solution, with the exception that the two sets of redox couples were more closely spaced and shifted to lower potentials. In the presence of NAD^+ (Figure 3.24, c), the two redox couples represent the conversion of the reduced leucoemeraldine form into the conducting intermediate oxidized emeraldine form and subsequent conversion to the fully oxidized pernigraniline form [41, 152]. However, in the case of NADH (Figure 3.24, b), the absence of a second redox couple suggests that the further oxidation of emeraldine to the pernigraniline form of PABA is not energetically favorable. These results are similar to the redox behavior of self-doped PABA in the presence of D-fructose as a function of pH of electrolyte solution (Figure 3.8) [85]. In that study, the redox behavior changed dramatically as a function of the pH of the solution due to the presence of different boronic acid species. Similarly, the difference in the redox behavior of PABA thin films in the presence of nucleotides was attributed to the difference in the structure of boronic acid-NAD^+ and -NADH complexes, which are known to form in solution [147–149]. These two distinct redox behaviors indicate that neither electron transfer between the polymer and the nucleotide nor exchange between the nucleotide complexed and in solution occurs on the timescale of the voltammetric experiment.

3.5.2.2 ^{11}B NMR

The ^{11}B NMR of monomer solutions in the presence of NAD^+ and NADH (Figure 3.25) reveals the nature of the complex of PABA with nucleotides. In the presence of NADH, a single resonance with a chemical shift of 28.8 ppm (Figure 3.25, a) indicates the presence of the neutral trigonal form of boronic acid species [6] as shown in Figure 3.26 (I or II). However, in the presence of NAD^+ (Figure 3.25, b), an additional resonance signal was observed approximately 20 ppm upfield from the trigonal boronic acid signal, indicating the formation of a tetrahedral anionic boronic acid [6] as shown in Figure 3.26 (III). The presence of two peaks suggests slow exchange between the two forms on the ^{11}B NMR timescale [148, 149]. These observed structural differences were attributed to the electrostatic stabilization of the tetrahedral anionic boron provided by the positively charged nicotinamide group in NAD^+ [148, 149].

Based on the ^{11}B NMR results, upon complexation with nucleotides, the nature of the boronate esters responsible for self-doping of PABA

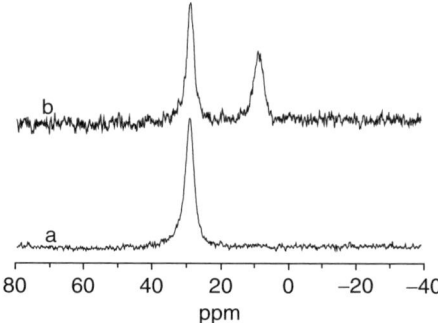

Figure 3.25 ^{11}B NMR spectra of 5 mM 3-aminophenylboronic acid in presence of 20 mM NADH (a) and NAD$^+$ (b) at pH 7.4 PBS. (Reprinted with permission from *Chemistry of Materials*, **17**, 2918. Copyright (2005) American Chemical Society.)

is different. In the presence of NAD$^+$, self-doping is thought to result from two sources – the anionic tetrahedral boronate ester as well as the nucleotide phosphate. The anionic tetrahedral boronic acid–NAD$^+$ adduct presumably supports the further oxidation of self-doped PABA (emeraldine to pernigraniline) due to the close proximity of the charged species to the polyaniline backbone. In contrast, the neutral trigonal boronic acid–NADH adduct only provides the anionic phosphates on the nucleotide which are further removed. Due to the relatively large separation of the charged phosphate group in the nucleotide from the cationic amine in the polymer backbone, oxidation from the emeraldine to the pernigraniline form of polyaniline is expected to be less energetically favorable.

3.5.2.3 Polarization Modulated Infrared Reflection Absorption Spectroscopy

The complexation of PABA with nucleotides was further studied using polarization modulated infrared reflection absorption spectroscopy (PM-IRRAS). PM-IRRAS spectra of PABA films exhibit all the characteristic vibrations of polyaniline and boronic acid (Figure 3.27) [93, 94]. After complexation with NAD$^+$ (Figure 3.27, b) and NADH (Figure 3.27, c), the disappearance of the free B–OH group vibration at 986 cm^{-1} and increase in the intensity of the asymmetric B–O bond vibration at 1330 cm^{-1} indicate the formation of boronate ester. The new vibrations at 1080 and 1470 cm^{-1} have been attributed to ribose and adenine moieties, respectively [153]. The vibrations at 1218 (NAD$^+$), 1245

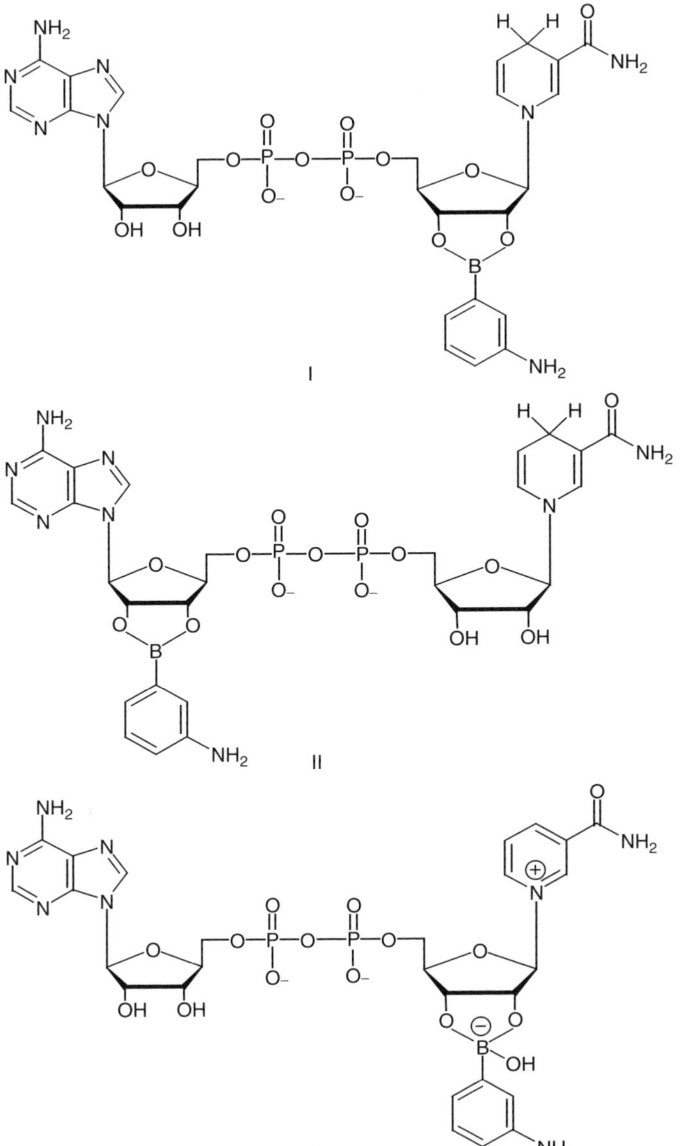

I

II

III

Figure 3.26 Proposed structures of boronate esters with NADH (I and II) and NAD^+ (III). (Reprinted with permission from *Chemistry of Materials*, **17**, 2918. Copyright (2005) American Chemical Society.)

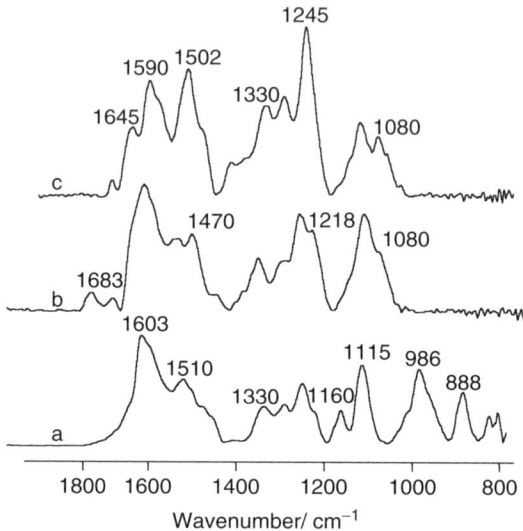

Figure 3.27 PM-IRRAS spectra of PABA film (a) reacted with 10 μM (b) NAD$^+$ and (c) NADH in pH 7.4 PBS. (Reprinted with permission from *Chemistry of Materials*, **17**, 2918. Copyright (2005) American Chemical Society.)

(NADH) and above 1603 cm^{-1} are assigned to the nicotinamide moiety [153]. The presence of these vibrations supports the complexation of nucleotides with PABA. After complexation with NAD$^+$ (Figure 3.27, b), the high relative intensity ratio of quinoid to benzenoid ring modes suggests that the PABA is in the oxidized state. These results suggest that complexation with NAD$^+$ results in self-doping through ester formation, in agreement with voltammetric and ^{11}B NMR studies. In contrast, following exposure to NADH (Figure 3.27, c), a red shift in the peaks corresponding to the quinoid and benzenoid ring modes as well as a decrease in their relative intensity ratio was observed. The decrease in the ratio of relative intensities of quinoid to benzenoid ring modes indicates that the polymer is reduced in the process. The reduction of PABA is probably due to hydride transfer from NADH to one of the nitrogens in the polyaniline backbone, consistent with previous reports [146].

3.5.2.4 Open-Circuit Potential Measurement

As discussed above, complexation changes the nature of the boronic acid substituent and in turn the electrochemical potential of polyaniline. Therefore, in addition to voltammetric, NMR and PM-IRRAS techniques, the complexation was monitored by measuring changes in the electrochemical potential of the polymer. The cyclic voltammetric behavior has shown a clear difference as a function of the redox state of the nucleotide and has suggested that the complexation reaction can be manipulated by altering the oxidation state of the polymer. The sensitivity of PABA films toward the redox state of the nucleotide was explored, using open circuit potential measurements, as a function of the oxidation state of the polymer (Figure 3.28). A significant difference in the open circuit potential sensitivity of PABA was observed upon complexation with nucleotides. A positive shift in electrochemical potential was observed for NAD^+ and a negative shift was observed for NADH. The positive shift in the potential of oxidized PABA for NAD^+ was presumably due to the formation of anionic boronate complex, similar to that observed for saccharides using PABA (see Figure 3.22) [7, 8]. However, the sensitivity was much higher in the case of NAD^+ than with saccharides, due to the higher binding constant associated with cyclopentane diol as well as the two sites available in the nucleotide. In contrast, in the case of NADH, the negative shift in the potential was attributed to the net effect of pK_a change and complexation, which is dominated by reduction of the polymer [146], consistent with PM-IRRAS results.

The sensitivity to NAD^+ and NADH was found to be dependent on the oxidation state of PABA. The maximum sensitivity was observed upon complexation with NAD^+ and NADH using oxidized and reduced PABA films, respectively. The magnitude of the change in electrochemical potential for NADH using a reduced PABA film (Figure 3.28, A; ○) was almost ten-times higher than that of NAD^+ using oxidized PABA film (Figure 3.28, B; ●). Also the potential reached a steady value within 10 min for NAD^+, while NADH required nearly 2 h to reach equilibrium. The observed high sensitivity and slow response for NADH, associated with the reduced polymer, can be attributed to the significant decrease in charging capacity and increased resistance to charge transport of the polymer in its nonconducting state. In other words, there is a larger potential change per unit charge when the polymer is in its reduced state compared with its oxidized conducting state. In order to determine a quantitative relationship between complexation and the open circuit potential of PABA, quartz crystal microbalance measurements were carried out. These results indicated that the amount of complexation is similar for NAD^+ and NADH using oxidized and reduced film, respectively. Therefore, the large change observed in the open circuit potential upon NADH complexation to the reduced form of PABA is associated with the oxidation state of the polymer. These results suggest

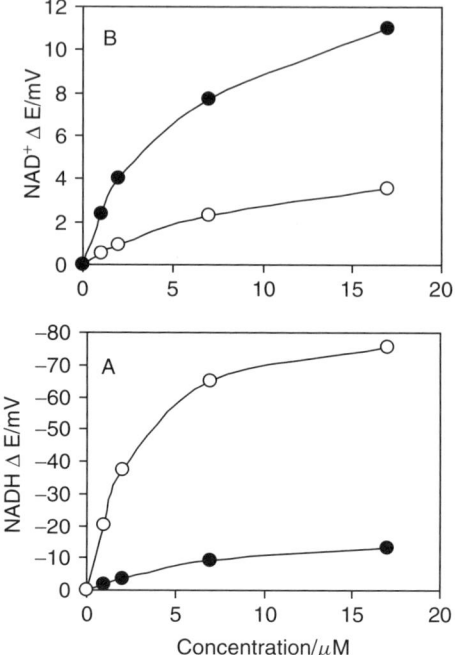

Figure 3.28 Sensitivity of PABA to NADH (A) and NAD$^+$ (B) in pH 7.4 PBS. PABA film was oxidized at 1.0 V (●) and reduced at −0.1 V (○) for 20 s. (Reprinted with permission from *Chemistry of Materials*, **17**, 2918. Copyright (2005) American Chemical Society.)

that this difference in degree of complexation as a function of oxidation state can form the basis of an extremely sensitive potentiometric sensor for detection of NAD$^+$ and NADH (down to the μM concentration range) and may prove an attractive alternative to amperometric-based sensors [144–146, 154].

3.5.3 Amine Sensors

The rapid detection of biogenic amines is of great importance in areas ranging from quality control of foodstuffs [155, 156], environmental and industrial monitoring applications [157, 158], to biomarkers of disease [159, 160]. Amine sensing has been investigated using various transduction modes such as electrochemical detection on various anode materials [161, 162], amperometric biosensors [163–165, 166, 167], ion selective electrodes, optical sensors [168, 169] and polymer based sensors [170–172]. Grubbs and Lewis *et al.* reported that chemiresistive amine vapor detectors consisting of polyaniline/carbon black composites display marked sensitivity and selectivity, and outperform the detection capabilities of human olfaction [172]. These composite sensors possess a detection threshold for one such amine, butylamine (BuNH$_2$), of 10 parts per trillion where human detection thresholds of this amine are reported to be 1–0.1 parts per million [173–175]. However, significant changes in the secondary structure of the film due to amine sorption reportedly results in quasireversibility of the response and a significant time for senor recovery. Recently, Freund *et al.* reported butylamine vapor sensing using PABA film in comparison with polyaniline deposited electrochemically on gold interdigitated electrodes [34]. Polyaniline has been used as an ammonia gas sensor because of its pH sensitivity, i.e., its unique doping/dedoping behavior. The conducting form of polyaniline (emeraldine salt) has been reported to deprotonate to emeraldine base and become insulating under alkaline conditions [150, 176]. In the case of PABA there is more complex sensitivity to amines due to boronate–alkyl ammonium interactions.

The typical relative differential resistance response, $\Delta R/R_b$, where ΔR is the change in resistance upon exposure to analyte and R_b is the baseline resistance before the exposure, of polyaniline and PABA film based detectors upon exposure to 100 ppb BuNH$_2$ is shown in Figure 3.29. Polyaniline exhibits a quasireversible response similar to that previously reported using polyaniline/carbon black composites [172]. In contrast, the PABA response was not reversible over the timescale of the exposure. However, the magnitudes of the response of polyaniline and PABA were similar. The dramatically different response/recovery rates allow for significant sensitivity of a differential response in time to exposures and essentially no sensitivity to the slow recovery process. The inset in Figure 3.29 shows the temporal differential resistance response, dR/dt, where dR is the difference between sequential resistance measurements and dt is the time between measurements of PABA obtained during the same exposure. Comparing dR/dt to $\Delta R/R_b$ values during the same exposure of BuNH$_2$ to the same PABA film illustrates the advantage that the slow recovery process gives to this detection method. A baseline dR/dt value is attained in 10–20 s after exposure. Comparing PABA to polyaniline in their usage as sensing material monitoring dR/dt values shows that baseline dR/dt values are reached more rapidly by PABA than by polyaniline on the basis that PABA recovers more slowly.

The absolute resistance and dR/dt response of PABA to exposures of BuNH$_2$ as a function of concentration is linear (Figure 3.30). The inset contains the average of peak dR/dt values ± 3 standard deviations

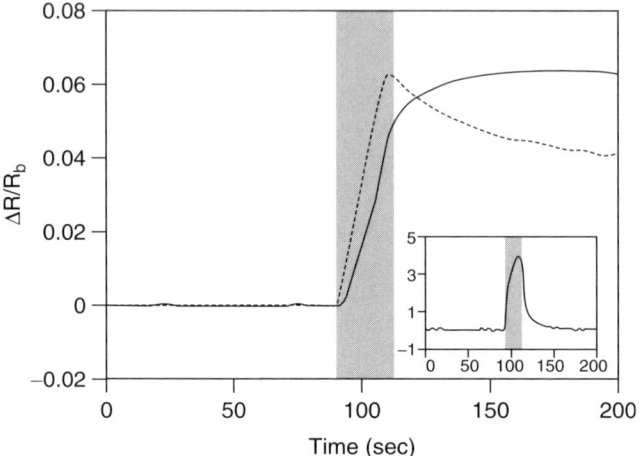

Figure 3.29 ΔR/Rb values of Polyaniline (dotted line) and PABA (solid line) in 100 ppb BuNH$_2$ (grey area). Inset are the dR/dt values (Ω/s) of PABA during the same exposure (Reprinted from *Sensors and Actuators*, **115**, J. T. English, B. A. Deore, M. S. Freund, 666. Copyright (2006), with permission from Elsevier.)

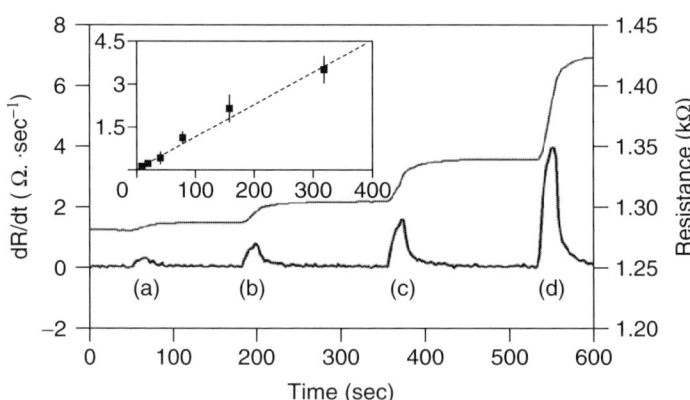

Figure 3.30 Absolute resistance (light top line) and dR/dt (heavy bottom line) values of PABA including exposures of 10 (a), 20 (b), 40 (c), and (d) 80 ppb BuNH$_2$. Inset are the peak dR/dt responses (y axis, Ω/s) during the same exposure to increasing concentrations of BuNH$_2$ (x axis, ppb) ±3 standard deviations. (Reprinted from *Sensors and Actuators*, **115**, J. T. English, B. A. Deore, M. S. Freund, 666. Copyright (2006), with permission from Elsevier.)

($n = 5$ in a single response) to 10, 20, 40, 80, 160, and 320 ppb $BuNH_2$. The responses display good linearity over this concentration range with a correlation coefficient of 0.98. The detection threshold ($>3\sigma$) was determined experimentally to be 2 ppb for $BuNH_2$ on this PABA film vapor detector. For a single PABA film detector exposed to a repeated concentration of $BuNH_2$ the peak dR/dt responses had a standard deviation of 5 %.

Upon exposure to amines, the reduction of polyaniline films occurs due to nucleophilic attack of amine on the quinoid ring of polyaniline. [177, 178]. Due to this reaction, the less stable quinoid ring reduces to the more stable benzenoid ring substituent. This reaction may contribute, in part, to the virtual irreversibility of PABA reaction with $BuNH_2$. The films are converted from the conducting emeraldine salt form to the insulating emeraldine base form, based on the spectroscopic evidence. However, assuming that the reactivity of $BuNH_2$ towards quinoid rings in polyaniline and PABA films is similar, it cannot explain the significantly different recoveries observed in both films. Therefore, the irreversibility of PABA is probably a result of electrostatic interactions between boronate and butyl ammonium ions as shown in Figure 3.31. As butyl ammonium accumulates and forms ion pairs with boronate, it stabilizes the dynamic equilibrium established between protonation and deprotonation of film by analyte. UV-Vis and FT-IR spectra of PABA after interaction with $BuNH_2$ suggest that the polymer is partially oxidized, implying that PABA can be partially self-doped.

3.5.4 Molecular Level Processing for Controlled Release of RNA

The processing of conducting polymers into thin films using a layer-by-layer deposition approach is an established technique. This approach is very simple and provides molecular level control over thickness and architecture of multilayer thin films. Using conducting and self-doped conducting polymers, the multilayer formation is carried out based on electrostatic interactions for various applications (for details see Chapter 1, section 1.6.1). In particular for biological applications, the use of polyaniline is restricted since the polymer is redox active only under acidic conditions, normally at pH < 4 [179, 180]. There are some reports on the use of polyaniline/gold nanoparticle [181] and polyaniline/carbon nanotube [182] multilayer films for biological applications, for example, detection of NADH and DNA hybridization. Self-doped polyaniline [44, 183, ,184] or polyaniline doped with negatively charged

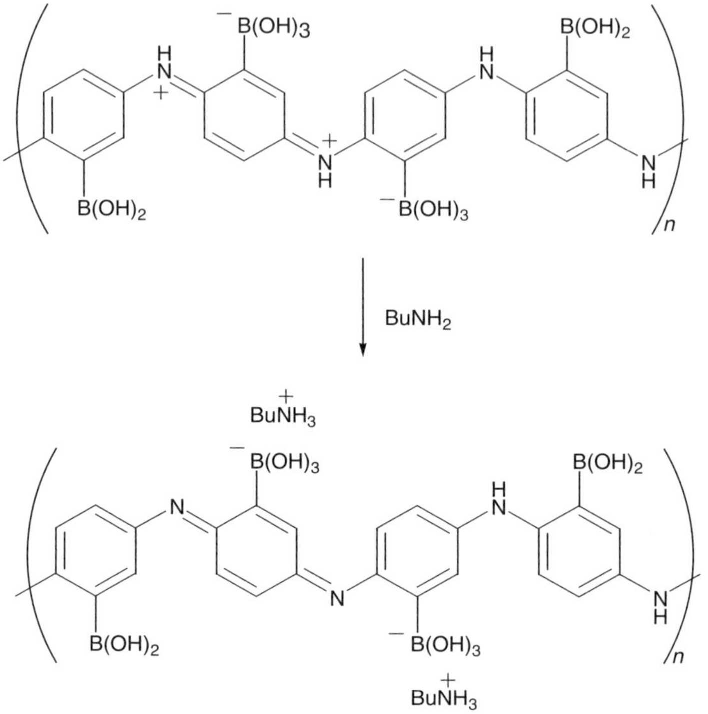

Figure 3.31 Proposed electrostatic interaction PABA films and BuNH$_2$ as caused by ion pairing between boronate sites and butyl ammonium ions. (Reprinted from *Sensors and Actuators*, **115**, J. T. English, B. A. Deore, M. S. Freund, 666. Copyright (2006), with permission from Elsevier.)

polyelectrolytes [144, 145, ,185] are however, redox active at neutral pH and have been successfully utilized for enzyme immobilization in multilayers [186, 187].

Recently, Freund *et al.* reported a novel approach for incorporating a biomolecule, namely RNA, into PABA to form a multilayer film using layer-by-layer deposition and subsequent release of RNA by applying potential under physiological conditions [188]. In particular, the interactions between PABA and RNA are exploited to prepare multilayer films. As mentioned above (Section 3.3.1), the complexation of boronic acid with D-fructose and the subsequent formation of a self-doped polymer extend the electroactivity of PABA to neutral and alkaline media (see Figure 3.8). Similarly, in this approach PABA interacts with RNA (Figure 3.32) through the formation of an anionic tetrahedral boronate ester and a boron–nitrogen dative bond, as well

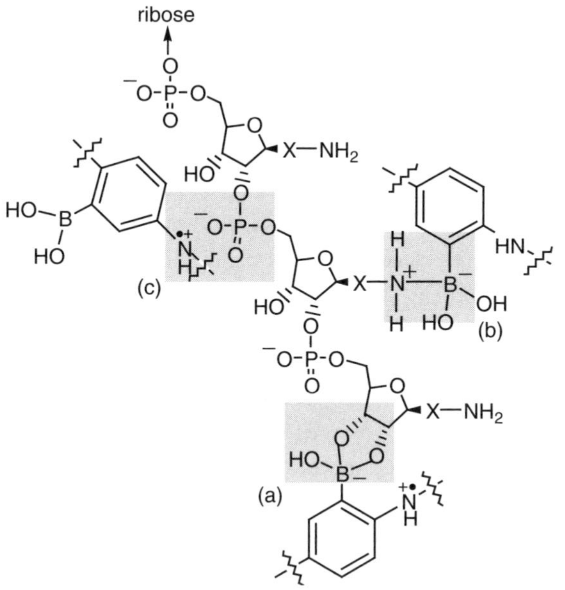

Figure 3.32 Proposed PABA/RNA bilayer interactions through (a) boronate-ester formation, (b) boron–nitrogen dative bond formation and/or (c) electrostatic interactions (Reprinted with permission from *Langmuir*, **22**, 2811. Copyright (2006) American Chemical Society.)

as electrostatic interactions of anionic phosphates with cationic amines, while maintaining the electroactivity under neutral conditions through self-doping for multilayer formation. Under these conditions, RNA acts as a polymeric counterion for multilayer formation as well as a dopant for PABA.

Multilayer films were prepared using water-soluble, self-doped, PABA prepared chemically in the presence of excess D-fructose and fluoride using a procedure discussed in Section 3.2.2. [37]. PABA solutions for dipping were typically 6 mg/mL and contained 3 M D-fructose and 40 mM NaF in water. The RNA solution for dipping contained 33 mg/mL prepared in PBS at pH 7.4. The following three step procedure resulted in the formation of one bilayer: (i) deposition of the first layer PABA by exposing the indium tin oxide (ITO) glass substrate to the PABA solution for 20 min to obtain nearly complete adsorption. The substrate was then allowed to stand for 15 min before it was rinsed in water for 1 min and dried under a stream of nitrogen gas. (ii) Fructose was removed from the PABA layer by exposing it to PBS for 20 min, after which time the substrate was removed, rinsed in water for 1 min, and dried again with nitrogen gas. (iii) The substrate with a PABA layer was then exposed to the RNA solution for 15 min, removed, rinsed in water for 15 s, and dried with nitrogen gas. These three steps were repeated to build multilayer films. The PABA/RNA multilayer films were characterized using various spectroscopic techniques and used for controlled release of RNA.

The interactions of PABA with RNA were investigated with UV-Vis and PM-IRRAS spectroscopy. Figure 3.33 shows the absorption spectra of a PABA layer (both salt and base form) and a PABA/RNA bilayer. The layer of PABA in the salt form (Figure 3.33, a) exhibits the characteristic absorption bands around 400 and 800 nm attributed to $\pi-\pi^*$ and bipolaron band transitions, respectively [53, 54]. The blue shift in the bipolaron band of the base form of the PABA layer (Figure 3.33, b) from 800 to 740 nm was observed upon exposure to PBS at pH 7.4 because of the removal of D-fructose and fluoride [37]. Subsequent complexation of the PABA layer in its base form with RNA resulted in a red shift in the bipolaron band from 740 to 800 nm, together with a small increase in the absorbance. These results reportedly confirmed the complexation of RNA with PABA under neutral conditions by the formation of the bilayer through anionic boronate esters, and subsequent conversion of the base form of PABA back to a self-doped salt form. The creation of the anionic tetrahedral boron forms the basis of multilayer formation. Further, the formation of boronate esters and a boron–nitrogen dative bond, as well as electrostatic interactions of anionic phosphates with cationic amines is supported by PM-IRRAS spectroscopy. After complexation

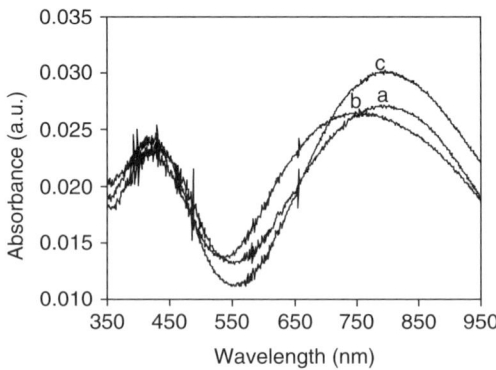

Figure 3.33 UV-Vis absorption spectra of (a) oxidized PABA first layer, (b) PABA base following removal of D-fructose, and (c) PABA/RNA bilayer. (Reprinted with permission from *Langmuir*, 22, 2811. Copyright (2006) American Chemical Society.)

with RNA, a peak was observed at 1065 cm^{-1}, which was attributed to C–O stretching and bending modes in the ribose moiety. The appearance of peaks at 1503 and 1670 cm^{-1} were assigned to the boron–nitrogen stretching mode and the carbonyl stretch of amides, which are known to be present in the nucleotide bases on RNA [153]. The peak at 1250 cm^{-1} was observed due to the electrostatic interaction of polyaniline with phosphate groups of the RNA [189].

The process of formation of a multilayer film on the ITO coated glass from sequential addition of PABA/RNA bilayers was observed with UV-Vis Spectroscopy as shown in Figure 3.34. The film growth observed with the deposition of additional bilayers suggests that the multilayer formation of PABA/RNA is reproducible with sequential deposition. All spectra exhibit an intense and sharp peak attributed to the $\pi-\pi^*$ and bipolaron band transitions. The bipolaron absorption band at 800 nm, associated with complexation of RNA with PABA, increases linearly with the number of PABA/RNA bilayers (Figure 3.34 inset). The linear relationship between absorbance and the number of deposited bilayers indicates that the deposition was reproducible from layer to layer, i.e., the amount of PABA adsorbed in each bilayer was the same. In addition to these results, multilayer formation was observed with ellipsometric and X-ray photoelectronic spectroscopy. The linear increase in film thickness with number of PABA/RNA bilayers was observed using ellipsometry. The average thickness of the PABA/RNA bilayer built up on a silicon substrate was approximately 10 nm. Additionally, X-ray photoelectron

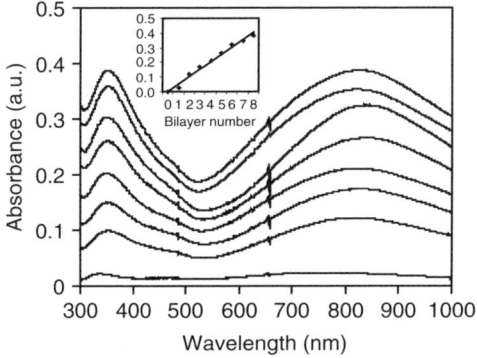

Figure 3.34 UV-Vis absorption spectra of PABA/RNA multilayers deposited on an ITO glass slide. The inset is a plot of λ_{max} at 800 nm vs the number of bilayers. (Reprinted with permission from *Langmuir*, 22, 2811. Copyright (2006) American Chemical Society.)

spectroscopy confirmed the linear increase in the atomic concentration of carbon to silicon with the number of bilayers.

PABA–PABA interactions influence multilayer film formation as demonstrated in control experiments without RNA. The comparison of the UV-Vis absorbance of the PABA/RNA multilayered film with PABA/PABA is shown in Figure 3.35. Without RNA, the multiple adsorptions of PABA layers show a linear increase in the absorbance at 800 nm. However, the magnitude of increase in the absorbance was significantly lower (75 %) than in the presence of RNA. The adsorption of PABA on PABA probably occurs because of intermolecular crosslinking between boronic acid groups and/or boronic acid and secondary amines on the polyaniline backbone. These results suggest that the macromolecular nature of RNA facilitates multilayer film growth by providing multiple binding sites for attachment to the previous layer of PABA and provides an anchor for the following layer of PABA.

In the context of controlled release, the layer-by-layer fabrication procedure offers potential advantages over conventional protein and nucleic acid encapsulation strategies, including the ability to control the order and location of multiple polymer layers with nanometer scale precision, and the ability to define the concentrations of incorporated materials simply by varying the number of polymer layers incorporated [190–192]. Although numerous reports describe the application of these materials to the sustained release of permeable small molecules [193–197], there are few examples of these assemblies designed to release macromolecular components. Several groups have performed

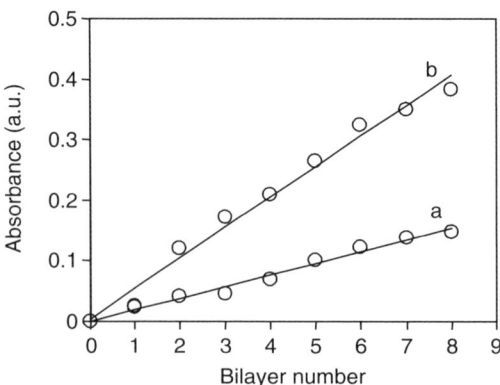

Figure 3.35 UV-Vis absorbance of (a) PABA/PABA and (b) PABA/RNA bilayers at 800 nm. (Reprinted with permission from *Langmuir*, **22**, 2811. Copyright (2006) American Chemical Society.)

controlled release of immobilized material by disrupting the multilayers usually by a change in pH or ionic strength in solution by using high salt concentrations (0.6–5.0 M) [198–201]. Neither of these approaches is particularly suitable for physiological conditions.

After successful preparation of PABA/RNA multilayer film, RNA was released from the multilayer film electrochemically under physiological conditions. When a conducting polymer is incorporated, whose properties can be altered electrochemically, it is possible to have more control over the release process under more suitable conditions. Figure 3.36 demonstrates this concept via cycling the potential applied to a multilayer between −0.2 and +1.4 V. The repetitive oxidation and reduction of the conducting polymer and the corresponding movement of counterions to

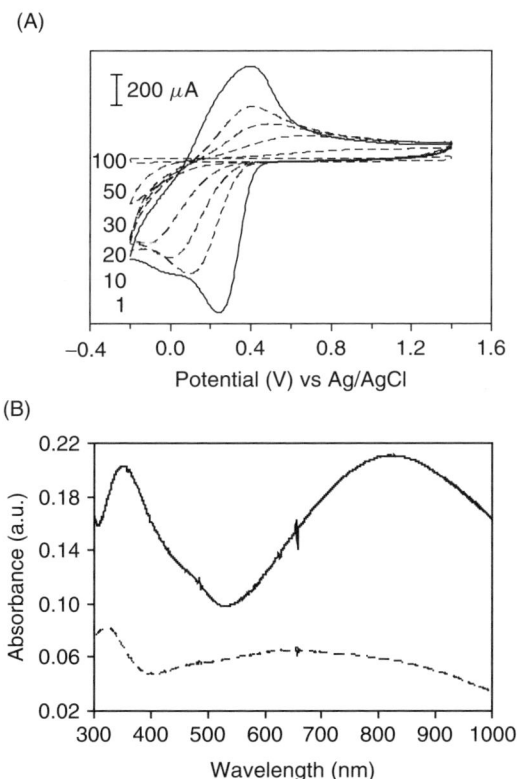

Figure 3.36 Potential induced release of RNA from multilayers (four bilayers) observed by (A) cyclic voltammetry and (B) UV-Vis absorption spectroscopy. PABA/RNA multilayers before (——) and after (- - -) disruption by applying a potential at a scan rate of 100 mV/s. (Reprinted with permission from *Langmuir*, **22**, 2811. Copyright (2006) American Chemical Society.)

maintain charge neutrality disrupts the interlayer bonding and, in turn, releases RNA. The PABA/RNA multilayer film is electroactive at neutral pH as shown in Figure 3.36A (——). The UV-Vis spectrum of the same film is shown in Figure 3.36B (——). However, when the film was reduced and oxidized with repetitive cycles, it lost its electroactivity because of the loss of RNA (see Figure 3.36, A; - - -). The significant decrease in the absorbance of the film and blue shift in the bipolaron band (see Figure 3.36, B; - - -) suggests that RNA was released after disrupting the multilayers. This shift in the bipolaron band was reversible upon exposing the remaining polymer to a solution containing RNA, and cycling polyaniline in PBS under identical conditions does not result in the degradation of the polymer, indicating that the polymer backbone does not experience degradation with repeated cycling.

REFERENCES

[1] J. Boeseken, 'The use of boric acid for the determination of the configuration of carbohydrates,' *Advances in Carbohydrate Chemistry* **1949**, *4*, 189.

[2] H. G. Kuivila, A. H. Keough, E. J. Soboczenski, 'Areneboronates from diols and polyols,' *Journal of Organic Chemistry* **1954**, *19*, 780.

[3] J. M. Sugihara, C. M. Bowman, 'Cyclic benzeneboronate esters,' *Journal of the American Chemical Society* **1958**, *80*, 2443.

[4] J. P. Lorand, J. O. Edwards, 'Polyol complexes and structure of the benzeneboronate ion,' *Journal of Organic Chemistry* **1959**, *24*, 769.

[5] T. D. James, K. Sandanayake, S. Shinkai, 'Saccharide sensing with molecular receptors based on boronic acid,' *Angewandte Chemie – International Edition in English* **1996**, *35*, 1911.

[6] G. Springsteen, B. H. Wang, 'A detailed examination of boronic acid–diol complexation,' *Tetrahedron* **2002**, *58*, 5291.

[7] E. Shoji, M. S. Freund, 'Potentiometric sensors based on the inductive effect on the pK_a of poly(aniline): a nonenzymatic glucose sensor,' *Journal of the American Chemical Society* **2001**, *123*, 3383.

[8] E. Shoji, M. S. Freund, 'Potentiometric saccharide detection based on the pK_a changes of poly(aniline boronic acid),' *Journal of the American Chemical Society* **2002**, *124*, 12486.

[9] T. D. James, K. Sandanayake, S. Shinkai, 'Chiral discrimination of monosaccharides using a fluorescent molecular sensor,' *Nature* **1995**, *374*, 345.

[10] H. Eggert, J. Frederiksen, C. Morin, J. C. Norrild, 'A new glucose-selective fluorescent bisboronic acid. First report of strong α-furanose complexation in aqueous solution at physiological pH,' *Journal of Organic Chemistry* **1999**, *64*, 3846.

[11] W. Yang, H. He, D. G. Drueckhammer, 'Computer-guided design in molecular recognition: design and synthesis of a glucopyranose receptor,' *Angewandte Chemie-International Edition* **2001**, *40*, 1714.

[12] W. Wang, S. Gao, B. Wang, 'Building fluorescent sensors by template polymerization: the preparation of a fluorescent sensor for fructose,' *Org. Lett.* **1999**, *1*, 1209.
[13] S. H. Gao, W. Wang, B. H. Wang, 'Building fluorescent sensors for carbohydrates using template-directed polymerizations,' *Bioorganic Chemistry* **2001**, *29*, 308.
[14] J. J. Lavigne, E. V. Anslyn, 'Teaching old indicators new tricks: a colorimetric chemosensing ensemble for tartrate/malate in beverages,' *Angewandte Chemie-International Edition* **1999**, *38*, 3666.
[15] B. D. Smith, S. J. Gardiner, T. A. Munro, M. F. Paugam, J. A. Riggs, 'Facilitated transport of carbohydrates, catecholamines and amino acids through liquid and plasticized organic membranes,' *Journal of Inclusion Phenomena and Molecular Recognition in Chemistry* **1998**, *32*, 121.
[16] V. Adamek, X. C. Liu, Y. A. Zhang, K. Adamkova, W. H. Scouten, 'New aliphatic boronate ligands for affinity-chromatography,' *Journal of Chromatography* **1992**, *625*, 91.
[17] A. Bergold, W. H. Scouten, 'Borate chromatography,' *Chemical Analysis* **1983**, *66*, 149.
[18] V. Bouriotis, I. J. Galpin, P. D. G. Dean, 'Applications of immobilized phenylboronic acids as supports for group-specific ligands in the affinity chromatography of enzymes,' *Journal of Chromatography* **1981**, *210*, 267.
[19] X.-C. Liu, J. L. Hubbard, W. H. Scouten, 'Synthesis and structural investigation of two potential boronate affinity chromatography ligands catechol [2-(diisopropylamino)carbonyl]phenylboronate and catechol [2-(diethylamino) carbonyl, 4-methyl]phenylboronate,' *Journal of Organometallic Chemistry* **1995**, *493*, 91.
[20] J. Psotova, O. Janiczek, 'Boronate affinity chromatography and the applications,' *Chemicke Listy* **1995**, *89*, 641.
[21] H. Seliger, V. Genrich, 'Synthesis of *m*-aminophenylboronic acid an effector in the affinity chromatography of nucleic acids,' *Experientia* **1974**, *30*, 1480.
[22] R. P. Singhal, S. DeSilva, M. Shyamali, 'Boronate affinity chromatography,' *Advances in Chromatography* **1992**, *31*, 293.
[23] P. R. Westmark, L. S. Valencia, B. D. Smith, 'Influence of eluent anions in boronate affinity chromatography,' *Journal of Chromatography A* **1994**, *664*, 123.
[24] S. Soundararajan, M. Badawi, C. M. Kohlrust, J. H. Hageman, 'Boronic acids for affinity chromatography – spectral methods for determinations of ionization and diol-binding constants,' *Analytical Biochemistry* **1989**, *178*, 125.
[25] R. E. London, S. A. Gabel, '^{19}F-NMR studies of fluorobenzeneboronic acids. 1. Interaction kinetics with biologically significant ligands,' *Journal of the American Chemical Society* **1994**, *116*, 2562.
[26] M. F. Paugam, B. D. Smith, 'Active-transport of uridine through a liquid organic membrane mediated by phenylboronic acid and driven by a fluoride-ion gradient,' *Tetrahedron Letters* **1993**, *34*, 3723.
[27] C. Dusemund, K. Sandanayake, S. Shinkai, 'Selective fluoride recognition with ferroceneboronic acid,' *Journal of the Chemical Society-Chemical Communications* **1995**, 333.

[28] T. D. James, K. Sandanayake, S. Shinkai, 'Novel photoinduced electron-transfer sensor for saccharides based on the interaction of boronic acid and amine,' *Chemical Communications* **1994**, 477.

[29] K. Sandanayake, S. Shinkai, 'Novel molecular sensors for saccharides based on the interaction of boronic acid and amines – saccharide sensing in neutral water,' *Journal of the Chemical Society-Chemical Communications* **1994**, 1083.

[30] E. Pringsheim, E. Terpetschnig, S. A. Piletsky, O. S. Wolfbeis, 'A polyaniline with near-infrared optical response to saccharides,' *Advanced Materials* **1999**, *11*, 865.

[31] M. Nicolas, B. Fabre, G. Marchand, J. Simonet, 'New boronic-acid- and boronate-substituted aromatic compounds as precursors of fluoride-responsive conjugated polymer films,' *European Journal of Organic Chemistry* **2000**, 1703.

[32] B. Fabre, L. Taillebois, 'Poly(aniline boronic acid)-based conductimetric sensor of dopamine,' *Chemical Communications* **2003**, 2982.

[33] J. Mathiyarasu, S. Senthilkumar, K. L. N. Phani, V. Yegnaraman, 'Selective detection of dopamine using a functionalized polyaniline composite electrode,' *Journal of Applied Electrochemistry* **2005**, *35*, 513.

[34] J. T. English, B. A. Deore, M. S. Freund, 'Biogenic amine vapour detection using poly(anilineboronic acid) films,' *Sensors and Actuators B: Chemical* **2006**, *115*, 666.

[35] B. A. Deore, M. S. Freund, 'Reactivity of poly(anilineboronic acid) with NAD^+ and NADH,' *Chemistry of Materials* **2005**, *17*, 2918.

[36] B. Deore, M. S. Freund, 'Saccharide imprinting of poly(aniline boronic acid) in the presence of fluoride,' *Analyst* **2003**, *128*, 803.

[37] B. A. Deore, I. Yu, M. S. Freund, 'A switchable self-doped polyaniline: Interconversion between self-doped and non-self-doped forms,' *Journal of the American Chemical Society* **2004**, *126*, 52.

[38] R. E. Mesmer, A. C. Rutenberg, '^{19}Fluorine nuclear magnetic resonance studies on fluoroborate species in aqueous solution,' *Inorg. Chem* **1973**, *12*, 699

[39] C. R. Cooper, N. Spencer, T. D. James, 'Selective fluorescence detection of fluoride using boronic acids,' *Chemical Communications* **1998**, 1365.

[40] H. Yamamoto, A. Ori, K. Ueda, C. Dusemund, S. Shinkai, 'Visual sensing of fluoride ion and saccharides utilizing a coupled redox reaction of ferrocenylboronic acids and dye molecules,' *Chemical Communications* **1996**, 407.

[41] W. S. Huang, B. D. Humphrey, A. G. MacDiarmid, 'Polyaniline, a novel conducting polymer – morphology and chemistry of its oxidation and reduction in aqueous electrolytes,' *Journal of the Chemical Society – Faraday Transactions I* **1986**, *82*, 2385.

[42] J. Yue, Z. H. Wang, K. R. Cromack, A. J. Epstein, A. G. MacDiarmid, 'Effect of sulfonic acid group on polyaniline backbone,' *Journal of the American Chemical Society* **1991**, *113*, 2665.

[43] Y. Wei, W. W. Focke, G. E. Wnek, A. Ray, A. G. MacDiarmid, 'Synthesis and electrochemistry of alkyl ring substituted polyanilines,' *The Journal of Physical Chemistry* **1989**, *93*, 495

[44] L. V. Lukachova, E. A. Shkerin, E. A. Puganova, E. E. Karyakina, S. G. Kiseleva, A. V. Orlov, G. P. Karpacheva, A. A. Karyakin, 'Electroactivity of chemically synthesized polyaniline in neutral and alkaline aqueous solutions – role of self-doping and external doping,' *Journal of Electroanalytical Chemistry* **2003**, *544*, 59.

[45] R. Mazeikiene, G. Niaura, A. Malinauskas, 'Voltammetric study of the redox processes of self-doped sulfonated polyaniline,' *Synthetic Metals* **2003**, *139*, 89.

[46] M. Vanduin, J. A. Peters, A. P. G. Kieboom, H. Vanbekkum, 'The pH-dependence of the stability of esters of boric acid and borate in aqueous medium as studied by ^{11}B NMR,' *Tetrahedron* **1984**, *40*, 2901.

[47] M. Vanduin, J. A. Peters, A. P. G. Kieboom, H. Vanbekkum, 'Studies on borate esters. 2. Structure and stability of borate esters of polyhydroxy-carboxylates and related polyols in aqueous alkaline media as studied by ^{11}B NMR,' *Tetrahedron* **1985**, *41*, 3411.

[48] J. G. Dawber, S. I. E. Green, J. C. Dawber, S. Gabrail, 'A polarimetric and ^{11}B and ^{13}C nuclear magnetic-resonance study of the reaction of the tetrahydroxyborate ion with polyols and carbohydrates,' *Journal of the Chemical Society – Faraday Transactions I* **1988**, *84*, 41.

[49] J. C. Norrild, H. Eggert, 'Boronic acids as fructose sensors. Structure determination of the complexes involved using (1) J(C–C) coupling constants,' *Journal of the Chemical Society – Perkin Transactions 2* **1996**, 2583.

[50] B. K. Shull, D. E. Spielvogel, G. Head, R. Gopalaswamy, S. Sankar, K. Devito, 'Studies on the structure of the complex of the boron neutron capture therapy drug, L-*p*-boronophenylalanine, with fructose and related carbohydrates: chemical and ^{13}C NMR evidence for the *β*-D-fructofuranose 2,3,6-(*p*-phenylalanylorthoboronate) structure,' *Journal of Pharmaceutical Sciences* **2000**, *89*, 215.

[51] C. A. Lindino, L. O. S. Bulhoes, 'The potentiometric response of chemically modified electrodes,' *Analytica Chimica Acta* **1996**, *334*, 317.

[52] A. A. Karyakin, M. Vuki, L. V. Lukachova, E. E. Karyakina, A. V. Orlov, G. P. Karpacheva, J. Wang, 'Processible polyaniline as an advanced potentiometric pH transducer. Application to biosensors,' *Analytical Chemistry* **1999**, *71*, 2534.

[53] S. Stafstrom, J. L. Bredas, A. J. Epstein, H. S. Woo, D. B. Tanner, W. S. Huang, A. G. MacDiarmid, 'Polaron lattice in highly conducting polyaniline – theoretical and optical studies,' *Physical Review Letters* **1987**, *59*, 1464.

[54] F. Wudl, R. O. Angus, F. L. Lu, P. M. Allemand, D. J. Vachon, M. Nowak, Z. X. Liu, A. J. Heeger, 'Poly(*para*-phenyleneamineimine) – synthesis and comparison to polyaniline,' *Journal of the American Chemical Society* **1987**, *109*, 3677.

[55] A. Kabumoto, K. Shinozaki, K. Watanabe, N. Nishikawa, 'Electrochemical degradation of polyaniline,' *Synthetic Metals* **1988**, *26*, 349.

[56] A. Abd-Elwahed, R. Holze, 'Ion size and size memory effects with electropolymerized polyaniline,' *Synthetic Metals* **2002**, *131*, 61.

[57] A. A. Nekrasov, V. F. Ivanov, A. V. Vannikov, 'Effect of pH on the structure of absorption spectra of highly protonated polyaniline analyzed by the Alentsev-Fock method,' *Electrochimica Acta* **2001**, *46*, 4051.

[58] D. E. Stilwell, S. M. Park, 'Electrochemistry of conductive polymers. 5. In situ spectroelectrochemical studies of polyaniline films,' *Journal of the Electrochemical Society* **1989**, *136*, 427.
[59] R. J. Cushman, P. M. McManus, S. C. Yang, 'Spectroelectrochemical study of polyaniline – the construction of a pH-potential phase-diagram,' *Journal of Electroanalytical Chemistry* **1987**, *219*, 335.
[60] J. Yue, G. Gordon, A. J. Epstein, 'Comparison of different synthetic routes for sulfonation of polyaniline,' *Polymer* **1992**, *33*, 4410.
[61] C. H. Hsu, P. M. Peacock, R. B. Flippen, J. Yue, A. J. Epstein, 'The molecular weight of sulfonic acid ring substituted polyaniline by laser light scattering,' *Synthetic Metals* **1993**, *60*, 223.
[62] M. Angelopoulos, Y. H. Liao, B. Furman, T. Graham, 'LiCl induced morphological changes in polyaniline base and their effect on the electronic properties of the doped form,' *Macromolecules* **1996**, *29*, 3046.
[63] D. Yang, P. N. Adams, R. Goering, B. R. Mattes, 'New methods for determining the molecular weight of polyaniline by size exclusion chromatography,' *Synthetic Metals* **2003**, *135*, 293.
[64] W. Zheng, M. Angelopoulos, A. J. Epstein, A. G. MacDiarmid, 'Experimental evidence for hydrogen bonding in polyaniline: mechanism of aggregate formation and dependency on oxidation state,' *Macromolecules* **1997**, *30*, 2953.
[65] X.-R. Zeng, T.-M. Ko, 'Structures and properties of chemically reduced polyanilines,' *Polymer* **1998**, *39*, 1187.
[66] Y. Wei, K. F. Hsueh, G. W. Jang, 'A study of leucoemeraldine and effect of redox reactions on molecular weight of chemically prepared polyaniline,' *Macromolecules* **1994**, *27*, 518.
[67] W. Zheng, M. Angelopoulos, A. J. Epstein, A. G. MacDiarmid, 'Concentration dependence of aggregation of polyaniline in NMP solution and properties of resulting cast films,' *Macromolecules* **1997**, *30*, 7634.
[68] A. Afzali, S. L. Buchwalter, L. P. Buchwalter, G. Hougham, 'Reaction of polyaniline with NMP at elevated temperatures,' *Polymer* **1997**, *38*, 4439.
[69] I. Yu, B. A. Deore, C. L. Recksiedler, T. C. Corkery, A. S. Abd-El-Aziz, M. S. Freund, 'Thermal stability of high molecular weight self-doped poly(aniline-boronic acid),' *Macromolecules* **2005**, *38*, 10022.
[70] J. Simon, S. Salzbrunn, G. K. S. Prakash, N. A. Petasis, G. A. Olah, 'Regioselective conversion of arylboronic acids to phenols and subsequent coupling to symmetrical diaryl ethers,' *Journal of Organic Chemistry* **2001**, *66*, 633.
[71] E. Shoji, M. S. Freund, 'Poly(aniline boronic acid): a new precursor to substituted poly(aniline)s,' *Langmuir* **2001**, *17*, 7183.
[72] C. L. Recksiedler, B. A. Deore, M. S. Freund, 'Substitution and condensation reactions with poly(anilineboronic acid): reactivity and characterization of thin films,' *Langmuir* **2005**, *21*, 3670.
[73] M. Abe, A. Ohtani, Y. Umemoto, S. Akizuki, M. Ezoe, H. Higuchi, K. Nakamoto, A. Okuno, Y. Noda, 'Soluble and high molecular weight polyaniline,' *Chemical Communications* **1989**, 1736.
[74] L. H. C. Mattoso, O. N. Oliveira, Jr, R. M. Faira, S. K. Manohar, A. J. Epstein, A. G. MacDiarmid, 'Synthesis of polyaniline/polytoluidine block copolymer via the pernigraniline oxidation state,' *Polymer International* **1994**, *35*, 89.

[75] P. N. Adams, P. J. Laughlin, A. P. Monkman, 'Synthesis of high molecular weight polyaniline at low temperatures,' *Synthetic Metals* **1996**, *76*, 157.
[76] J. Stejskal, A. Riede, D. Hlavata, J. Prokes, M. Helmstedt, P. Holler, 'The effect of polymerization temperature on molecular weight, crystallinity and electrical conductivity of polyaniline,' *Synthetic Metals* **1998**, *96*, 55.
[77] P. M. Beadle, Y. F. Nicolau, E. Banka, P. Rannou, D. Djurado, 'Controlled polymerization of aniline at sub-zero temperatures,' *Synthetic Metals* **1998**, *95*, 29.
[78] D. Yang, P. N. Adams, B. R. Mattes, 'Intrinsic viscosity measurement of dilute emeraldine base solutions for estimating the weight average molecular weight of polyaniline,' *Synthetic Metals* **2001**, *119*, 301.
[79] R. Gangopadhyay, A. De, 'Conducting semi-IPN based on polyaniline and crosslinked poly(vinyl alcohol),' *Synthetic Metals* **2002**, *132*, 21.
[80] G. Abbati, E. Carone, L. D'Ilario, A. Martinelli, 'Polyurethane-polyaniline conducting graft copolymer with improved mechanical properties,' *Journal of Applied Polymer Science* **2003**, *89*, 2516.
[81] H. S. O. Chan, S. C. Ng, P. K. H. Ho, 'Polyanilines doped with phosphonic acids – their preparation and characterization,' *Macromolecules* **1994**, *27*, 2159.
[82] F. T. Liu, K. G. Neoh, E. T. Kang, S. Li, H. S. Han, K. L. Tan, 'Effects of crosslinking on polyaniline films' doping behavior and degradation under weathering,' *Polymer* **1999**, *40*, 5285.
[83] C. Cruz, E. A. Ticianelli, 'Electrochemical and ellipsometric studies of polyaniline films grown under cycling conditions,' *Journal of Electroanalytical Chemistry* **1997**, *428*, 185.
[84] J. Yue, A. J. Epstein, 'Synthesis of self-doped conducting polyaniline,' *Journal of the American Chemical Society* **1990**, *112*, 2800.
[85] B. A. Deore, S. Hachey, M. S. Freund, 'Electroactivity of electrochemically synthesized poly(aniline boronic acid) as a function of pH: role of self-doping,' *Chemistry of Materials* **2004**, *16*, 1427.
[86] Y. Tomata, M. Sasaki, K. Tanino, M. Miyashita, 'The first C2 selective halide substitution reaction of 2,3-epoxy alcohols by the use of $(CH_3O)(3)B-MX$ (X = I, Br, Cl) system,' *Tetrahedron Letters* **2003**, *44*, 8975.
[87] L. K. Mohler, A. W. Czarnik, 'Alpha-amino-acid chelative complexation by an arylboronic acid,' *Journal of the American Chemical Society* **1993**, *115*, 7037.
[88] W. Kliegel, G. Lubkowitz, J. O. Pokriefke, S. J. Rettig, J. Trotter, 'Structural studies of organoboron compounds, part 72 – nitrones and oximes of bifunctional carbonyl compounds and their reaction products with diarylborinic acids. Crystal and molecular structure of examples of five-, six-, and seven-membered boron chelates,' *Canadian Journal of Chemistry – Revue Canadienne De Chimie* **2000**, *78*, 1325.
[89] H. Hopfl, A. Farfan, 'Synthesis and structural characterization of (2′-hydroxy-acetophenoneazine)mono(diphenylboron) chelate,' *Canadian Journal of Chemistry – Revue Canadienne De Chimie* **1998**, *76*, 1853.
[90] E. Vedejs, R. W. Chapman, S. Lin, M. Muller, D. R. Powell, 'Crystallization-induced asymmetric transformation vs quasi-racemate formation in tetravalent boron complexes,' *Journal of the American Chemical Society* **2000**, *122*, 3047.

[91] M. P. Groziak, A. D. Ganguly, P. D. Robinson, 'Boron heterocycles bearing a peripheral resemblance to naturally occurring purines – design, syntheses, structures, and properties,' *Journal of the American Chemical Society* **1994**, *116*, 7597.

[92] B. A. Deore, I. S. Yu, P. M. Aguiar, C. Recksiedler, S. Kroeker, M. S. Freund, 'Highly crosslinked, self-doped polyaniline exhibiting unprecedented hardness,' *Chemistry of Materials* **2005**, *17*, 3803.

[93] A. J. Epstein, R. P. McCall, J. M. Ginder, A. G. MacDiarmid, 'Spectroscopy and photoexcitation spectroscopy of polyaniline: a model system for new phenomena,' *Advances in Spectroscopy* **1991**, *19*, 355.

[94] G. Socrates, *Infrared Characteristic Group Frequencies: Tables and Charts*, 2nd edn, Wiley, Chichester, UK, **1994**.

[95] N. Colthup, L. H. Daly, S. E. Wiberley, *Introduction to Infrared and Raman Spectroscopy*, 2nd ed., Academic, New York, N. Y., **1975**.

[96] X. N. Chen, G. Y. Liang, D. Whitmire, J. P. Bowen, '*Ab initio* and molecular mechanics (MM3) calculations on alkyl- and arylboronic acids,' *Journal of Physical Organic Chemistry* **1998**, *11*, 378.

[97] G. Zamfirova, V. Lorenzo, R. Benavente, J. M. Perena, 'On the relationship between modulus of elasticity and microhardness,' *Journal of Applied Polymer Science* **2003**, *88*, 1794.

[98] R. J. Crawford, 'Microhardness testing of plastics,' *Polymer Testing* **1982**, *3*, 37.

[99] M. Irigoyen, P. Bartolomeo, F. X. Perrin, E. Aragon, J. L. Vernet, 'UV ageing characterisation of organic anticorrosion coatings by dynamic mechanical analysis, Vickers microhardness, and infrared analysis,' *Polymer Degradation and Stability* **2001**, *74*, 59.

[100] M. Krumova, C. Klingshirn, F. Haupert, K. Friedrich, 'Microhardness studies on functionally graded polymer composites,' *Composites Science and Technology* **2001**, *61*, 557.

[101] E. T. Kang, Z. H. Ma, K. L. Tani, O. N. Tretinnikov, Y. Uyama, Y. Ikada, 'Surface hardness of pristine and modified polyaniline films,' *Langmuir* **1999**, *15*, 5389.

[102] E. T. Kang, K. G. Neoh, Y. Q. Dong, Z. H. Ma, K. L. Tan, 'Super-hard-surfaced polyaniline films by bulk and surface modifications,' *Synthetic Metals* **1999**, *101*, 696.

[103] B. Wrackmeyer, 'Nuclear magnetic resonance spectroscopy of boron compounds containing two-, three- and four-coordinate boron,' *Annual Reports on NMR Spectroscopy* **1988**, *20*, 61.

[104] S. Kroeker, P. S. Neuhoff, J. F. Stebbins, 'Enhanced resolution and quantitation from "ultrahigh" field NMR spectroscopy of glasses,' *Journal of Non-Crystalline Solids* **2001**, *293*, 440.

[105] A. Samoson, E. Lippmaa, 'Excitation phenomena and line-intensities in high-resolution NMR powder spectra of half-integer quadrupolar nuclei,' *Physical Review B* **1983**, *28*, 6567.

[106] D. Massiot, C. Bessada, J. P. Coutures, F. Taulelle, 'A quantitative study of Al-27 MAS NMR in crystalline YAG,' *Journal of Magnetic Resonance* **1990**, *90*, 231.

[107] L. Frydman, J. S. Harwood, 'Isotropic spectra of half-integer quadrupolar spins from bidimensional magic-angle-spinning NMR,' *Journal of the American Chemical Society* **1995**, *117*, 5367.
[108] J. Yue, A. J. Epstein, Z. Zhong, P. K. Gallagher, A. G. MacDiarmid, 'Thermal stabilities of polyanilines,' *Synthetic Metals* **1991**, *41*, 765.
[109] S. A. Chen, G. W. Hwang, 'Structure characterization of self-acid-doped sulfonic acid ring-substituted polyaniline in its aqueous solutions and as solid film,' *Macromolecules* **1996**, *29*, 3950.
[110] C. C. Han, C. H. Lu, S. P. Hong, K. F. Yang, 'Highly conductive and thermally stable self-doping propylthiosulfonated polyanilines,' *Macromolecules* **2003**, *36*, 7908.
[111] A. G. MacDiarmid, C. K. Chiang, A. F. Richter, N. L. D. Somasiri, A. J. Epstein, *Conducting Polymers: Special applications*, 1st edn, D. Reidel Publishing Co., Dordrecht, **1987**.
[112] F. Cataldo, P. Maltese, 'Synthesis of alkyl and n-alkyl-substituted polyanilines – a study on their spectral properties and thermal stability,' *European Polymer Journal* **2002**, *38*, 1791.
[113] Y. Wei, K. F. Hsueh, 'Thermal analysis of chemically synthesized polyaniline and effects of thermal aging on conductivity,' *Journal of Polymer Science, Part A – Polymer Chemistry* **1989**, *27*, 4351.
[114] Y. D. Wang, M. F. Rubner, 'An investigation of the conductivity stability of acid-doped polyanilines,' *Synthetic Metals* **1992**, *47*, 255.
[115] S. Kim, I. J. Chung, 'Annealing effect on the electrochemical property of polyaniline complexed with various acids,' *Synthetic Metals* **1998**, *97*, 127.
[116] S. A. Chen, G. W. Hwang, 'Water-soluble self-acid-doped conducting polyaniline – structure and properties,' *Journal of the American Chemical Society* **1995**, *117*, 10055.
[117] A. Gok, B. Sari, M. Talu, 'Synthesis and characterization of conducting substituted polyanilines,' *Synthetic Metals* **2004**, *142*, 41.
[118] N. A. Zaidi, J. P. Foreman, G. Tzamalis, S. C. Monkman, A. P. Monkman, 'Alkyl substituent effects on the conductivity of polyaniline,' *Advanced Functional Materials* **2004**, *14*, 479.
[119] D. Vogna, R. Marotta, A. Napolitano, M. d'Ischia, 'Advanced oxidation chemistry of paracetamol. UV/H_2O_2-induced hydroxylation/degradation pathways and N-15-aided inventory of nitrogenous breakdown products,' *Journal of Organic Chemistry* **2002**, *67*, 6143.
[120] S. Q. Liu, L. Bakovic, A. C. Chen, 'Specific binding of glycoproteins with poly(aniline boronic acid) thin film,' *Journal of Electroanalytical Chemistry* **2006**, *591*, 210.
[121] G. S. Wilson, Y. B. Hu, 'Enzyme based biosensors for *in vivo* measurements,' *Chemical Reviews* **2000**, *100*, 2693.
[122] D. A. Gough, J. C. Armour, 'Development of the implantable glucose sensor – what are the prospects and why is it taking so long,' *Diabetes* **1995**, *44*, 1005.
[123] S. Vaidyanathan, G. Macaloney, J. Vaughn, B. McNeil, L. M. Harvey, 'Monitoring of submerged bioprocesses,' *Critical Reviews in Biotechnology* **1999**, *19*, 277.

[124] L. C. Clark, Jr, C. Lyons, 'Electrode systems for continuous monitoring in cardiovascular surgery,' *Annals of the New York Academy of Sciences* **1962**, *102*, 29.
[125] G. S. Wilson, Y. Zhang, G. Reach, D. Moattisirat, V. Poitout, D. R. Thevenot, F. Lemonnier, J. C. Klein, 'Progress toward the development of an implantable sensor for glucose,' *Clinical Chemistry* **1992**, *38*, 1613.
[126] J. Pickup, 'Developing glucose sensors for *in vivo* use,' *Trends in Biotechnology* **1993**, *11*, 285.
[127] C. Meyerhoff, F. J. Mennel, F. Sternberg, E. F. Pfeiffer, 'Current status of the glucose sensor,' *Endocrinologist* **1996**, *6*, 51.
[128] E. Wilkins, P. Atanasov, 'Glucose monitoring: state of the art and future possibilities,' *Medical Engineering & Physics* **1996**, *18*, 273.
[129] W. H. Wu, C. Greene, 'Interaction of boric acid with diol compounds and its application for the measurement of sugar in urine,' *Clinical Chemistry* **1986**, *32*, 1193.
[130] J. Yoon, A. W. Czarnik, 'Fluorescent chemosensors of carbohydrates – a means of chemically communicating the binding of polyols in water based on chelation enhanced quenching,' *Journal of the American Chemical Society* **1992**, *114*, 5874.
[131] T. D. James, P. Linnane, S. Shinkai, 'Fluorescent saccharide receptors: a sweet solution to the design, assembly and evaluation of boronic acid derived pet sensors,' *Chemical Communications* **1996**, 281.
[132] H. Shinmori, M. Takeuchi, S. Shinkai, 'Spectroscopic sugar sensing by a stilbene derivative with push (Me2N)-pull ((Ho)2B-)-type substituents,' *Tetrahedron* **1995**, *51*, 1893.
[133] K. Tsukagoshi, S. Shinkai, 'Specific complexation with monosaccharides and disaccharides that can be detected by circular dichroism,' *Journal of Organic Chemistry* **1991**, *56*, 4089.
[134] M. Takeuchi, T. Imada, S. Shinkai, 'Highly selective and sensitive' sugar tweezer" designed from a boronic-acid-appended μ-oxobis[porphinato-iron(III)],' *Journal of the American Chemical Society* **1996**, *118*, 10658.
[135] R. Gabai, N. Sallacan, V. Chegel, T. Bourenko, E. Katz, I. Willner, 'Characterization of the swelling of acrylamidophenylboronic acid-acrylamide hydrogels upon interaction with glucose by faradaic impedance spectroscopy, chronopotentiometry, quartz crystal microbalance (QCM), and surface plasmon resonance (SPR) experiments,' *Journal of Physical Chemistry B* **2001**, *105*, 8196.
[136] A. Kikuchi, K. Suzuki, O. Okabayashi, H. Hoshino, K. Kataoka, Y. Sakurai, T. Okano, 'Glucose-sensing electrode coated with polymer complex gel containing phenylboronic acid,' *Analytical Chemistry* **1996**, *68*, 823.
[137] A. N. J. Moore, D. D. M. Wayner, 'Redox switching of carbohydrate binding to ferrocene boronic acid,' *Canadian Journal of Chemistry – Revue Canadienne De Chimie* **1999**, *77*, 681.
[138] F. H. Arnold, W. Zheng, A. S. Michaels, 'A membrane-moderated, conductimetric sensor for the detection and measurement of specific organic solutes in aqueous solutions,' *Journal of Membrane Science* **2000**, *167*, 227.
[139] B. A. Deore, M. D. Braun, M. S. Freund, 'pH dependent equilibria of poly(anilineboronic acid) – saccharide complexation in thin films,' *Macromolecular Chemistry and Physics* **2006**, *207*, 660.

[140] H. B. White, III, *Evolution of Coenzymes and the Origin of Pyridine Nucleotides*, Academic, New York, NY, **1982**.
[141] J. D. Rawn, *Biochemistry*, International edn, Neil Patterson, Burlington, NC, **1989**.
[142] W. T. Bresnahan, P. J. Elving, 'The role of adsorption in the initial one-electron electrochemical reduction of nicotinamide adenine-dinucleotide (NAD^+),' *Journal of the American Chemical Society* **1981**, *103*, 2379.
[143] L. Gorton, 'Chemically modified electrodes for the electrocatalytic oxidation of nicotinamide coenzymes,' *Journal of the Chemical Society – Faraday Transactions I* **1986**, *82*, 1245.
[144] P. N. Bartlett, P. R. Birkin, E. N. K. Wallace, 'Oxidation of β-nicotinamide adenine dinucleotide (NADH) at poly(aniline)-coated electrodes,' *Journal of the Chemical Society – Faraday Transactions* **1997**, *93*, 1951.
[145] P. N. Bartlett, E. N. K. Wallace, 'The oxidation of beta-nicotinamide adenine dinucleotide (NADH) at poly(aniline)-coated electrodes: Part II. Kinetics of reaction at poly(aniline)-poly(styrenesulfonate) composites,' *Journal of Electroanalytical Chemistry* **2000**, *486*, 23.
[146] P. N. Bartlett, E. Simon, 'Measurement of the kinetic isotope effect for the oxidation of NADH at a poly(aniline)-modified electrode,' *Journal of the American Chemical Society* **2003**, *125*, 4014.
[147] N. V. C. Ralston, C. D. Hunt, 'Diadenosine phosphates and S-adenosylmethionine: novel boron binding biomolecules detected by capillary electrophoresis,' *Biochimica Et Biophysica Acta – General Subjects* **2001**, *1527*, 20.
[148] D. H. Kim, B. N. Marbois, K. F. Faull, C. D. Eckhert, 'Esterification of borate with NAD^+ and NADH as studied by electrospray ionization mass spectrometry and B-11 NMR spectroscopy,' *Journal of Mass Spectrometry* **2003**, *38*, 632.
[149] D. H. Kim, K. F. Faull, A. J. Norris, C. D. Eckhert, 'Borate–nucleotide complex formation depends on charge and phosphorylation state,' *Journal of Mass Spectrometry* **2004**, *39*, 743.
[150] A. Ray, A. F. Richter, A. G. MacDiarmid, A. J. Epstein, 'Polyaniline – protonation deprotonation of amine and imine sites,' *Synthetic Metals* **1989**, *29*, E151.
[151] A. G. MacDiarmid, L. S. Yang, W. S. Huang, B. D. Humphrey, 'Polyaniline – electrochemistry and application to rechargeable batteries,' *Synthetic Metals* **1987**, *18*, 393.
[152] N. S. Sariciftci, H. Kuzmany, H. Neugebauer, A. Neckel, 'Structural and electronic transitions in polyaniline – a Fourier transform infrared spectroscopic study,' *Journal of Chemical Physics* **1990**, *92*, 4530.
[153] K. T. Yue, C. L. Martin, D. Chen, P. Nelson, D. L. Sloan, R. Callender, 'Raman-spectroscopy of oxidized and reduced nicotinamide adenine dinucleotides,' *Biochemistry* **1986**, *25*, 4941.
[154] C. R. Raj, T. Ohsaka, 'Electrocatalytic sensing of NADH at an *in situ* functionalized self-assembled monolayer on gold electrode,' *Electrochemistry Communications* **2001**, *3*, 633.
[155] H. D. Belitz, W. Grosch, *Chemistry of Foods*, Acribia, Zaragoza, Spain, **1987**.

[156] E. K. Paleologos, M. G. Kontominas, 'On-line solid phase extraction with surfactant accelerated on-column derivatization and micellar liquid chromatographic separation as a tool for the determination of biogenic amines in various food substrates,' *Analytical Chemistry* **2004**, *76*, 1289.
[157] C. S. Koupris, J. Northcott, in *Encyclopedia of Chemical Technology, Vol. 2*, 2nd edn (Eds. R. E. Kirk, D. E. Othmer), Wiley, New York, **1978**, pp. 411.
[158] A. C. Stern *Air Quality Management, Vol. 5*, Academic, New York, NY, **1977**.
[159] G. Preti, J. N. Labows, J. G. Kostelc, S. Aldinger, R. Daniele, 'Analysis of lung air from patients with bronchogenic-carcinoma and controls using gas chromatography mass spectrometry,' *Journal of Chromatography – Biomedical Applications* **1988**, *432*, 1.
[160] M. L. Simenhoff, J. F. Burke, J. J. Saukkonen, A. T. Ordinario, R. Doty, 'Biochemical profile or uremic breath,' *The New England Journal of Medicine* **1977**, *297*, 132.
[161] M. D. Koppang, M. Witek, J. Blau, G. M. Swain, 'Electrochemical oxidation of polyamines at diamond thin film electrodes,' *Analytical Chemistry* **1999**, *71*, 1188.
[162] I. G. Casella, S. Rosa, E. Desimoni, 'Electrooxidation of aliphatic amines and their amperometric detection in flow injection and liquid chromatography at a nickel based glassy carbon electrode,' *Electroanalysis (New York)* **1998**, *10*, 1005.
[163] M. Niculescu, C. Nistor, I. Frebort, P. Pec, B. Mattiasson, E. Csoregi, 'Redox hydrogel based amperometric bienzyme electrodes for fish freshness monitoring,' *Analytical Chemistry* **2000**, *72*, 1591.
[164] M. Niculescu, T. Ruzgas, C. Nistor, I. Frebort, M. Sebela, P. Pec, E. Csoregi, 'Electrooxidation mechanism of biogenic amines at amine oxidase modified graphite electrode,' *Analytical Chemistry* **2000**, *72*, 5988.
[165] S. Iwaki, M. Ogasawara, R. Kurita, O. Niwa, K. Tanizawa, Y. Ohashi, K. Maeyama, 'Real time monitoring of histamine released from rat basophilic leukemia (RBL-2H3) cells with a histamine microsensor using recombinant histamine oxidase,' *Analytical Biochemistry* **2002**, *304*, 236.
[166] K. Aoki, Y. Teragishi, M. Tokieda, 'Light scattering of polyaniline films responding to electrochemical switching,' *Journal of Electroanalytical Chemistry* **1999**, *460*, 254.
[167] S. Amemiya, Y. Umezawa, 'Chemical sensing based on molecular recognition at membrane surfaces,' *Journal of Synthetic Organic Chemistry Japan* **1997**, *55*, 436.
[168] W. Qin, P. Parzuchowski, W. Zhang, M. E. Meyerhoff, 'Optical sensor for amine vapors based on dimer–monomer equilibrium of indium(iii) octaethylporphyrin in a polymeric film,' *Analytical Chemistry* **2003**, *75*, 332.
[169] Z. Loukou, A. Zotou, 'A comparative survey of the simultaneous ultraviolet and fluorescence detection in the RP-HPLC determination of dansylated biogenic amines in alcoholic beverages,' *Chromatographia* **2003**, *58*, 579.
[170] K. Zeng, H. Tachikawa, Z. Y. Zhu, V. L. Davidson, 'Amperometric detection of histamine with a methylamine dehydrogenase polypyrrole based sensor,' *Analytical Chemistry* **2000**, *72*, 2211.
[171] Y. B. Wang, G. A. Sotzing, R. A. Weiss, 'Conductive polymer foams as sensors for volatile amines,' *Chemistry of Materials* **2003**, *15*, 375.

REFERENCES

[172] G. A. Sotzing, J. N. Phend, R. H. Grubbs, N. S. Lewis, 'Highly sensitive detection and discrimination of biogenic amines utilizing arrays of polyaniline/carbon black composite vapor detectors,' *Chemistry of Materials* **2000**, *12*, 593.

[173] T. M. Hellman, F. H. Small, 'Characterization of petrochemical odors,' *Chemical Engineering Progress* **1973**, *69*, 75.

[174] T. M. Hellman, F. H. Small, 'Characterization of the odor properties of 101 petrochemicals using sensory methods,' *Journal of the Air Pollution Control Association* **1974**, *24*, 979.

[175] D. G. Laing, H. Panhuber, R. I. Baxter, 'Olfactory properties of amines and n-butanol,' *Chemical Senses & Flavour* **1978**, *3*, 149.

[176] A. G. MacDiarmid, '"Synthetic metals": a novel role for organic polymers (Nobel lecture),' *Angewandte Chemie – International Edition* **2001**, *40*, 2581.

[177] C. C. Han, R. C. Jeng, 'Concurrent reduction and modification of polyaniline emeraldine base with pyrrolidine and other nucleophiles,' *Chemical Communications* **1997**, 553.

[178] D. Yang, G. Zuccarello, B. R. Mattes, 'Physical stabilization or chemical degradation of concentrated solutions of polyaniline emeraldine base containing secondary amine additives,' *Macromolecules* **2002**, *35*, 5304.

[179] A. F. Diaz, J. A. Logan, 'Electroactive polyaniline films,' *Journal of Electroanalytical Chemistry* **1980**, *111*, 111.

[180] J. Yue, A. J. Epstein, A. G. MacDiarmid, 'Sulfonic acid ring substituted polyaniline, a self-doped conducting polymer,' *Molecular Crystals and Liquid Crystals* **1990**, *189*, 255.

[181] S. J. Tian, J. Y. Liu, T. Zhu, W. Knoll, 'Polyaniline/gold nanoparticle multilayer films: assembly, properties and biological applications,' *Chemistry of Materials* **2004**, *16*, 4103.

[182] J. Y. Liu, S. J. Tian, W. Knoll, 'In neutral solution and their application for stable low potential detection of reduced β-nicotinamide adenine dinucleotide,' *Langmuir* **2005**, *21*, 5596.

[183] H. Tang, A. Kitani, T. Yamashita, S. Ito, 'Highly sulfonated polyaniline electrochemically synthesized by polymerizing aniline-2,5-disulfonic acid and copolymerizing it with aniline,' *Synthetic Metals* **1998**, *96*, 43.

[184] A. A. Karyakin, A. K. Strakhova, A. K. Yatsimirsky, 'Self-doped polyanilines electrochemically active in neutral and basic aqueous solutions – electropolymerization of substituted anilines,' *Journal of Electroanalytical Chemistry* **1994**, *371*, 259.

[185] P. N. Bartlett, E. Simon, 'Poly(aniline)-poly(acrylate) composite films as modified electrodes for the oxidation of NADH', *Physical Chemistry and Chemical Physics* **2000**, *2*, 2599.

[186] O. A. Raitman, E. Katz, A. F. Buckmann, I. Willner, 'Integration of polyaniline/poly(acrylic acid) films and redox enzymes on electrode supports: an *in situ* electrochemical/surface plasmon resonance study of the bioelectrocatalyzed oxidation of glucose or lactate in the integrated bioelectrocatalytic systems,' *Journal of the American Chemical Society* **2002**, *124*, 6487.

[187] O. Ngamna, A. Morrin, S. E. Moulton, A. J. Killard, M. R. Smyth, G. G. Wallace, 'An HRP based biosensor using sulphonated polyaniline,' *Synthetic Metals* **2005**, *153*, 185.

[188] C. L. Recksiedler, B. A. Deore, M. S. Freund, 'A novel layer-by-layer approach for the fabrication of conducting polymer/RNA multilayer films for controlled release,' *Langmuir* **2006**, *22*, 2811.
[189] R. Nagarajan, W. Liu, J. Kumar, S. K. Tripathy, F. F. Bruno, L. A. Samuelson, 'Manipulating DNA conformation using intertwined conducting polymer chains,' *Macromolecules* **2001**, *34*, 3921.
[190] J. T. Zhang, L. S. Chua, D. M. Lynn, 'Multilayered thin films that sustain the release of functional DNA under physiological conditions,' *Langmuir* **2004**, *20*, 8015.
[191] G. Decher, 'Fuzzy nanoassemblies: toward layered polymeric multicomposites,' *Science* **1997**, *277*, 1232.
[192] P. Bertrand, A. Jonas, A. Laschewsky, R. Legras, 'Ultrathin polymer coatings by complexation of polyelectrolytes at interfaces: suitable materials, structure and properties,' *Macromolecular Rapid Communications* **2000**, *21*, 319.
[193] W. Jin, X. Y. Shi, F. Caruso, 'High activity enzyme microcrystal multilayer films,' *Journal of the American Chemical Society* **2001**, *123*, 8121.
[194] O. P. Tiourina, A. A. Antipov, G. B. Sukhorukov, N. L. Larionova, Y. Lvov, H. Mohwald, 'Entrapment of α-chymotrypsin into hollow polyelectrolyte microcapsules,' *Macromolecular Bioscience* **2001**, *1*, 209.
[195] Y. Lvov, A. A. Antipov, A. Mamedov, H. Mohwald, G. B. Sukhorukov, 'Urease encapsulation in nanoorganized microshells,' *Nano Letters* **2001**, *1*, 125.
[196] X. P. Qiu, S. Leporatti, E. Donath, H. Mohwald, 'Studies on the drug release properties of polysaccharide multilayers encapsulated ibuprofen microparticles,' *Langmuir* **2001**, *17*, 5375.
[197] X. Y. Shi, F. Caruso, 'Release behavior of thin walled microcapsules composed of polyelectrolyte multilayers,' *Langmuir* **2001**, *17*, 2036.
[198] C. Schuler, F. Caruso, 'Decomposable hollow biopolymer based capsules,' *Biomacromolecules* **2001**, *2*, 921.
[199] S. A. Sukhishvili, S. Granick, 'Layered, erasable, ultrathin polymer films,' *Journal of the American Chemical Society* **2000**, *122*, 9550.
[200] S. T. Dubas, T. R. Farhat, J. B. Schlenoff, 'Multiple membranes from 'true' polyelectrolyte multilayers,' *Journal of the American Chemical Society* **2001**, *123*, 5368.
[201] S. T. Dubas, J. B. Schlenoff, 'Polyelectrolyte multilayers containing a weak polyacid: construction and deconstruction,' *Macromolecules* **2001**, *34*, 3736.

4
Self-Doped Polythiophenes

Polythiophenes and their derivatives have been studied intensively and utilized in various applications such as electrical conductors, nonlinear optical devices, antistatic coatings, supercapacitors, transparent electrodes, electrochromic displays, light emitting diodes, photovoltaic cells, batteries, field effect transistors, etc. [1–5]. The low solubility and/or fusibility of polythiophene have motivated chemists to introduce various substituents along the backbone. This can not only improve the processability but can also modulate its electronic and physical properties; it can even lead to physical phenomena that are not found in the parent unsubstituted polythiophene. Polythiophenes can be prepared by chemical synthesis such as metal catalyzed polycondensation [6, 7] and using ferric chloride as oxidizing agent [8], and by electrochemical polymerization [9, 10]. The first environmentally stable, organic solvent soluble, 3-alkyl substituted polythiophene was chemically synthesized by Elsenbaumer *et al.* [11–13]. After this, various groups reported the chemical and electrochemical synthesis of substituted polythiophenes [14–16].

Following these studies, the synthesis of water-soluble derivatives, poly(ω-(3'-thienyl)alkanesulfonates) were reported by Wudl and Heeger *et al.* [17, 18]. In addition to providing water solubility, the covalently bound sulfonate groups serve as charge balancing counterions when the polymers are oxidatively- or p-doped (see Chapter 1 Figure 1.13). This led to the concept of 'self-doped' conducting polymers. Furthermore, the sulfonic acid form of the polymers was found to protonate the π-conjugated backbone, thus resulting in 'auto-doped' conducting polymers [19–21]. Since this discovery, many analogous polymers,

Self-Doped Conducting Polymers   M.S. Freund and B.A. Deore
© 2007 John Wiley & Sons, Ltd

such as self-doped polythiophenes and other conducting polymers in general, have been synthesized. The detailed electrochemical and chemical synthesis, properties and applications are discussed in the following sections.

4.1 SULFONIC ACID DERIVATIVES

4.1.1 Electrochemical Polymerization

Wudl and Heeger *et al.* [17] electrochemically synthesized the sodium salts and acid forms of poly(3-thiophene ethanesulfonate) and poly(3-thiophene butanesulfonate). The monomers 3-thiophene ethanesulfonate and 3-thiophene butanesulfonate were prepared by the route shown in Figure 4.1. However, attempts to electropolymerize these monomers or their sulfonic acid derivatives were not successful. Therefore, the monomers, 3-thiophene alkanesulfonate methyl esters, were polymerized first, followed by conversion of the ester to the sodium salt via the sulfonyl chloride derivative. The sodium salts of poly(3-thiophene ethanesulfonate) and poly(3-thiophene butanesulfonate) are reportedly soluble in water in their neutral (insulating) and doped (conducting)

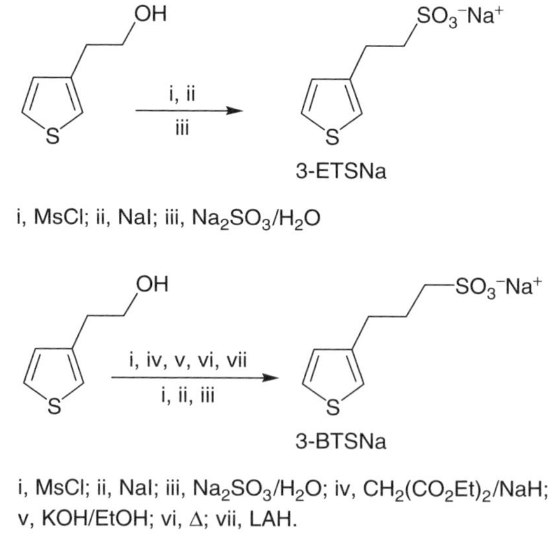

Figure 4.1 Synthesis of monomer sodium (3-thiophene-β-ethanesulfonate) (3-ETSNa) and sodium (3-thiophene-δ-butanesulfonate) (3-BTSNa). (Reprinted with permission from *Journal of the American Chemical Society*, **109**, 1858. Copyright (1987) American Chemical Society.)

SULFONIC ACID DERIVATIVES

states. The water-soluble acid form was prepared by passing the polymer through an ion-exchange resin consisting of the acid form of sulfonated polystyrene. The films of these polymers, cast from water solutions, exhibited electrical conductivity in the range 10^{-7}–10^{-2} S/cm depending on the relative humidity. A conductivity of ~ 10 S/cm was observed when exposed to bromine vapors. The molecular weights of these polymers were not reported.

The UV-Vis absorption spectra of the sodium salts and acid forms of poly(3-thiophene ethanesulfonate) and poly(3-thiophene butanesulfonate) films show a $\pi-\pi^*$ transition at 425 nm as well as transitions associated with the charge carrying bipolarons at 800 nm. The absorption peak attributed to bipolarons is more pronounced in the case of poly(3-thiophene butanesulfonic acid) solid film as shown in Figure 4.2. [18]. A relatively small shift in the absorption spectrum of poly(3-thiophene butanesulfonic acid) polymer is observed upon dissolution in water. In contrast, a significant blue shift has been reported between

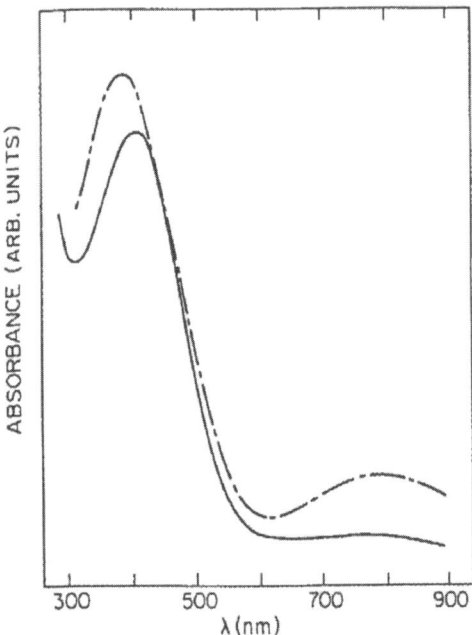

Figure 4.2 UV-Vis absorption spectra of poly(3-thiophene butanesulfonic acid) (cast from water; dash curve) and for a solution of the same polymer in water (solid curve). (Reprinted from *Synthetic Metals*, **20**, A. O. Patil, Y. Ikenoue, N. Basescu, N. Colaneri, J. Chen, F. Wudl, A. J. Heeger, 151. Copyright (1987), with permission from Elsevier.)

the solid film and solution spectra of 3-alkylthienyl substituted polythiophene. Electron spin resonance measurements showed a narrow line shape ($\Delta H \sim 4.5$) and a g-value corresponding to delocalized electrons. The magnetic spin susceptibility values of poly(3-thiophene ethanesulfonic acid) and poly(3-thiophene butanesulfonic acid) were approximately $8-9 \times 10^{-7}$ emu/mol, suggesting the presence of one spin per 1500 monomer units. The electron spin resonance followed Curie's law ($\chi = C/T$) at room temperature. It was suggested that, similarly to polythiophene and the 3-alkylthienyl substituted polythiophene, charge storage via bipolarons is the dominant mechanism in the self-doped sodium salt and acid forms of poly(3-thiophene alkanesulfonate)s [18]. However, in these preliminary studies, the self-doping mechanism, i.e., expulsion of cations associated with sulfonate groups or insertion of additional anions upon electrochemical oxidation, was not explicitly mentioned.

In further studies, the self-doping mechanism of these polymers was verified by cyclic voltammetry, pH measurements and atomic absorption spectroscopy [17, 20]. The cyclic voltammograms of the sodium salt and acid forms of poly(3-thiophene butanesulfonate) cast films are shown in

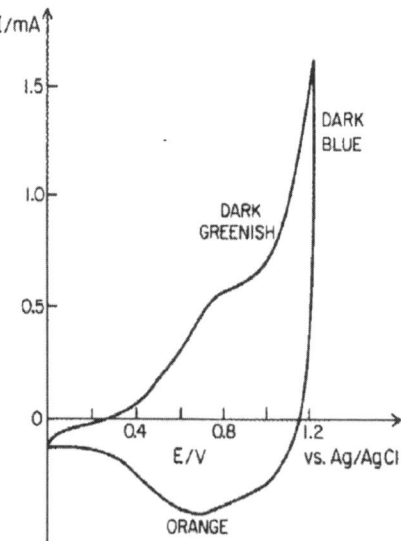

Figure 4.3 A typical cyclic voltammogram of sodium salts of poly(3-thiophene butanesulfonate) cast film on Pt foil obtained after several 'break-in cycles' in acetonitrile with 0.1 N Bu$_4$NClO$_4$. Scan rate 10 mV/s. (Reprinted from *Synthetic Metals*, **30**, Y. Ikenoue, N. Uotani, A. O. Patil, F. Wudl, A. J. Heeger, 305. Copyright (1989), with permission from Elsevier.)

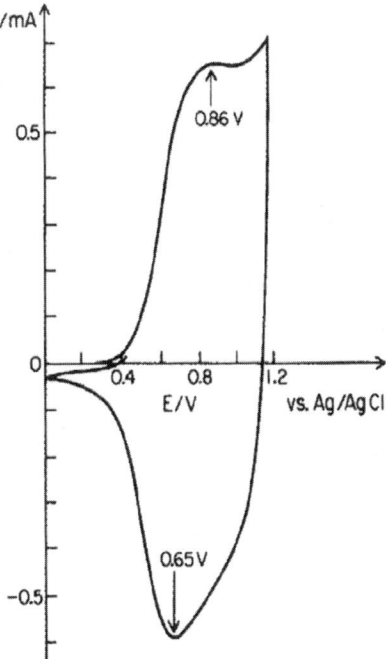

Figure 4.4 A typical cyclic voltammogram of poly(3-thiophene butanesulfonic acid) cast film on Pt foil obtained after several 'break-in cycles' in acetonitrile with 0.1 N Bu$_4$NClO$_4$. Scan rate 100 mV/s. (Reprinted from *Synthetic Metals*, 30, Y. Ikenoue, N. Uotani, A. O. Patil, F. Wudl, A. J. Heeger, 305. Copyright (1989), with permission from Elsevier.)

Figures 4.3 and 4.4, respectively. The polymer films are electroactive in nonaqueous media and show color changes as a function of potential. At a reduction potential of 0.6 V vs Ag/AgCl, polymer films were insulating and orange colored. At oxidation potentials around 0.6 to 0.8 V, the polymer films were dark green and finally turned a dark blue at higher potentials. It was reported that these color changes are particularly evident in case of sodium salts of poly(3-thiophene butanesulfonate) films in the presence of small amount of water. Poly(3-thiophene butanesulfonic acid) films were found to be hygroscopic, i.e., coated with a water layer, based on FT-IR results. The expulsion of H$^+$ ions associated with the mechanism of self-doping was confirmed by an *in situ* pH measurement during cyclic voltammetry of poly(3-thiophene butanesulfonic acid) in acetonitrile with 0.1 M Bu$_4$NClO$_4$. Similarly, sodium ion concentration was determined by atomic absorption spectroscopy as a function of potential. Upon oxidation, expulsion of Na$^+$ from sodium salt of poly(3-thiophene butanesulfonate) film was observed. The auto-doping or self-acid-doping mechanism of the acid form of poly(3-thiophene alkanesulfonate)s is described in detail in Chapter 1 (Section 1.4.3).

Following the initial reports of Wudl and Heeger et al. [17, 18], Havinga et al. prepared self-doped polythiophene by direct electropolymerization of monomer, the potassium salt of 3'-propylsulfonate 2,2':5',5''-terthienyl, in the absence of added electrolyte [22]. The monomer was synthesized by alkylation of diketones, followed by functional group transformations as shown in Figure 4.5. In this case, the monomer itself acts as the electrolyte during electrochemical polymerization, taken as evidence of self-doping. The UV-Vis spectra of self-doped poly(3'-propylsulfonate 2,2':5',5''-terthienyl) film is shown in Figure 4.6, a. The self-doped polymer was water soluble; however, the polymer solutions were reported to be unstable (Figure 4.6, b). The spectrum of the polymer methanol solution (Figure 4.6, c) suggested that the polymer is dedoped while going into solution. The conductivity of polymer pellets was reported to be 0.01 S/cm. Wallace et al. [23] electrochemically polymerized the sodium salt of poly(2-(3-thienyl)ethyl sulfonate in an aqueous solution without supporting electrolyte. The polymer was synthesized at a constant potential of 1.8 V (vs Ag/AgCl) using 0.1 M 2-(3'-thienyl)ethyl sulfonate. The polymer was not deposited on the electrode substrate due to its solubility in water.

Chen et al. [24] studied the structure and effect of the side chain length on the doping level of self-doped poly(n-(3'-thienyl)alkanesulfonic acid)s with alkanes of carbon numbers 2, 6 and 10. They suggested that self-doping of poly(n-(3'-thienyl)alkanesulfonic acid)s is dependent on the side chain length (for details see Chapter 1 Section 1.4.3). In subsequent work, Chen et al. reported the irreversible thermal dedoping

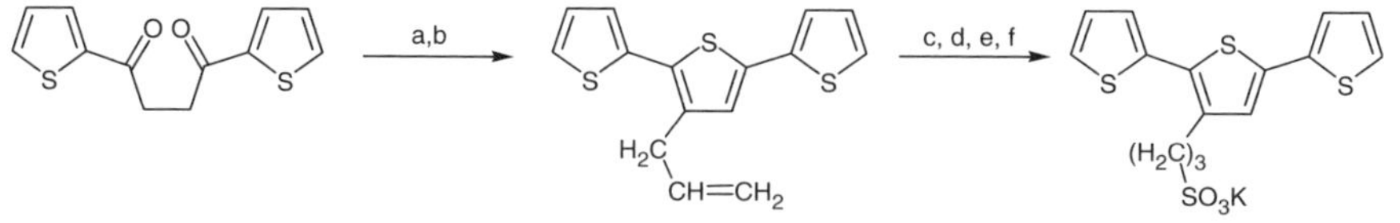

a, ClCH$_2$CH=CH$_2$, KOH, DMSO;   b, P$_2$S$_5$, NaHCO$_3$, ether;   c, BH$_3$, THF, H$_2$O$_2$, NaOH;

d, PBr$_3$, CH$_2$Cl$_2$;   e, Na$_2$SO$_3$, BuNBr, EtOH, water;   f, HCl, water; KOH, water

Figure 4.5 Synthesis of monomer potassium salt of 3'-propylsulfonate 2,2':5',5''-terthienyl. (Reproduced from *Polymer Bulletin*, 1987, 18, 277, E. E. Havinga, L. W. Van Horssen, W. Tenhoeve, H. Wynberg, E. W. Meijer, with kind permission of Springer Science and Business Media.)

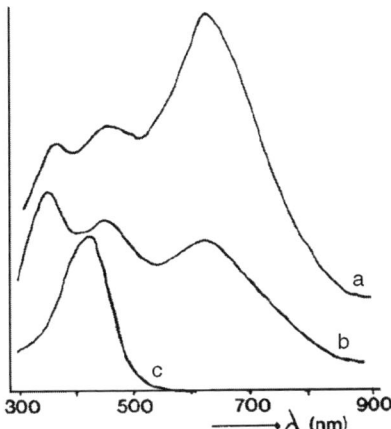

Figure 4.6 UV-Vis spectra of poly(3'-propylsulfonate 2,2':5',5"-terthienyl) (a) solid film grown on ITO in acetonitrile, (b) water solutions obtained from these films and (c) methanol solution of the polymer. (Reproduced from *Polymer Bulletin*, 1987, 18, 277, E. E. Havinga, L. W. van Horssen, H. Wynberg, E. W. Meijer, with kind permission of Springer Science and Business Media.)

characteristics and accompanying structural changes of self-acid-doped poly(2-(3'-thienyl)ethanesulfonic acid) (P3TESH) [25]. At a temperature of 40 °C or above, the P3TESH thin films are susceptible to thermal dedoping, resulting in a change of color from brownish green to brownish orange. The UV-Vis near infrared spectra of P3TESH thin films measured under vacuum at different temperatures are shown in Figure 4.7, A. The P3TESH films show a $\pi-\pi^*$ transition at 440 nm as well as polaron and bipolaron band transitions at 800 and 1900 nm, respectively. The intensity of the $\pi-\pi^*$ transition increases while polaron and bipolaron bands decrease with temperature. At 70 °C, polaron and bipolaron bands disappear completely. The color of the films changes irreversibly from brownish green to brownish orange at 45 °C and to yellowish orange at 70 °C. These results suggest the dedoping of P3TESH films with temperature. The dedoping of P3TESH films is significant even at temperatures below 40 °C. The spectral changes with time at 40 °C showed the decrease in the intensity of the polaron and bipolaron bands (Figure 4.7, B). The color change from brownish green to brownish orange was reported after 240 min.

Similarly to the UV-Vis near infrared results, irreversible changes in the conductivity of P3TESH films are observed with temperature (Figure 4.8, A). The conductivity decreases by a factor of 10^5 (10^{-2} to 10^{-7} S/cm) from the doped to the dedoped state with temperatures between 20

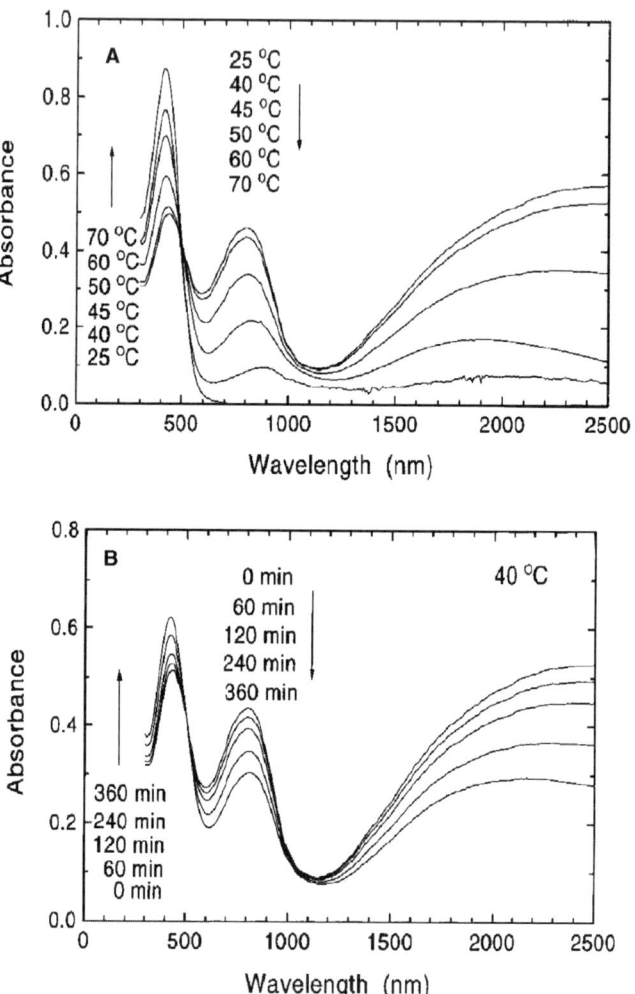

Figure 4.7 (A) UV-Vis near infrared spectra of P3TESH thin films measured under vacuum at different temperatures; (B) variations of UV-Vis near infrared spectra of P3TESH thin film with time measured under vacuum and 40 °C. (Reprinted with permission from *Chemistry of Materials*, **9**, 2750. Copyright (1997) American Chemical Society.)

and 70 °C. A significant decrease in conductivity was observed even at the relatively low temperatures of 30 and 40 °C after 5 min heating (Figure 4.8, B). Based on NMR results, Chen *et al.* [25] suggested that the thermal dedoping results from the nucleophilic attack of the $-CH_2CH_2SO_3^-$ side chain at the carbocation to yield C–O bonding and a coupling of the unpaired electrons in polarons on the main chain. As a

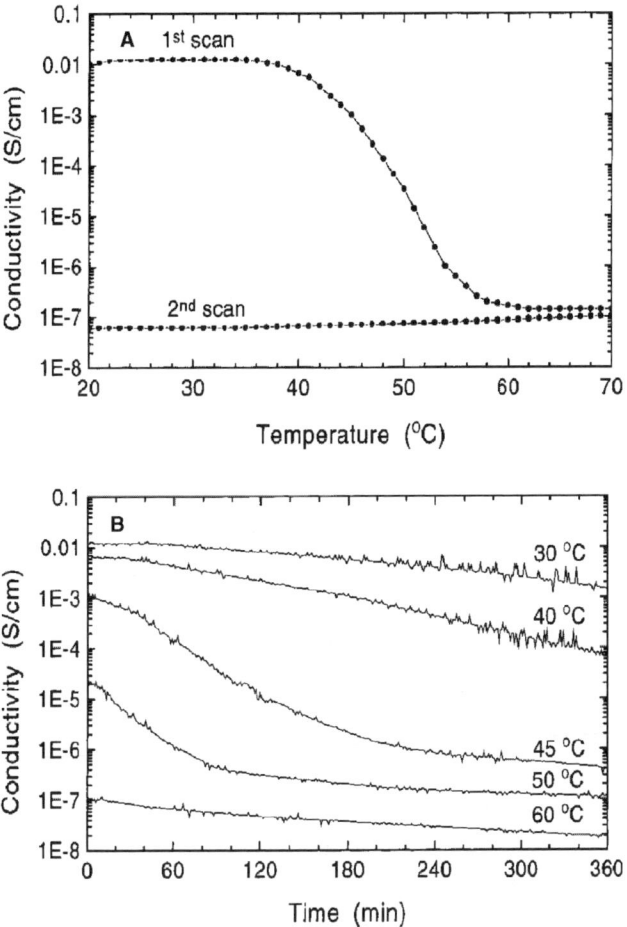

Figure 4.8 (A) Conductivity of P3TESH film vs temperature scanning from 20 to 70 °C under vacuum; (B) variations of conductivity of P3TESH films with time at different temperatures under vacuum. (Reprinted with permission from *Chemistry of Materials*, 9, 2750. Copyright (1997) American Chemical Society.)

result, P3TESH films are more susceptible to thermal dedoping than are other conjugated polymers such as alkyl substituted polythiophene [26, 27] and sulfonated polyanilines [28], where thermal dedoping is observed at temperatures above 100 °C. Therefore, the thermal dedoping characteristics of P3TESH films at low temperatures are reported to be useful for the permanent recording of temperature increases at relatively low temperatures. For example, it was suggested that these films can be used in temperature indicators for foods and drugs that need to be kept fresh at low temperature.

Zotti et al. [29] prepared self-doped polythiophene by electropolymerization of the monomer, tetrabutyl ammonium 4-(4H-cyclopenta[2,1-b:3,4-b']-dithienyl)-butanesulfonate (Figure 4.9). The use of this monomer was reported to be advantageous due to its lower oxidation potential, functional group remote from the coupling site, elimination of adverse inductive effects through the use of a spacer methylene group, and high solubility in acetonitrile. The polymer was electrochemically synthesized in acetonitrile with 0.1 M tetrabutyl ammonium perchlorate. The oxidation potential of the monomer (Figure 4.9) was the same as 4-alkyl substituted cyclopentadithiophene (0.6 V vs Ag/Ag$^+$), indicating that substitution of the sulfonate group did not affect the electronic properties of the monomer. The cyclic voltammogram of poly(4-(4H-cyclopenta[2,1-b:3,4-b']-dithienyl)butanesulfonate) in acetonitrile with 0.1 M tetrabutyl ammonium perchlorate exhibited two redox waves at −0.3 and 0.1 V vs Ag/Ag$^+$. The UV-Vis spectra of neutral and oxidized polymer (film and aqueous solution) showed absorption peaks at around 550 and 870 nm, respectively (Figure 4.10). The neutral polymer has been reported to be moderately soluble in water, however, higher solubility (10 g/L) is observed for bulk oxidized polymer produced by oxidative electrolysis. The maximum in situ conductivity of polymer in acetonitrile with 0.1 M tetrabutyl ammonium perchlorate was found to be 0.6 S/cm (Figure 4.11). The higher conductivity is attributed to the sulfonate–tetrabutyl ammonium ion pair, which reportedly increases the interchain distance. Zotti et al. [29] suggested that lower conductivity of alkyl sulfonated polymer compared with alkylated polymer is due to ion aggregation or ion pairing along the chain, which introduces disorder in

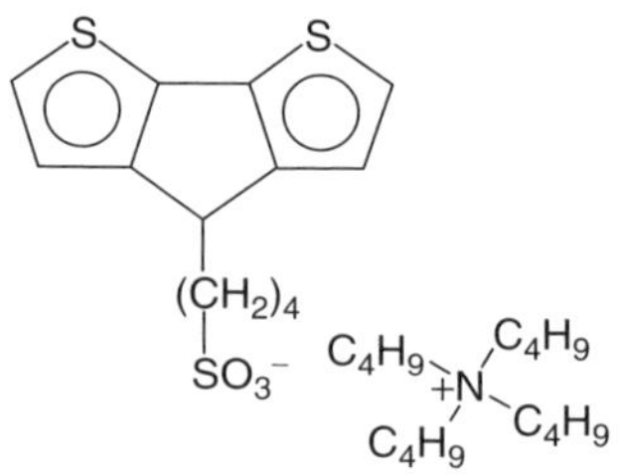

Figure 4.9 Structure of tetrabutyl ammonium 4-(4H-cyclopenta[2,1-b:3,4-b']-dithienyl)butanesulfonate. (Reprinted with permission from *Chemistry of Materials*, 9, 2940. Copyright (1997) American Chemical Society.)

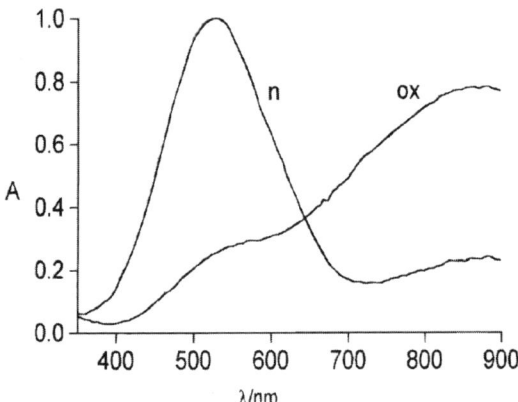

Figure 4.10 UV-Vis absorption spectra of neutral (n) and oxidized (ox) poly(4-(4H-cyclopenta[2,1-b:3,4-b']dithienyl)butanesulfonate) in water solution. (Reprinted with permission from *Chemistry of Materials*, **9**, 2940. Copyright (1997) American Chemical Society.)

the structure that is associated with decoplanarization of the thiophene ring. The four-point-probe conductivity of films cast from an oxidized polymer water solution was 0.03 S/cm. This conductivity value is lower than *in situ* conductivity in acetonitrile and is attributed to the partial dedoping of the polymer in water.

The self-doping mechanism of poly(4-(4H-cyclopenta[2,1-b:3,4-b'] dithienyl)butanesulfonate) has been confirmed by electrochemical quartz-crystal microbalance measurements in acetonitrile containing 0.1 M perchlorate salts of various cations. The mass change during the cyclic voltammograms of the polymer indicated the expulsion of cations during oxidation doping. It is also suggested that the cation expulsion process is accompanied by a pronounced anion insertion (1:1 ratio). The high-cation flux indicates the self-doping ability of the polymer. The cation–anion flux during positive doping of polymer is outlined in Figure 4.12.

The electrochemical synthesis of self-doped poly(3,4-ethylenedioxythiophene)-sulfonate in water without supporting electrolyte has been reported by Chevort *et al.* [30]. The monomer, sulfonated 3,4-ethylenedioxythiophene, was synthesized by various steps as shown in Figure 4.13. The electrochemical polymerization in water resulted in the formation of soluble oligomers. However, the copolymer films of poly(3,4-ethylenedioxythiophene)sulfonate could be prepared in water using an equimolar amount of 3,4-ethylenedioxythiophene without supporting electrolyte in the potential range of −1.4 to 1.2 V (vs

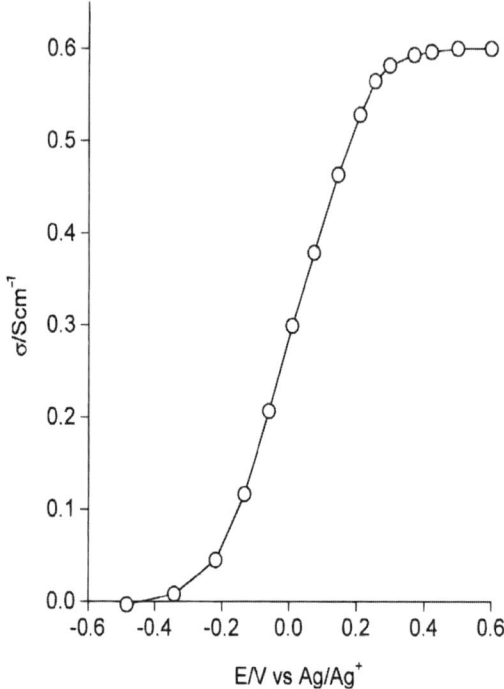

Figure 4.11 *In situ* conductivity of poly(4-(4H-cyclopenta[2,1-b:3,4-b']dithienyl)butanesulfonate) in acetonitrile + 0.1 M Bu$_4$NClO$_4$. (Reprinted with permission from *Chemistry of Materials*, 9, 2940. Copyright (1997) American Chemical Society.)

SCE). The ion exchange properties of these sulfonated copolymer films were evaluated for binding of cations like Ru(NH$_3$)$_6^{3+}$ or UO$_2^{2-}$. The cyclic voltammetry and radiochemical measurements clearly showed the permanent cation exchange properties allowing efficient incorporation of cations like Ru(NH$_3$)$_6^{3+}$ or UO$_2^{2-}$.

The homopolymer films of sulfonated poly(3,4-ethylenedioxythiophene) were electrochemically polymerized using a mixture of acetonitrile and water (5:1 ratio), as reported by Kumar *et al.* [31]. The UV-Vis spectra and the molecular weight of 15 800 g/mol confirmed the polymeric nature of the deposited films. The polymer was found to be soluble in water. The electrochemical and chemical synthesis of water-soluble, highly conductive oligomers of sulfonated poly(3,4-ethylenedioxythiophene) in aqueous acid and nonaqueous media has been reported by Zotti *et al.* [32]. In aqueous HClO$_4$ solution, the constant potential electrolysis of the sodium salt of sulfonated

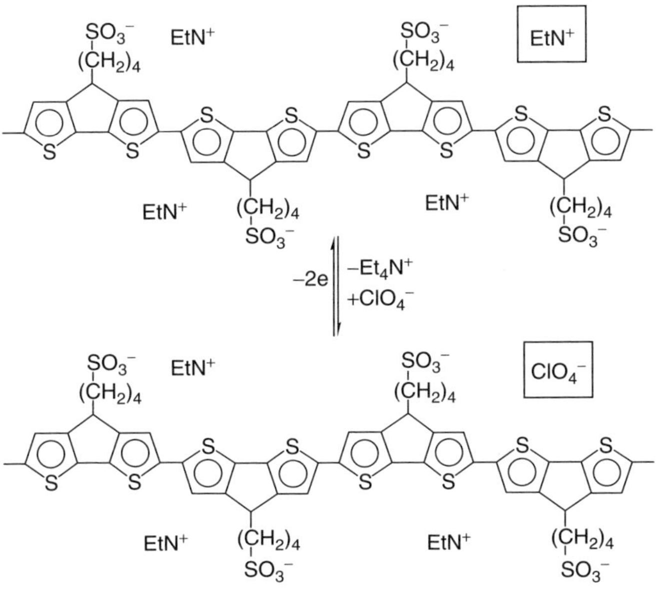

Figure 4.12 Cation–anion flux during positive doping of poly(4-(4H-cyclopenta-[2,1-b:3,4-b']dithienyl)butanesulfonate). (Reprinted with permission from *Chemistry of Materials*, **9**, 2940. Copyright (1997) American Chemical Society.)

3,4-ethylenedioxythiophene at 1.3 V vs SCE resulted in a blue solution, indicating the presence of the oxidized conducting form of the polymer in solution. The acid solution was neutralized with LiOH and evaporated to dryness. Finally a dark solid polymer was obtained by adding acetonitrile to dissolve the lithium perchlorate. The isolated polymer was soluble in water with a molecular weight of 2500 g/mol as observed with gel permeation chromatography. These results indicate that the degree of polymerization is, on average, eight monomer units and is therefore an oligomer. The conductivity of the oligomer was reported to be 10 S/cm. The structure of self-doped poly(3,4-ethylenedioxythiophene) is shown in Figure 4.14. Sulfonated poly(3,4-ethylenedioxythiophene) can also be polymerized in acetonitrile. The electropolymerization of the tetrabutyl ammonium salt of 3,4-ethylenedioxythiophene in acetonitrile with and without the addition of tetrabutyl ammonium perchlorate electrolyte reportedly results in the production of soluble oligomers. However, polymer films can be obtained by polymerization of the monomer, protonated-3,4-ethylenedioxythiophene, in acetonitrile with tetrabutyl ammonium perchlorate electrolyte. The dry and *in situ* conductivity in

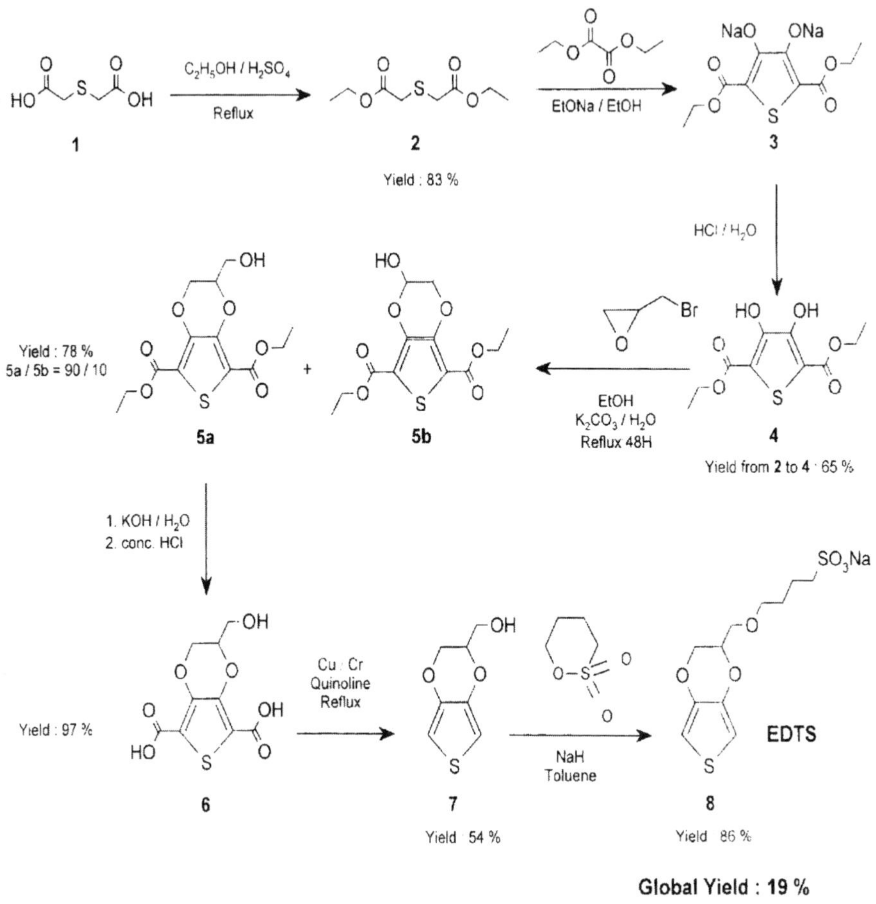

Figure 4.13 Synthesis of monomer sulfonated 3,4-ethyledioxythiophene. **1**: thioglycolic acid, **2**: diethyl thiodiglycolate, **3**: diethyl 3,4-dihydroxythiophene-2,5-dicarboxylate disodium salt, **4**: diethyl 3,4-dihydroxythiophene-2-5-dicarboxylate, **5**: diethyl 2-(hydroxymethyl)-2,3-dihydrothieno[3,4-b][1,4]dioxin-5,7-dicarboxylate, **6**: diethyl 2-(hydroxymethyl)-2,3-dihydrothieno[3,4-b][1,4]dioxin-5,7-dicarboxylic acid, **7**: 2,3-dihydrothieno[3,4-b][1,4]dioxin-2-yl methanol, **8**: sulfonated 3,4-ethylenedioxythiophene. (Reprinted from *Journal of Electroanalytical Chemistry*, 443, O. Stephan, P. Schottland, P. Y. Le Gall, C. Chevrot, C. Mariet, M. J. Carrier, 217. Copyright (1998), with permission from Elsevier.)

acetonitrile of the tetrabutyl ammonium perchlorate salt of poly(3,4-ethylenedioxythiophene) sulfonic acid was approximately 2 S/cm. Quartz-crystal microbalance results suggested 60 % self-doping and 40 % external doping within the polymer films. The chemical synthesis of water-soluble, sulfonated, poly(3,4-ethylenedioxythiophene) can also

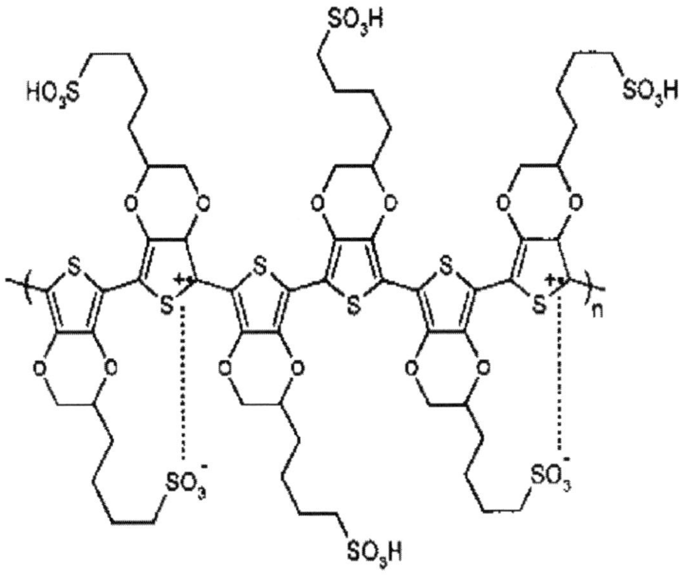

Poly(EDTS)

Figure 4.14 Structure of self-doped poly(3,4-ethylenedioxythiophene). (Reproduced from *Macromolecular Chemistry and Physics*, 2002, 203, 1958, G. Zotti, S. Zecchin, G. Schiavon, L. B. Groenendaal, with permission from Wiley-VCH.)

be achieved in aqueous solution using iron(III) tosylate at 80 °C [32]. The acid form of sulfonated poly(3,4-ethylenedioxythiophene) was obtained by ion exchange. The molecular weight and conductivity of the polymer is reported to be around 5000 g/mol and 1–5 S/cm.

Yildiz et al. [33, 34] synthesized self-doped sulfonated-poly(3-methyl thiophene) (MTy) and -poly(thiophene) by *in situ* electrophilic substitution of sulfonic acid on the polythiophene backbone. The electrochemical synthesis of sulfonated poly(3-methyl thiophene) was carried out in acetonitrile/0.1 M tetrabutyl ammonium perchlorate and acetonitrile/ 0.1 M lithium perchlorate, containing anhydrous fluorosulfonic acid (HSO_3F) by potential cycling in the range of −0.2 to 1.8 V (vs Ag/AgCl) at a sweep rate of 100 mV/s [34]. HSO_3F was used as a sulfonation reagent for thiophene. The sulfonation of polymer was confirmed by FT-IR spectroscopy. The degree of sulfonation of the polymer was controlled by varying HSO_3F and 3-methyl thiophene concentrations in the solutions used for polymerization. A decrease in growth rate and electroactivity of the sulfonated-poly(3-methyl thiophene) film with increasing concentration of HSO_3F was observed. Similarly, the

Table 4.1 The degree of sulfonation ratios and the dry conductivity values of the polymer films obtained from the acetonitrile/0.1 M tetrabutyl ammonium perchlorate solutions containing 200 mM 3-MTy and different concentrations of HSO_3F. (Reprinted from *European Polymer Journal*, 40, Y. A. Udum, K. Pekmez, A. Yildiz, 1057. Copyright (2004), with permission from Elsevier.)

Polymer composition	C/S	Conductivity (S/cm)
200 mM 3-MTy	5.2	72
200 mM 3-MTy + 25 mM HSO_3F	6.4	56
200 mM 3-MTy + 50 mM HSO_3F	6.1	38
200 mM 3-MTy + 75 mM HSO_3F	5.8	32
200 mM 3-MTy + 100 mM HSO_3F	5.8	26
200 mM 3-MTy + 150 mM HSO_3F	5.8	22
200 mM 3-MTy + 200 mM HSO_3F	5.7	19

conductivity decreased with an increase in concentration of HSO_3F (Table 4.1) as expected. The morphology of poly(3-methyl thiophene) (Figure 4.15, A) in comparison with sulfonated poly(3-methyl thiophene) (Figure 4.15, B) clearly shows the influence of HSO_3F on the polymerization of 3-methyl thiophene and hence on its conductivity. In the absence of the sulfonate group, poly(3-methyl thiophene) films are uniform and dense (Figure 4.15, A). However, a cauliflower-like morphology is observed for the sulfonated polymer (Figure 4.15, B). The molecular weight of the polymer was reported to be in the range 6000–8000 g/mol. The solubility of sulfonated polymer in 0.1 M KOH increases from 2–13 mg/mL with increase in concentration of HSO_3F from 25 to 250 mM. Similarly, electrochemical synthesis of sulfonated poly(thiophene) can be carried out in acetonitrile/0.1 M lithium perchlorate containing anhydrous HSO_3F [33]. The polymer conductivity decreases from 1.6 to 0.0025 S/cm and solubility in 0.1 M KOH increases from 1.1 to 9.6 mg/mL with increasing degree of sulfonation. The molecular weight of the polymer is reported to be 30 000 to 40 000 g/mol. Recently, Yildiz *et al.* [35, 36] electrochemically synthesized sulfonated conducting copolymers using aniline and thiophene monomers in a similar manner.

4.1.2 Chemical Polymerization

The chemical synthesis of self-doped polythiophene was first reported by Ikenoue *et al.* [21]. They synthesized self-doped poly(3-(3'-thienyl)propanesulfonate) in aqueous media using ferric chloride as an oxidizing agent. The electrochemical polymerization of the monomer, 3-(3'-

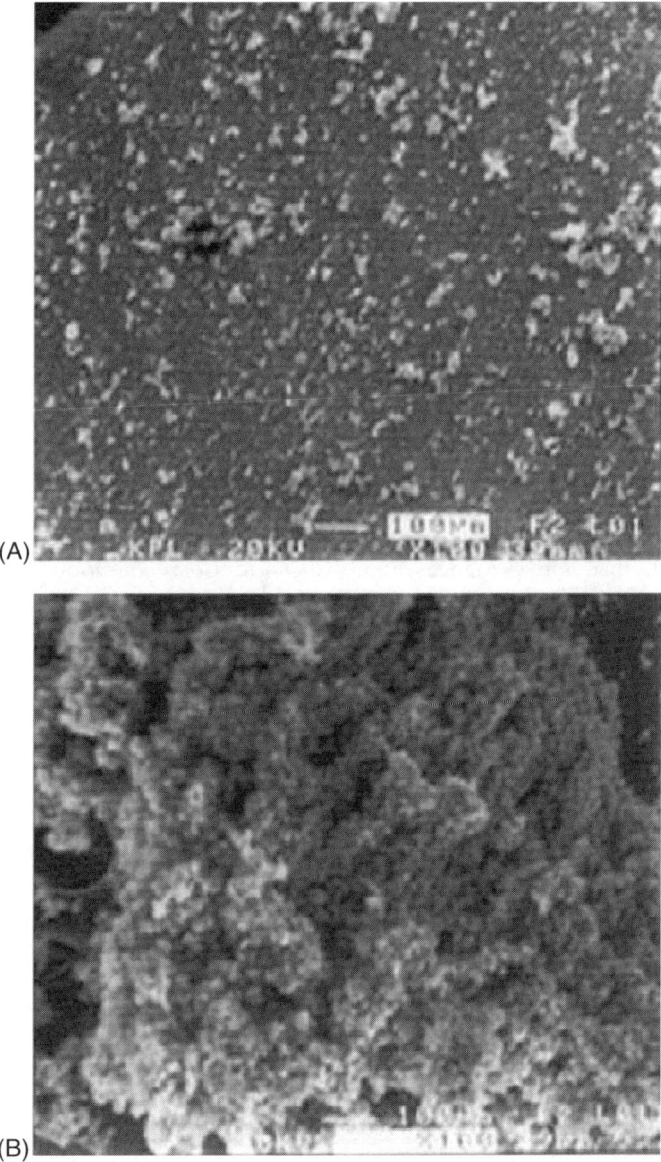

Figure 4.15 Scanning electron micrographs of (A) poly(3-methyl thiophene) and (B) sulfonated poly(3-methyl thiophene). (Reprinted from *European Polymer Journal*, 40, Y. A. Udum, K. Pekmez, A. Yildiz, 1057. Copyright (2004), with permission from Elsevier.)

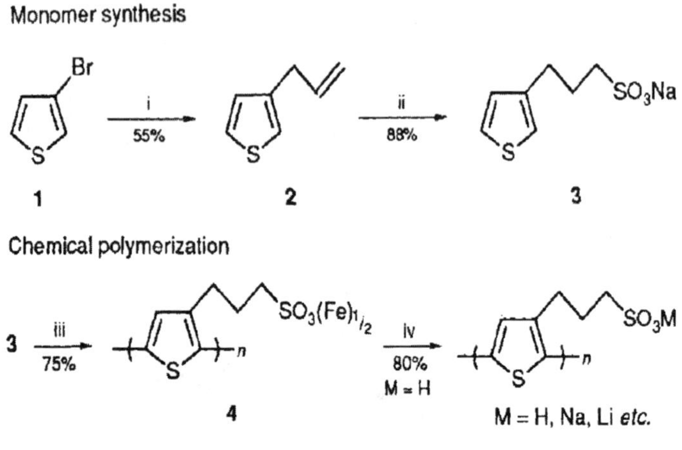

Figure 4.16 Synthesis of sulfonated monomer and polymer. Reagents and conditions: i, BuLi, Et$_2$O, $-73\,°$C, then allyl bromide, 2 h, $-73\,°$C; ii, NaHSO$_3$, azoisobutyronitrile in MeOH/H$_2$O, 5 h, 80 °C; iii, FeCl$_3$ in H$_2$O, room temp.; iv, NaOH, then ion exchange resin. (*Chemical Communications* 1990, 1694, Y. Ikenoue, Y. Saida, M. Kira, H. Tomozawa, H. Yashima, M. Kobayashi. Reproduced by permission of The Royal Society of Chemistry.)

thienyl)propanesulfonate (Figure 4.16, 3) in methanol or water and chemical polymerization in chloroform using ferric chloride as a oxidizing agent were unsuccessful. The monomer synthesis and chemical polymerization is schematically represented in Figure 4.16. The chemical polymerization of the monomer, 3-(3'-thienyl)propanesulfonate (Figure 4.16, 3) resulted in an intermediate polymer, poly(3-(3'-thienyl) propanesulfonate)$_2$Fe (Figure 4.16, 4). The polymer was insoluble in water; however, after treatment with NaOH, soluble poly(3-(3'-thienyl) propanesulfonate) (Figure 4.16, 5) was formed. This polymer was passed through an ion exchange resin column (H$^+$-type) to obtain self-doped poly(3-(3'-thienyl)propanesulfonic acid). The UV-Vis spectra of self-doped neutral polymer (sodium salt of poly(3-(3'-thienyl)propanesulfonate)) in water and the poly(3-(3'-thienyl)propanesulfonic acid) cast film are shown in Figure 4.17. The UV-Vis near infrared spectra of the poly(3-(3'-thienyl)propanesulfonic acid) cast film shows highly doped behavior. The four-point probe conductivity of self-doped polymer films without external dopant is 0.1 S/cm. The molecular weight measured using gel permeation chromatography was approximately 100 000 g/mol and the average degree of polymerization was found to be 440.

The electrochemical behavior of cast films of poly(3-(3'-thienyl)-propanesulfonic acid) in acetonitrile/HBF$_4$ containing 6 % water is

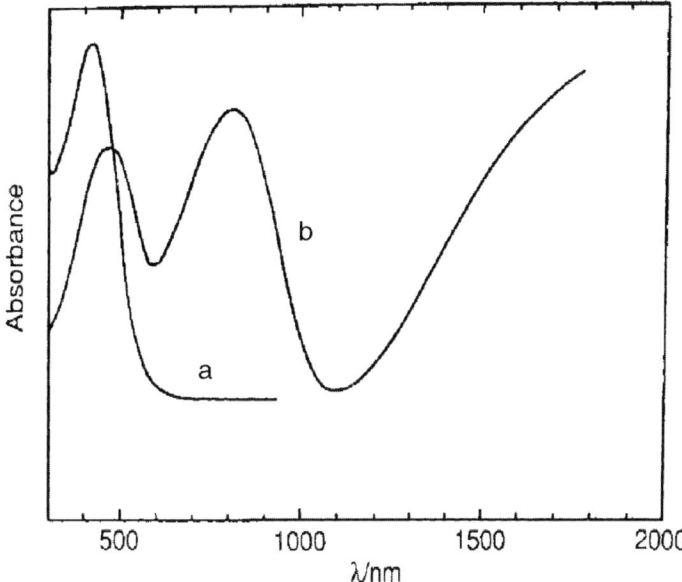

Figure 4.17 UV-Vis and near infrared absorption spectra of self-doped poly(3-(3'-thienyl)propanesulfonic acid): (a) is neutral 5 (Figure 4.16)(M = Na) in water; (b) cast film of self-doped 5 (Figure 4.16) (M = H) (acid form). (*Chemical Communications* 1990, 1694, Y. Ikenoue, Y. Saida, M. Kira, H. Tomozawa, H. Yashima, M. Kobayashi. Reproduced by permission of The Royal Society of Chemistry.)

shown in Figure 4.18. The polymer exhibited stable and reversible fast redox switching over a range of scan rates. These results suggest that the electrochemical behavior is not dominated by the diffusion limited process associated with counterions. The electrochemical kinetics were found to be faster than poly(3-hexylthiophene) [37]. Time dependent UV-Vis near infrared optical spectra and the normalized time dependent change in optical density at 500 nm (λ_{max}) of a poly(3-(3'-thienyl)-propanesulfonic acid) film in 0.5 M HBF_4/acetonitrile are shown in Figure 4.19 a and b, respectively. The polymer shows the characteristic $\pi-\pi^*$ transition, polaron and bipolaron bands at approximately 500, 800 and 1100 nm, respectively. The $\pi-\pi^*$ transition (500 nm) of the self-doped polymer exhibited an electrochemical switching time of 50 ms (Figure 4.19 b). The redox switching of these films was stable over 10 000 cycles.

Holdcroft *et al.* [38] prepared the sodium salt and acid forms of poly(n-(3'-thienyl)alkanesulfonates) following the Ikenoue *et al.* chemical synthesis method mentioned above [21]. The aim of this study was to

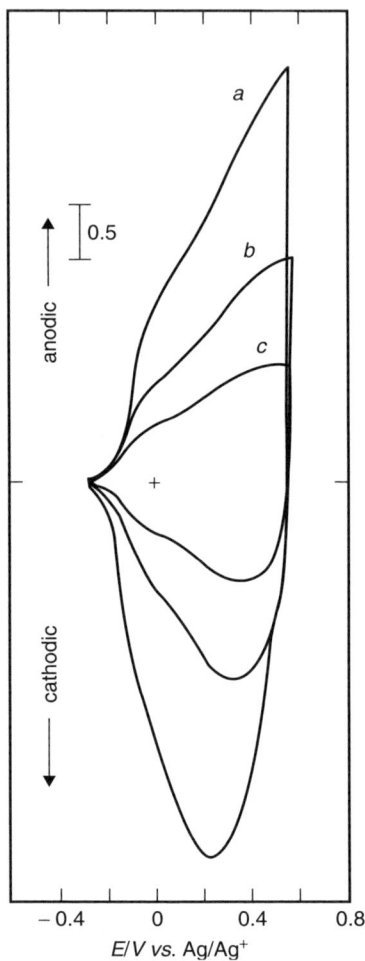

Figure 4.18 Electrochemical behavior of a cast film of poly(3-(3′-thienyl)propane sulfonic acid) at several scan rates in acidic electrolyte solution, 0.5 M HBF$_4$ (6 % H$_2$O)–acetonitrile where the working electrode was a 3100 Å thick film of poly(3-(3′-thienyl)propanesulfonic acid) (M = H) cast on ITO (indium tin oxide), the counter electrode was a Pt mesh and the reference electrode was Ag/Ag$^+$. Cyclic voltammograms were obtained at (a) 200, (b) 100 and (c) 50 mV/s respectively. (Reprinted from *Synthetic Metals*, 40, Y. Ikenoue, H. Tomozawa, Y. Saida, M. Kira, H. Yashima, 333. Copyright (1991), with permission from Elsevier.)

develop a water based photoresist that could be used in photolithography. Poly(n-(3′-thienyl)alkanesulfonates containing propanesulfonate, hexanesulfonate and octanesulfonate substituents were synthesized in order to study the effect of chain length on the rate of dissolution of thin films and photolithographic contrast. One very interesting

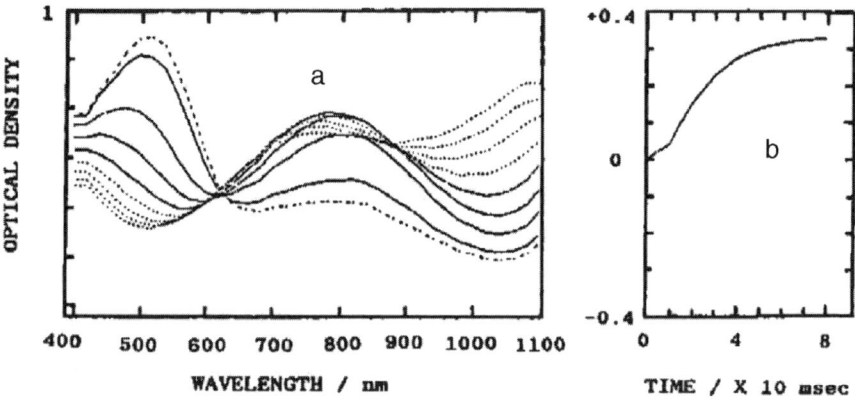

Figure 4.19 (a) Time dependent optical changes in UV-Vis-near infrared spectra of poly(3-(3′-thienyl)propanesulfonic acid) film in 0.5 M HBF$_4$/acetonitrile. Spectra were taken every 10 ms. (b) Normalized time dependent change in optical density at 500 nm (λ_{max}) on poly(3-(3′-thienyl)propanesulfonic acid) film. (Reprinted from *Synthetic Metals*, 40, Y. Ikenoue, H. Tomozawa, Y. Saida, M. Kira, H. Yashima, 333. Copyright (1991), with permission from Elsevier.)

observation was that the sodium salt and acid forms of poly(n-(3′-thienyl)alkanesulfonate) films gave a featureless X-ray diffraction spectra, implying a completely amorphous polymer, whereas irregular poly(3-hexylthiophene) shows some crystallinity. The featureless diffraction spectra of the acid forms of poly(n-(3′-thienyl)alkanesulfonates) suggest that self-doping does not lead to a significant improvement in self-organization. The coupling of 3-substituted thiophenes via the 2- and 5-positions leads to polymers with varying degrees of regioregularity, due to formation of head-to-head (HH), head-to-tail (HT) and tail-to-tail (TT) configurational isomers as shown in Figure 4.20. Holdcroft *et al.* [38] postulated that head-to-head (HH) linkage prevents individual chains from achieving coplanar conformation and chain packing, leading to amorphous character. The ratio of head-to-tail (HT) dyads to head-to-head (HH) dyads was found to be 4:1.

The conductivities of the self-doped acid forms of poly(n-(3′-thienyl)-alkanesulfonates) were in the range of 5×10^{-2}–10^{-1} S/cm. The sodium salt and acid forms of poly(n-(3-thienyl)alkanesulfonates are reportedly completely water soluble. The absorption spectra of aqueous solution of sodium salt of poly(n-(3′-thienyl)alkanesulfonates) containing propanesulfonate, hexanesulfonate and octanesulfonate substituents show $\pi-\pi^*$ transitions at 436, 446, and 466 nm, respectively. It was suggested that the red shift in absorption spectra with increasing side-chain length is

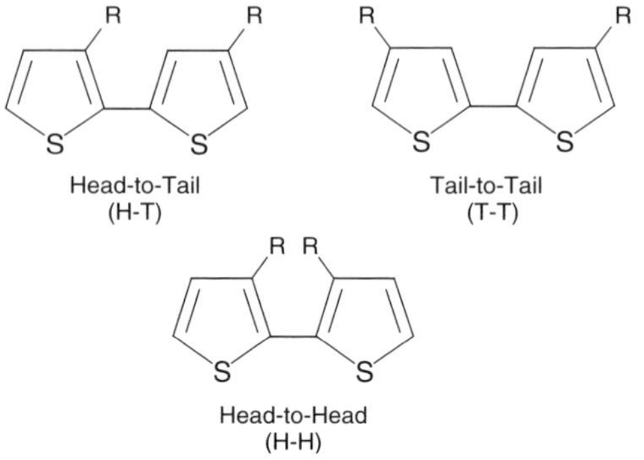

Figure 4.20 Regioisomers of poly(n-(3'-thienyl)alkanesulfonate) dyads. (Reprinted with permission from *Macromolecules*, 28, 975. Copyright (1995) American Chemical Society.)

due to the fact that longer alkanesulfonate substituents assume a more rod-like conformation in solution. However, the absorption spectra of thin films of sodium salt of poly(n-(3'-thienyl)alkanesulfonates) exhibited a blue shift with increasing side chain length and also blue shifted in the solid state compared with nonsulfonated analogs. A model for aggregation in poly(n-(3'-thienyl)alkanesulfonate), wherein the polymer in solution is depicted as a random coil, is shown in Figure 4.21. Upon removal of solvent the polymers are prevented from self-organizing by ion aggregation. In absorption spectra of thin films of the acid form of poly(n-(3'-thienyl)alkanesulfonates) containing propanesulfonate, hexanesulfonate and octanesulfonate, $\pi-\pi^*$ transitions are observed at 452, 438 and 460 nm, respectively, and bipolaron band transitions are observed at around 800 nm, consistent with self-doping in the polymer.

The chemical and physical changes that result from the interaction with UV-Vis radiation were studied for potential applications of water-soluble π-conjugated polymers in photolithography and electroluminescence. Solutions of poly(n-(3'-thienyl)alkanesulfonate) reportedly undergo a small degree of photobleaching upon exposure to UV or visible light for long periods of time. FT-IR spectra of photolyzed polymers show characteristic peaks of hydroxyl groups, keto/or aldehydic groups and the formation of sulfine residues. These observations are consistent with a proposed photochemical mechanism of degradation for nonsulfonated analogs [39]. It is suggested that photobleaching occurs by singlet oxygen sensitization and subsequent 1,4 Diels–Alder addition to thienyl units, wherein intermediate endoperoxide residues are unstable

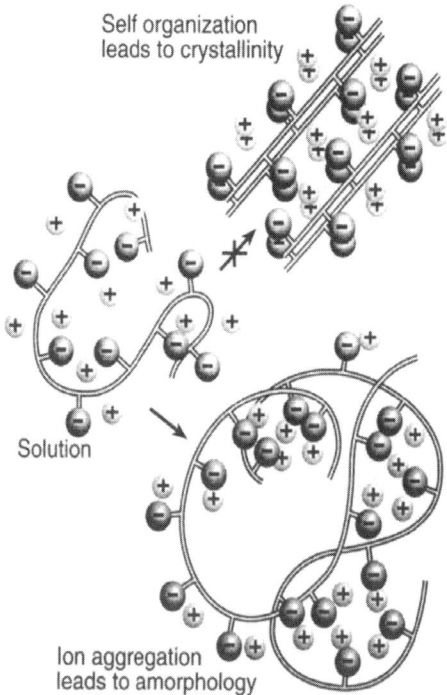

Figure 4.21 Diagram illustrating how ion aggregation prevents self-organization and crystallization of poly(n-(3'-thienyl)alkanesulfonate). (Reprinted with permission from *Macromolecules*, 28, 975. Copyright (1995) American Chemical Society.)

and rearrange with retention of sulfur to produce a sulfine, or elimination of sulfur to yield a *trans*-diketone. The blue shift in λ_{max} observed upon prolonged photolysis is consistent with endoperoxide formation and shortening of the π-conjugation length.

The rates of photobleaching of aqueous solutions of the sodium salt of poly(n-(3'-thienyl)alkanesulfonate) are found to be more photostable than their nonaqueous analogs, due to the lifetime of photosensitized singlet oxygen being much shorter in aqueous solution. Photobleaching of polymer films exposed to ambient atmosphere was significantly slower compared with polymer solutions, and much slower than films of their nonsulfonated analogs, as shown in Figure 4.22. Holdcroft *et al*. [38] speculated that the sodium salts of poly(n-(3'-thienyl)alkanesulfonate) might take up sufficient moisture from the atmosphere, due to their hygroscopic nature, to quench photosensitized singlet oxygen and lead to lower rates of photobleaching. Anhydrous films of sodium salt of poly(n-(3'-thienyl)hexane sulfonates) photobleached much faster,

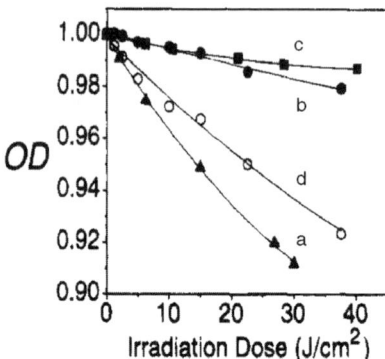

Figure 4.22 Change in optical density (at λ_{max}) of polymer films during photolysis in ambient air (unless stated): (a) poly(3-hexylthiophene)(P3HT); (b) sodium salt of poly(n-(3-thienyl)hexane sulfonate)s, P3THSNa; (c) acid form of poly(n-(3'-thienyl)hexanesulfonates), P3THSH; (d) anhydrous sodium salt of poly(n-(3-thienyl)hexanesulfonate)s, anhydrous P3THSNa in dry air. Incident intensity = 2.5 mW/cm^2 at 313 nm. Initial optical density = 1.0 at 313 nm. (Reprinted with permission from *Macromolecules*, **28**, 975. Copyright (1995) American Chemical Society.)

confirming that residual moisture is responsible for increasing photostability (Figure 4.22, d). The acid form of poly(n-(3'-thienyl)alkanesulfonate) films exposed to ambient atmosphere showed even higher photostability towards disruption of the π-conjugated backbone than did their sodium salt analogs (Figure 4.22, c). This is reportedly due to a combination of moisture uptake and the enhanced rigidity of the polymer backbone, which leads to efficient internal quenching of excitation via lattice relaxation. It was suggested that the presence of polaronic and bipolaronic charges on the polymer backbone originating from self-doping may serve as quenching centers for deactivation of excitation. Photoimaging experiments found that water-soluble poly(n-(3'-thienyl)alkanesulfonates) were able to form both negative and positive images using irradiation, depending on the media conditions of the irradiation, the oxidation state of the polymer and the length of alkanesulfonate chain.

Leclerc et al. [40] chemically synthesized the water-soluble sodium salts of poly(2-(3'-thienyloxy)ethanesulfonate) and poly(2-(4-methyl-3'-thienyloxy)ethanesulfonate) and their self-acid-doped forms (Figure 4.23). The monomer, sodium 2-(4-methyl-3'-thienyloxy)ethanesulfonate was synthesized in three steps from 3-bromo-4-methylthiophene (Figure 4.24) [41]. Similarly, sodium 2-(3'-thienyloxy)ethanesulfonate, was synthesized from 3-methyloxythiophene. The synthesis of the sodium

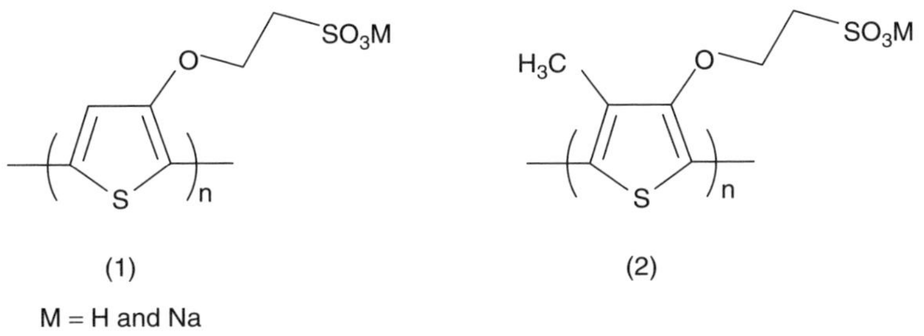

SULFONIC ACID DERIVATIVES

(1) M = H and Na

(2)

Figure 4.23 Structure of sodium salt and acid form of (1) poly(2-(3′-thienyloxy)ethanesulfonate), and (2) poly(2-(4-methyl-3′-thienyloxy)ethanesulfonate). (Reprinted with permission from *Chemistry of Materials*, 9, 2902. Copyright (1997) American Chemical Society.)

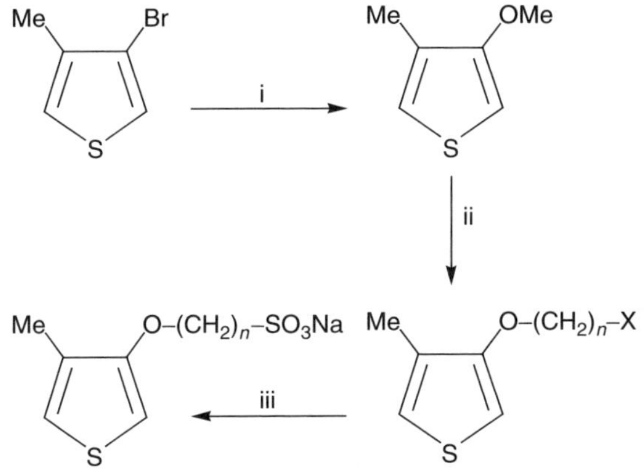

Figure 4.24 Synthesis of sodium 2-(4-methyl-3′-thienyloxy)ethanesulfonate. Reagents and conditions: i, CuBr, NaOMe, NMP, 90 %; ii, NaHSO$_4$, HO(CH$_2$)$_n$X (X = Br, Cl, OH, SH, NH$_2$, etc., n = 2, 3, 6), PhMe, 65 %; iii, Na$_2$SO$_2$, H$_2$O, Me$_2$CO, 60 %; (*Chemical Communications* 1996, 2761, K. Faid, M. Leclerc. Reproduced by permission of The Royal Society of Chemistry.)

salt of the polymer was carried out in chloroform using dry ferric chloride as an oxidizing agent. The aqueous solutions of the sodium salts of the polymers were passed through a cation (H$^+$) resin column to obtain the self-acid-doped forms of the polymer. The UV-Vis absorption spectra of aqueous solutions of the sodium salt forms of the polymers showed an absorption maximum near 550 nm, suggesting the neutral forms of polymer. After passing these solutions through a cation (H$^+$) resin column, i.e. protonation, a new absorption appeared near 800 nm, characteristic of the oxidized conducting

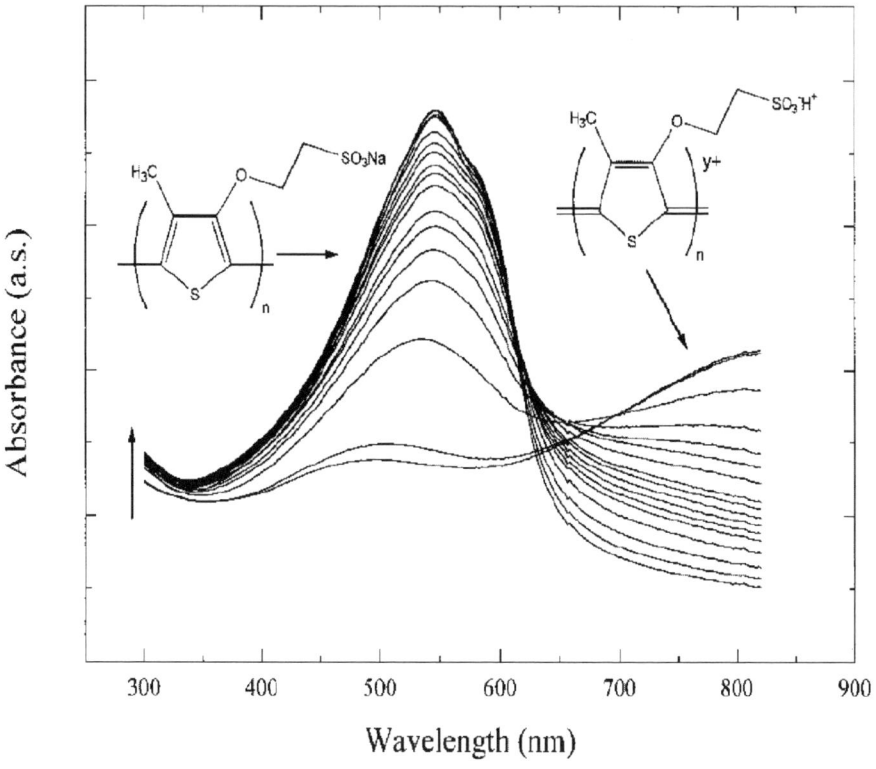

Figure 4.25 UV-Vis absorption spectrum of an aqueous solution of poly(2-(4-methyl-3'-thienyloxy)ethanesulfonic acid) upon addition of NaOH, at room temperature. (Reprinted with permission from *Chemistry of Materials*, 9, 2902. Copyright (1997) American Chemical Society.)

state of polythiophene. This process is reported to be reversible; upon deprotonation with sodium hydroxide, insulating, dark polymers are obtained (Figure 4.25). The number average molecular weight and polydispersity of both the polymers measured using size exclusion chromatography were 6000–8000 and 1.2, respectively. The conductivities of self-acid-doped poly(2-(3'-thienyloxy)ethanesulfonic acid) and poly(2-(4-methyl-3'-thienyloxy)ethanesulfonic acid) were found to be 0.5 and 5 S/cm, respectively. The conductivity of the neutral, insulating polymer was reported to be less than 10^{-6} S/cm. The oxidation potentials of the sodium salts of poly(2-(3'-thienyloxy)ethanesulfonate) and poly(2-(4-methyl-3'-thienyloxy)ethanesulfonate) were found to be lower (0.5 and 0.44 V vs Ag/AgCl) compared with poly(ω-(3'-thienyl)alkanesulfonates) (0.8 V vs Ag/AgCl).

SULFONIC ACID DERIVATIVES

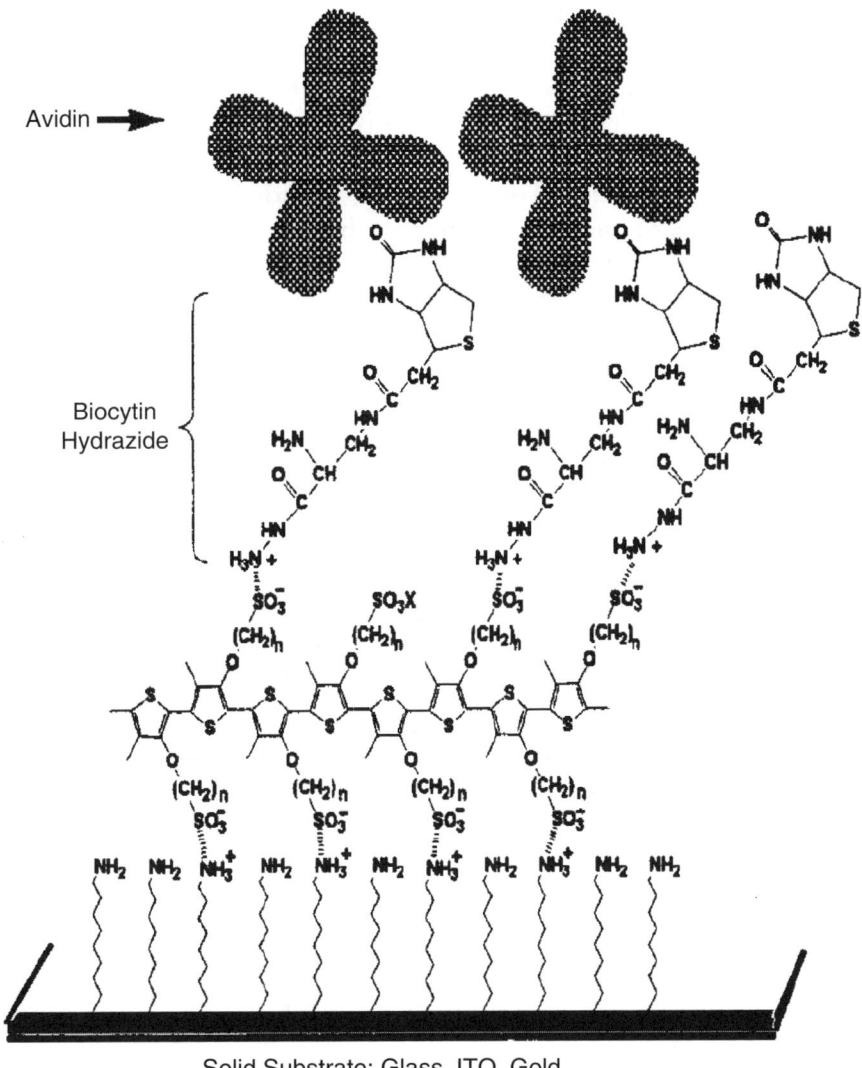

Solid Substrate: Glass, ITO, Gold

Figure 4.26 Schematic description of the modified electrode. (Reprinted with permission from *Journal of the American Chemical Society*, **120**, 5274. Copyright (1998) American Chemical Society.)

In a subsequent study Leclerc *et al.* [42] prepared water-soluble biochromic poly-(2-(4-methyl-3′-thienyloxy)ethanesulfonic acid) based on biotin–avidin interactions as shown in Figure 4.26. They suggest that the self-acid-doped poly-(2-(4-methyl-3′-thienyloxy)ethanesulfonic acid) can react with various basic molecules, such as n-butylamine and biocytin hydrazine, to yield tunable, functionalized (through electrostatic interactions) and chromic polymers both in solution and in the solid state. Upon complexation of poly-(2-(4-methyl-3'-thienyloxy)ethanesulfonic acid) with amine derivatives, the absorption peak at 800 nm disappears and a new peak appears at around 540 nm, due to neutralization of the acid form and subsequent formation of a neutral amine bearing polymeric material. Using biochromic poly-(2-(4-methyl-3'-thienyloxy)ethanesulfonic acid) (Figure 4.26), the optical detection of the protein avidin in water was achieved down to 10^{-11} mol as shown in Figure 4.27 (inset). After addition of avidin to form a 1:1 complex of biocytin hydrazide and poly(2-(4-methyl-3'-thienyloxy)ethanesulfonic acid), a decrease in the absorption band at 400 nm and an increase in the absorption band at 500–550 nm was observed (Figure 4.27). It was suggested that the addition of avidin probably induces rigidity of the

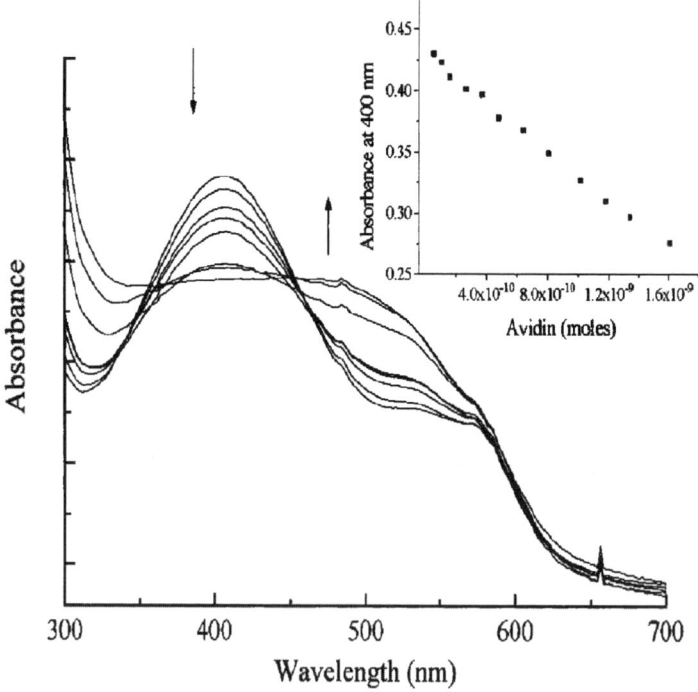

Figure 4.27 UV-Vis absorption spectra of the 1:1 complex between biocytin hydrazide and poly(2-(4-methyl-3'-thienyloxy)ethanesulfonic acid) ($\sim 5 \times 10^{-7}$ mol) as a function of the addition of avidin, in water/methanol (98:2, v/v). (Reprinted with permission from *Journal of the American Chemical Society*, **120**, 5274. Copyright (1998) American Chemical Society.)

polymeric chains that could be related to the fact that four binding sites are present on each avidin molecule, each capable of interacting with a nearby biotin, which could possibly result in cooperative bindings. The interactions between protein avidin and the biotinylated poly(2-(4-methyl-3′-thienyloxy)ethanesulfonic acid) were also detected electrochemically. Upon complexation, a 200 mV increase in the oxidation potential of the polymer was observed. The shift in potential was reportedly due to transition of planar-to-nonplanar conformations of the polymer backbone induced by incorporation of the very large avidin molecule. The self-doped sulfonated poly(alkoxythiophene) derivatives were also used for preparation of multilayer films using layer-by-layer deposition in aqueous media [43, 44].

Reynolds *et al.* [45] have prepared a water-soluble, self-doped poly(4-(2,3-dihydrothieno[3,4-*b*]-[1,4]dioxin-2-yl-methoxy)-1-butanesulfonic acid, sodium salt) (PEDOT-S) (Figure 4.28) by chemical synthesis using ferric chloride in chloroform. The monomer, sulfonated 3,4-ethylenedioxythiophene, was synthesized by following Chevrot *et al.*'s synthesis method as shown in Figure 4.13 [30]. The electrochromic and hole-transporting multilayer films of PEDOT-S and poly(allylamine

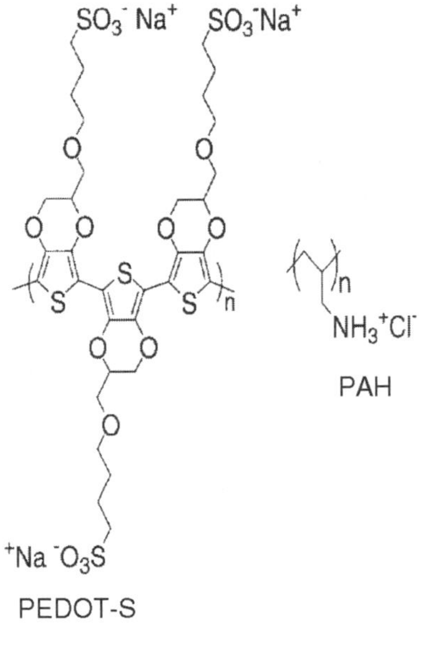

Figure 4.28 Structure of poly(4-(2,3-dihydrothieno[3,4-*b*]-[1,4]dioxin-2-yl-methoxy)-1-butanesulfonic acid, sodium salt) (PEDOT-S) and poly(allylamine hydrochloride)(PAH). (Reproduced from *Advanced Materials* 2002, 14, 684, C. A. Cutler, M. Bouguettaya, T. S. Kang, J. R. Reynolds, with permission from Wiley-VCH.)

hydrochloride)(PAH) (Figure 4.28) were prepared based on electrostatic interactions [45, 46]. The PAH-PEDOT-S multilayer films are redox active and show a reversible redox behavior, despite the insulating nature of the interstitial layer. Reynolds *et al.* suggest that the multilayer film has an interpenetrating polymer network as opposed to discrete bilayers within the film. The scan rate dependence of PAH-PEDOT-S multilayer films with 10–50 bilayers exhibited a rapid redox switching as shown in Figure 4.29. A linear increase in the current with scan rate up to 50 bilayers was observed, indicating that the redox process is not diffusion limited for scan rates as high as 1 V/s. The fast redox switching is attributed to the microporous open film morphology of the bilayer network derived from the use of high ionic strength salt/polyelectrolyte solution and low pH. In the case of self-doped PEDOT-S, oxidation and reduction of the polymer backbone is coupled with cation expulsion from (during oxidation) and insertion into (during reduction) the polymer film. The rapid ion switching at high scan rate is attributed to an open morphology of the bilayer network.

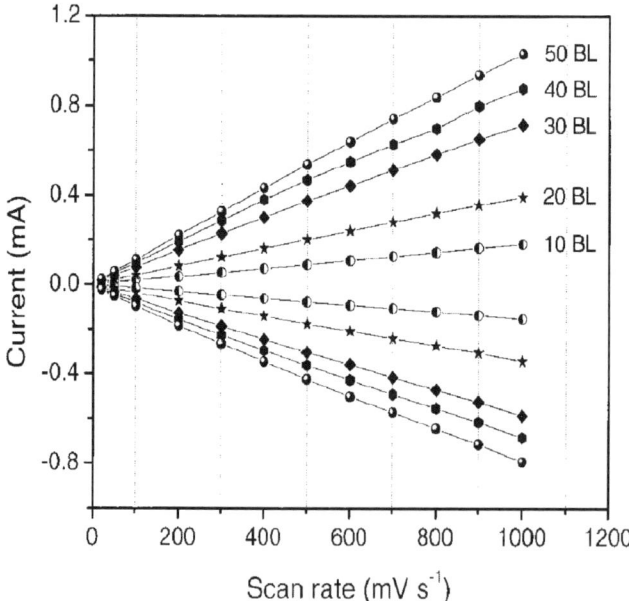

Figure 4.29 Scan rate dependence of PAH-PEDOT-S multilayer films of 10–50 bilayers. Cyclic voltammograms were scanned between −0.7 V and 0.6 V (vs Ag/AgCl) at scan rates ranging from 20–1000 mV/s in $NaSO_4$ (0.2 M/H_2O). (Reproduced from *Advanced Materials*, 2002, 14, 684, C. A. Cutler, M. Bouguettaya, T. S. Kang, J. R. Reynolds, with permission from Wiley-VCH.)

CARBOXYLIC ACID DERIVATIVES

Figure 4.30 Chemical structure of thieno[3,4-*b*]thiophene. (Reproduced from *Advanced Materials*, 2002, **17**, 1792, B. S. Lee, V. Seshadri, H. Palko, G. A. Sotzing, with permission from Wiley-VCH.)

4.1.3 Post Polymerization Modification

Recently, the ring-sulfonated, water-soluble, low band gap poly(thienothiophene) was synthesized by Sotzing *et al.* [47]. They first synthesized neutral poly(thieno[3,4-*b*]thiophene) by oxidation of thieno[3,4-*b*]thiophene (Figure 4.30) in ferric chloride/chloroform followed by reduction of oxidized poly(thieno[3,4-*b*]thiophene) with hydrazine hydrate. Electrophilic substitution, i.e., sulfonation, of poly(thieno[3,4-*b*]thiophene) was carried out with fuming sulfuric acid. By controlling sulfonation level, 56 and 65 % ring sulfonated poly(thieno[3,4-*b*]thiophene) were prepared, the conductivities reported to be 5×10^{-3} and 6.5×10^{-4} S/cm, respectively. The conductivity of 56 % ring sulfonated poly(thieno[3,4-*b*]thiophene) was increased to 0.02 S/cm upon exposure to iodine vapors. The molecular weights of 56 and 65 % ring sulfonated poly(thieno[3,4-*b*]thiophene) were 360 000 and 260 000 g/mol and polydispersities of 1.7 and 1.3, respectively. The molecular weights were measured by gel permeation chromatography using dimethyl formamide as the solvent. These polymers are reported to be soluble in water up to 25 mg/mL. Single UV-Vis absorption peaks are observed at 764 nm and 620 nm for 56 and 65 % ring sulfonated poly(thieno[3,4-*b*]thiophene)s, suggesting that polymers are partially oxidized.

4.2 CARBOXYLIC ACID DERIVATIVES

The synthesis of carboxyl substituted polythiophenes has been investigated to a lesser degree in comparison with sulfonated polythiophenes. The self-doping in carboxyl substituted polythiophenes was first reported by Bauerle *et al.* [48]. They prepared a series of carboxylic acid substituted poly(3-alkylthienylenes). The monomers, 3-(ω-carboxyalkyl)thiophenes, were prepared by reaction

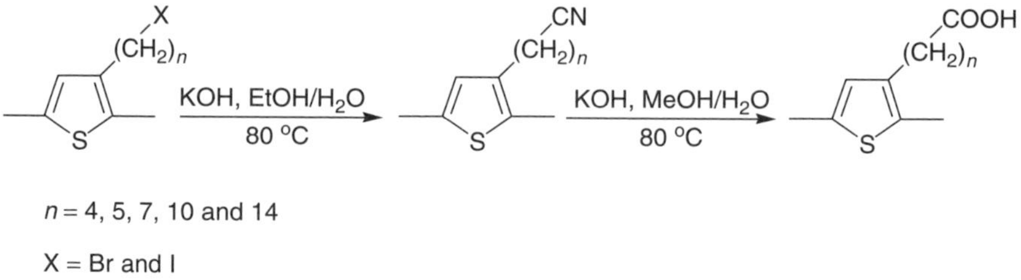

n = 4, 5, 7, 10 and 14

X = Br and I

Figure 4.31 Synthesis of monomer 3-(ω-carboxyalkyl)thiophene. (Reproduced from *Advanced Materials*, 1990, 2, 490. P. G. Bauerle, K. U. Gaudl, F. Wuerthner, N. S. Sariciftci, H. Neugebauer, M. Mehring, C. Zhong, K. Doblhofer, with permission of Wiley-VCH.)

of ω-haloalkylthiophenes with potassium cyanide to obtain 3-(ω-cyanoalkyl)thiophenes. Further saponification resulted in the desired monomers as shown in Figure 4.31. These monomers can be electrochemically polymerized in acetonitrile to poly(3-(ω-carboxyalkyl)thiophene)s. The oxidation potentials of the monomers ($\sim$1.43 to 1.47 V vs Ag/Ag$^+$) are similar to nonfunctionalized 3-alkylthiophenes, indicating that they are unaffected by either the presence of the carboxylic acid group or the length of the alkyl chains. In addition, the redox potentials of poly(3-(ω-carboxyalkyl)thiophene) films (0.44 to 0.48 V vs Ag/Ag$^+$) are similar to nonfunctionalized 3-alkylthiophenes. However, the conductivity of these polymer films are approximately 10^{-3} S/cm, lower than nonfunctionalized 3-alkylthiophenes. The absorption spectra of the neutral polymer films show a $\pi-\pi^*$ transition between 450 and 470 nm. An additional bipolaron band is observed in the range 785 to 804 nm for the oxidized polymer films. In a spectroelectrochemistry study of (8-(3′-thienyl)octanoic acid) films in acetonitrile/tetrabutyl ammonium perchlorate, a bipolaron band is observed only at higher electrode potentials (0.2–0.5 V vs Ag/Ag$^+$) for oxidized polymer. This band is absent in neutral polymer at low electrode potentials, indicating that the polymer is not self-doped in nonaqueous media. X-ray fluorescence measurements and elemental analysis of polymer films electrochemically oxidized in aqueous KClO$_4$/HClO$_4$ solutions, at pH values below the pK_a of the carboxylic acid, suggest the absence of perchlorate anions in the films. These results confirm the self-doping in poly(8-(3′-thienyl)octanoic acid) films, i.e., the oxidized polymer backbone is charge compensated by the covalently bound carboxylate anions. Based on these results, Bauerle *et al.* [48] proposed

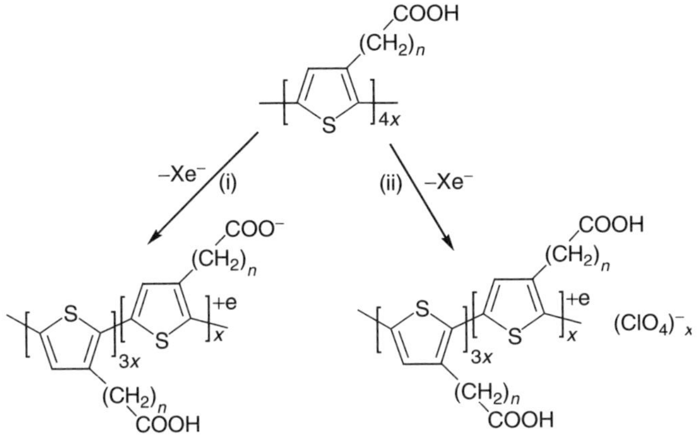

CARBOXYLIC ACID DERIVATIVES

$n = 4, 5, 7, 10$ and 14

Figure 4.32 Doping mechanism of poly(3-(ω-carboxyalkyl)thiophene) in aqueous (i) and nonaqueous (ii) electrolytes. (i) H$_2$O/KClO$_4$/HClO$_4$/pH 4 and (ii) CH$_3$CN/tetrabutyl ammonium perchlorate. (Reproduced from *Advanced Materials*, 1990, 2, 490, P. Bauerle, K. U. Gaudl, F. Wurthner, N. S. Sariciftci, H. Neugebauer, M. Mehring, C. Zhong, K. Doblhofer, with permission from Wiley-VCH.)

a doping mechanism in aqueous and nonaqueous media as shown in Figure 4.32.

Albery *et al.* [39, 49] prepared poly(3-thiopheneacetic acid) and its copolymer with thiophene by electrochemical polymerization. Bartlett *et al.* [50] electrochemically synthesized conducting poly(3-thiopheneacetic acid) films in dry acetonitrile containing tetraethyl ammonium tetrafluoroborate. These films are redox active in acetonitrile, however, stability was reportedly poor in comparison with poly(3-methylthiophene) and poly(methyl 3-thiopheneacetate) due to traces of water. In dry acetonitrile, the polymer can be electrochemically oxidized and reduced. Upon oxidation in water and methanol, poly(3-thiopheneacetic acid) film converted into a passive film. Based on the electrochemistry and an FT-IR study, Bartlett *et al.* postulate the mechanism for the electrochemical passivation shown in the Figure 4.33. In the mechanism, passivation of the polymer involves the formation of an intermediate cyclic lactone and subsequent breakdown by reaction with solvent. This process does not destroy the conductivity of the polymer so the process can continue until all the monomer units within the film are converted to a lactone form (Figure 4.33, IV). The electrochemical passivation is not observed

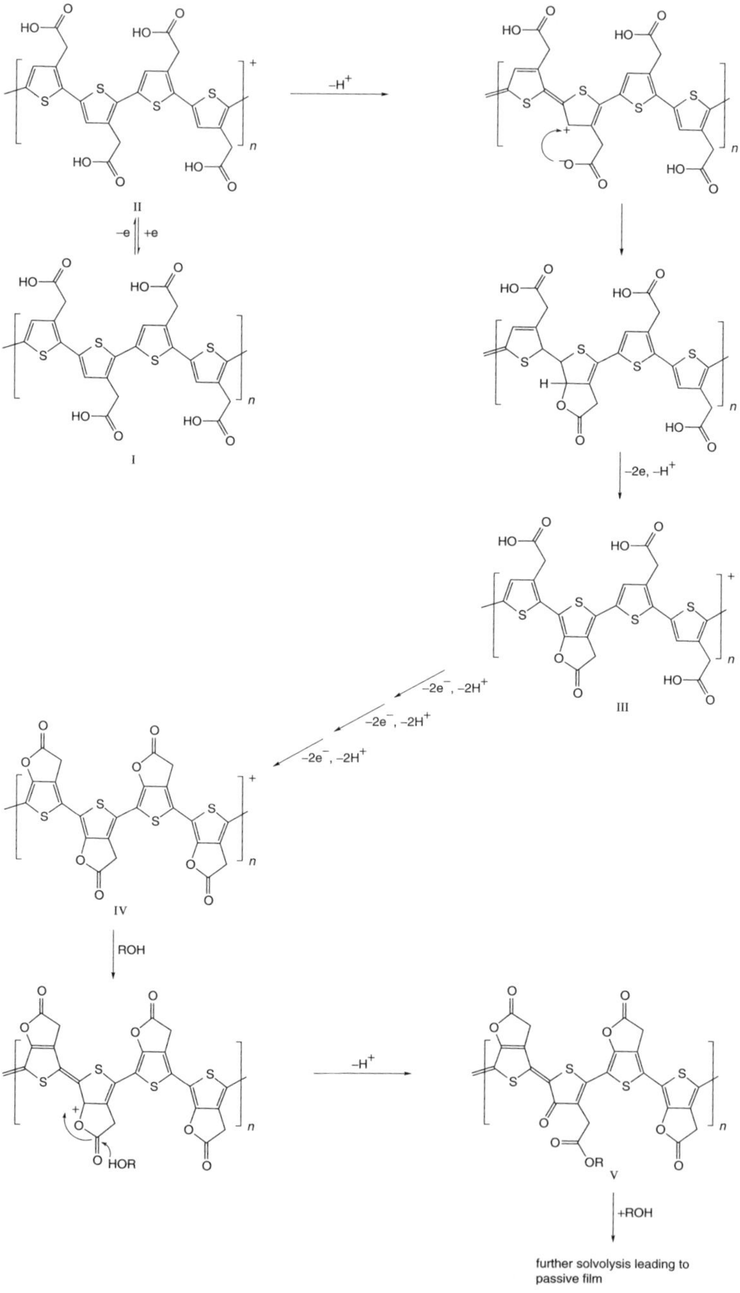

Figure 4.33 Proposed mechanism of the electrochemical passivation of poly(3-thiopheneacetic acid) in water and methanol. (*Journal of Material Chemistry*, 1994, **4**, 1805, P. N. Bartlett, D. H. Dawson. Reproduced by permission of The Royal Society of Chemistry.)

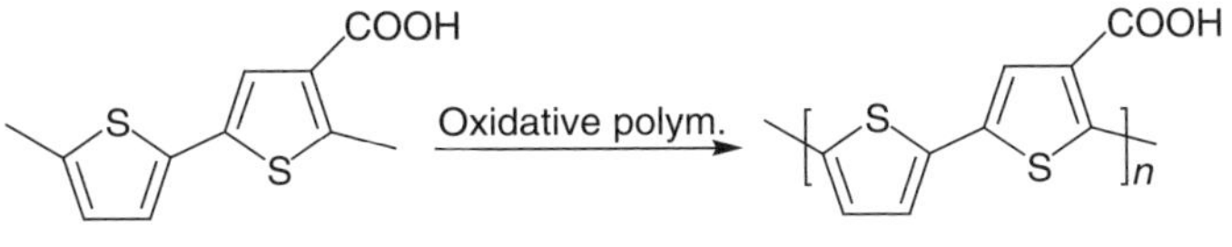

Figure 4.34 Synthesis of poly(4-carboxy-2,2′-bithiophene). (Reprinted with permission from *Macromolecules*, 31, 933. Copyright (1998) American Chemical Society.)

for the electrochemically polymerized films of the corresponding methyl ester or methyl 3-thiopheneacetate.

The poly(4-carboxy-2,2′-bithiophene) can be prepared by direct chemical or electrochemical polymerization of 4-carboxy-2,2′-bithiophene (Figure 4.34) [51]. Hutchison et al. have suggested that this approach reduces undesirable side reactions and allows facile incorporation of strong electron-withdrawing side chains. Chemical synthesis was carried out using anhydrous ferric chloride in chloroform. The electrochemical polymerization was carried out in propylene-carbonate-containing tetrabutyl ammonium hexafluorophosphate. The conductivity of a pressed pellet of chemically synthesized polymer was approximately 10^{-4} S/cm and the polymer was water soluble in its neutral form.

The synthesis of water soluble, regioregular, head-to-tail 2,5-poly(thiophene-3-propionic acid) has been reported by McCullough et al. [52]. The polymer synthesis is outlined in Figure 4.35. It was reported that the regioregular water soluble polythiophenes with a propionic acid at the 3-position of the thiophene ring can switch from purple to yellow by increasing the amount of NH_4OH in aqueous solution; in addition, a cation size dependent chromaticity due to a sterically induced disruption of the aggregated phase was reported. Osada et al. [53] prepared water soluble poly(3-thiophene acetic acid) (Figure 4.36) and its copolymers with 3-n-methylthiophene or with 3-n-octadecylthiophene. The solution properties of these polymers were studied by potentiometric titration, viscosity measurements and UV-Vis spectroscopy. No information regarding self-doping was reported in either report. Following these reports, Liaw et al. [54] electrochemically synthesized poly(3-thiopheneacetic acid)-crosslinked film using crosslinker 3,3′-bithiophene or hexamethylene diisocyanate-bithiophene (Figure 4.37). The brittle properties of poly(3-thiopheneacetic acid) films were improved via network formation, without loss of conductivity and self-doping properties. The conductivities of poly(3-thiopheneacetic acid) network films were found to be around 10^{-4} S/cm. However, after iodine doping,

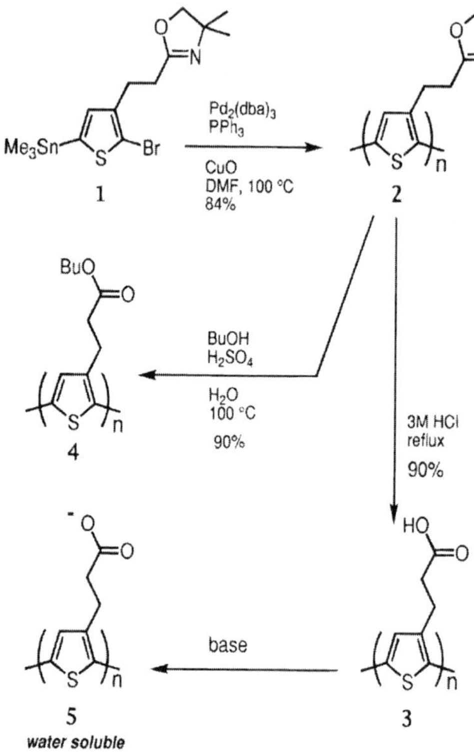

Figure 4.35 Synthesis of regioregular, head-to-tail polythiophenes. (Reprinted with permission from *Journal of the American Chemical Society*, **119**, 633. Copyright (1997) American Chemical Society.)

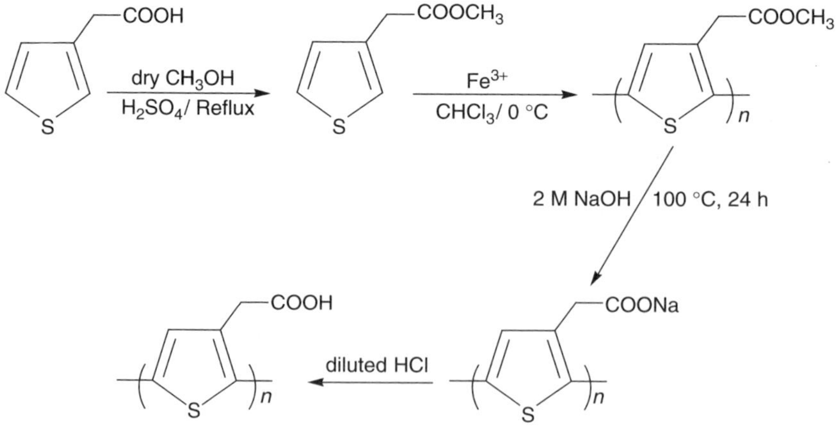

Figure 4.36 Synthesis of poly(3-thiopheneacetic acid). (Reprinted with permission from *Macromolecules*, **32**, 3964. Copyright (1999) American Chemical Society.)

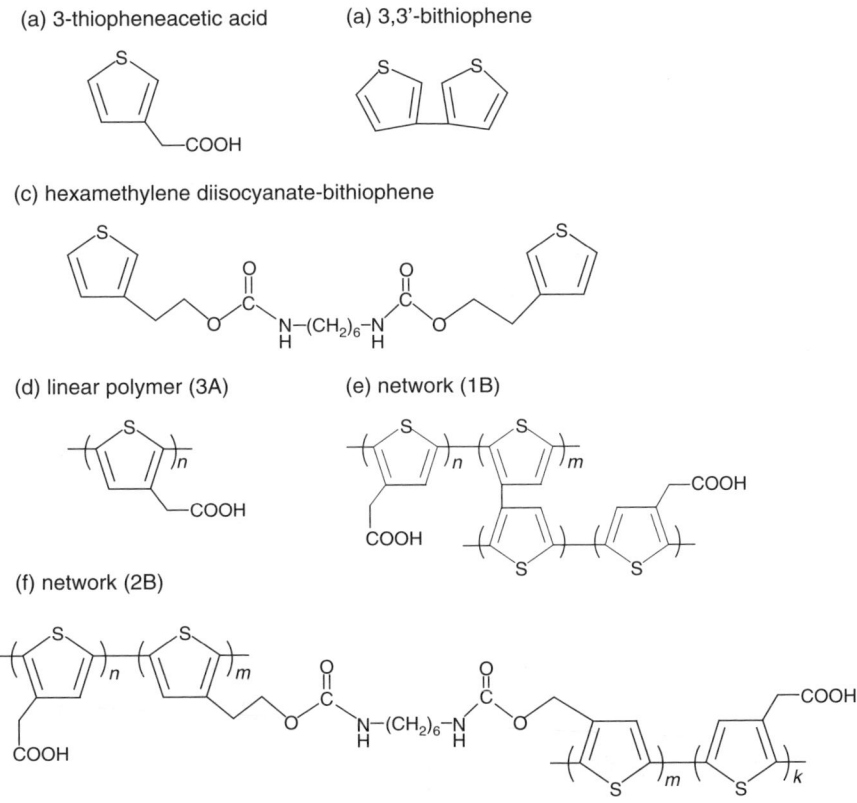

Figure 4.37 Synthesis of poly(3-thiopheneacetic acid) network. (Reprinted from *Synthetic Materials*, 99, D. J. Liaw, B. Y. Liaw, J. P. Gong, Y. Osada, 53. Copyright (1999), with permission from Elsevier.)

conductivities dropped almost one order of magnitude, reportedly due to overoxidation and degradation of poly(3-thiopheneacetic acid) by iodine [50].

4.3 PHOSPHONIC ACID DERIVATIVES

Synthesis of a water-soluble, phosphonic acid derivatized, polythiophene, poly(3-(3'-thienyloxy)propanephosphonate) has been reported by Viinikanoja *et al.* [55]. The polymerizations of both 3-(3'-thienyloxy) propanephosphonic acid and its sodium salt were unsuccessful; however an ethyl ester form was electropolymerized in 0.1 M lithium perchlorate/acetonitrile:dichloromethane (1:1) at 1.8 V vs sodium SCE.

Further, silyldealkylation of poly(3-(3′-thienyloxy)propanephosphonic acid diethyl ester) in dichloromethane was carried out with addition of bromotrimethylsilane in an inert atmosphere to obtain poly(3-(3′-thienyloxy)propanephosphonate. Matrix assisted laser desorption ionization (MALDI) measurements showed that poly(3-(3′-thienyloxy) propanephosphonate) is an oligomer with an average chain length of 10 units. It was reported that poly(3-(3′-thienyloxy)propanephosphonate) is a self-doped polymer. The pK_a values of 3-(3′-thienyloxy)propanephosphonate are close to 2.5 and 8.2, suggesting that within that pH range the pendent phosphonic acid groups are dissociated and exist mainly in the form of monoanions. The polymer is reportedly soluble in weakly

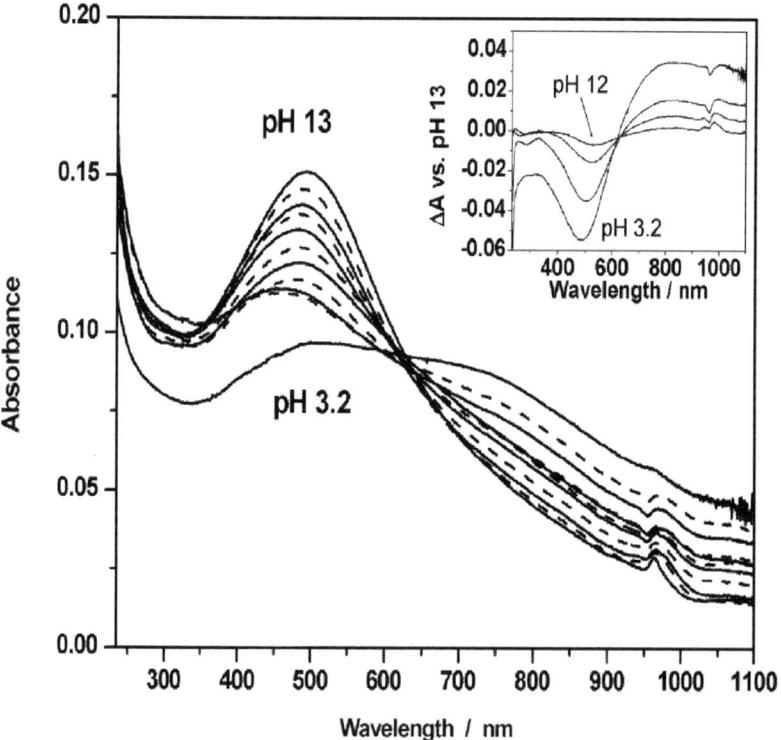

Figure 4.38 Absorption spectra of poly(3-(3′-thienyloxy)propanephosphonate) solutions (40 μM) at different pH: 13.0, 12.1, 11.0, 10.0, 9.0, 8.1, 7.1, 6.1, 5.0, 4.0, and 3.2. Inset shows the difference spectra at selected pH (12.1, 10.0, 6.1 and 3.2) vs pH 13.0. (Reprinted with permission from *Langmuir*, **19**, 2768. Copyright (2003) American Chemical Society.)

acidic, neutral and alkaline aqueous solution, below pH 2.6, the polymer precipitates from the solution.

The UV-Vis absorption spectra of poly(3-(3'-thienyloxy)propanephosphonate) solutions at different pH values are shown in Figure 4.38. Under alkaline conditions, the $\pi-\pi^*$ transition of the dedoped form is observed at 496 nm. The spectra of the polymer under alkaline conditions exhibit a noticeable absorbance at higher wavelengths, which has been attributed to the presence of oxidized polymer and/or aggregation. After addition of acid, a decrease in the $\pi-\pi^*$ transition (496 nm) and an increase in the polaron band (750 nm) is observed due to oxidation (Figure 4.38). These changes are attributed to partial protonation of the polymer. A proposed mechanism for the self-acid-doping of poly(3-(3'-thienyloxy)propanephosphonate) involving a tautomeric proton transfer from phosphonate to the thiophene backbone is shown in Figure 4.39. Multilayer films can be prepared using layer-by-layer deposition of polyelectrolytes, poly(3-(3'-thienyloxy)propanephosphonate)

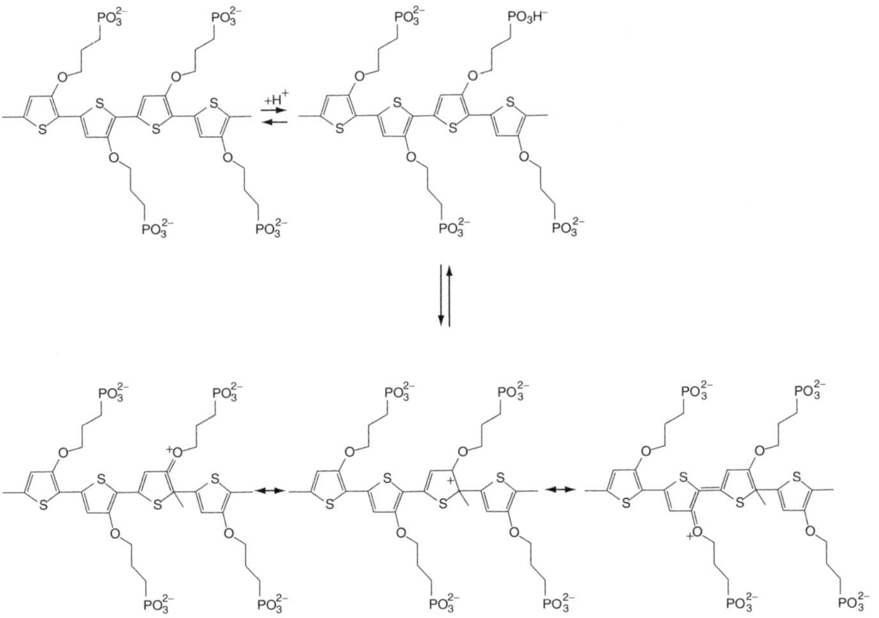

Figure 4.39 A proposed mechanism for the self-acid-doping of poly(3-(3'-thienyloxy)propanephosphonate). (Reprinted with permission from *Langmuir*, **19**, 2768. Copyright (2003) American Chemical Society.)

with poly(diallyldimethyl ammonium chloride) based on electrostatic interactions and metal phosphate linkage with Zr^+ ions. The multilayers showed both electrochromism and pH induced halochromism.

REFERENCES

[1] J. Roncali, 'Conjugated poly(thiophenes) – synthesis, functionalization and applications,' *Chemical Reviews* **1992**, *92*, 711.
[2] T. Yamamoto, 'Electrically conducting and thermally stable p-conjugated polyarylenes prepared by organometallic processes,' *Progress in Polymer Science* **1992**, *17*, 1153.
[3] G. Tourillon, *Handbook of Conducting Polymers*, Dekker, New York, **1986**.
[4] G. I. Gustafsson, O. Inganas, W. R. Salaneck, J. Laakso, M. Loponen, T. Taka, J. E. Osterholm, H. Stubb, T. Hjertberg, *Processable Conducting Poly(3-alkylthiophenes)*. Kluwer, Dordrecht, **1991**.
[5] R. D. McCullough, 'The chemistry of conducting polythiophenes,' *Advanced Materials* **1998**, *10*, 93.
[6] T. S. Yamamoto, K. Sanechika, A. Yamamoto, 'Preparation of thermostable and electric conducting poly(2,5-thienylene)' *Journal of Polymer Science Polymer Letters Ed.* **1980**, *18*, 9.
[7] J. W. D. Lin, L. P. Dudek 'Synthesis and properties of poly(2,5-thienylene)' *Journal of Polymer Science Polymer Letters Ed.* **1980**, *18*, 2869.
[8] K. H. Yoshino, S. Hayashi, R. Sugimoto 'Preparation and properties of conducting heterocyclic polymer films by chemical method,' *Japan Journal of Applied Physics Part 2* **1984**, *23*, 899.
[9] A. Diaz, 'Electrochemical preparation and characterization of conducting polymers,' *Chemica Scripta* **1981**, *17*, 145.
[10] G. Tourillon, F. Garnier, 'New electrochemically generated organic conducting polymers,' *Journal of Electroanalytical Chemistry* **1982**, *135*, 173.
[11] K. Y. Jen, R. Oboodi, R. L. Elsenbaumer, 'Processible and environmentally stable conducting polymers,' *Polymeric Materials Science and Engineering* **1985**, *53*, 79.
[12] K. Y. Jen, G. G. Miller, R. L. Elsenbaumer, 'Highly conducting, soluble and environmentally-stable poly(3-alkylthiophenes),' *Chemical Communications* **1986**, 1346.
[13] R. L. J. Elsenbaumer, K. Y. Jen, R. Oboodi, 'Processible and environmentally stable conducting polymers,' *Synthetic Metals* **1986**, *15*, 169.
[14] R. T. Sugimoto, S. Takeda, H. B. Gu, K. Yoshino, 'Preparation of soluble polythiophene derivatives utilizing transition metal halides as catalysts and their property,' *Chemistry Express* **1986**, *1*, 635.
[15] M. Sato, S. Tanaka, K. Kaeriyama, 'Soluble conducting polythiophenes,' *Chemical Communications* **1986**, 873.
[16] S. Hotta, S. D. D. V. Rughooputh, A. J. Heeger, F. Wudl, 'Spectroscopic studies of soluble poly(3-alkylthienylenes)' *Macromolecules* **1987**, *20*, 212.
[17] A. O. Patil, Y. Ikenoue, F. Wudl, A. J. Heeger, 'Water-soluble conducting polymers,' *Journal of the American Chemical Society* **1987**, *109*, 1858.

[18] A. O. Patil, Y. Ikenoue, N. Basescu, N. Colaneri, J. Chen, F. Wudl, A. J. Heeger, 'Self-doped conducting polymers,' *Synthetic Metals* **1987**, *20*, 151.
[19] Y. Ikenoue, J. Chiang, A. O. Patil, F. Wudl, A. J. Heeger, 'Verification of the cation-popping doping mechanism of self-doped polymers,' *Journal of the American Chemical Society* **1988**, *110*, 2983.
[20] Y. Ikenoue, N. Uotani, A. O. Patil, F. Wudl, A. J. Heeger, 'Electrochemical studies of self-doped conducting polymers – verification of the cation-popping doping mechanism,' *Synthetic Metals* **1989**, *30*, 305.
[21] Y. Ikenoue, Y. Saida, M. Kira, H. Tomozawa, H. Yashima, M. Kobayashi, 'A facile preparation of a self-doped conducting polymer,' *Chemical Communications* **1990**, 1694.
[22] E. E. Havinga, L. W. Vanhorssen, W. Tenhoeve, H. Wynberg, E. W. Meijer, 'Self-doped water soluble conducting polymers,' *Polymer Bulletin* **1987**, *18*, 277.
[23] H. C. Zhao, F. Chen, T. W. Lewis, W. E. Price, G. G. Wallace, 'Studies of electropolymerization of sodium 2-(3'-thienyl)ethyl sulfonate,' *Reactive and Functional Polymers* **1997**, *34*, 27.
[24] S. A. Chen, M. Y. Hua, 'Structure and doping level of the self-acid-doped conjugated conducting polymers – poly[n-(3'-thienyl)alkanesulfonic acid]s,' *Macromolecules* **1993**, *26*, 7108.
[25] M. Y. Y. Hua, S. W. Yang, A. A.Chen, 'Sensitive thermal-undoping characteristics of the self-acid-doped conjugated conducting polymer poly[2-(3'-thienyl)ethanesulfonic acid]' *Chemistry of Materials* **1997**, *9*, 2750.
[26] J. E. P. Osterholm, P. Passiniemi, H. Isotalo, H. Stubb, 'Synthesis and properties of tetrachloroferrate-anion-doped polythiophene,' *Synthetic Metals* **1987**, *18*, 213.
[27] O. G. Inganaes, G. Gustafsson, 'Thermochromism in poly(3-alkylthiophenes) and their polymer blends,' *Synthetic Metals* **1990**, *37*, 195.
[28] J. Yue, A. J. Epstein, Z. Zhong, P. K. Gallagher, A. G. MacDiarmid, 'Thermal stabilities of polyanilines,' *Synthetic Metals* **1991**, *41*, 765.
[29] G. Zotti, S. Zecchin, G. Schiavon, G. Berlin, G. Pagani, A. Canavesi, 'Doping-induced ion exchange in the highly conjugated self-doped polythiophene from anodic coupling of 4-(4H-cyclopentadithien-4-yl)butanesulfonate,' *Chemistry of Materials* **1997**, *9*, 2940.
[30] O. Stephan, P. Schottland, P. Y. Le Gall, C. Chevrot, C. Mariet, M. Carrier, 'Electrochemical behaviour of 3,4-ethylenedioxythiophene functionalized by a sulphonate group. Application to the preparation of poly(3,4-ethylenedioxythiophene) having permanent cation-exchange properties,' *Journal of Electroanalytical Chemistry* **1998**, *443*, 217.
[31] K. Krishnamoorthy, M. Kanungo, A. V. Ambade, A. Q. Contractor, A. Kumar, 'Electrochemically polymerized electroactive poly(3,4-ethylenedioxythiophene) containing covalently bound dopant ions: poly {2-(3-sodiumsulfinopropyl)-2,3-dihydrothieno 3,4-b 1,4 dioxin},' *Synthetic Metals* **2001**, *125*, 441.
[32] G. Zotti, S. Zecchin, G. Schiavon, L. Groenendaal, 'Electrochemical and chemical synthesis and characterization of sulfonated poly(3,4-ethylenedioxythiophene): a novel water soluble and highly conductive conjugated oligomer,' *Macromolecular Chemistry and Physics* **2002**, *203*, 1958.

[33] Y. A. Udum, K. Pekmez, A. Yildiz, 'Electropolymerization of self-doped polythiophene in acetonitrile containing FSO$_3$H,' *Synthetic Metals* **2004**, *142*, 7.
[34] Y. A. Udum, K. Pekmez, A. Yildiz, 'Electrochemical synthesis of soluble sulfonated poly(3-methyl thiophene),' *European Polymer Journal* **2004**, *40*, 1057.
[35] Y. A. Udum, K. Pekmez, A. Yildiz, 'Electrochemical preparation of a soluble conducting aniline-thiophene copolymer,' *European Polymer Journal* **2005**, *41*, 1136.
[36] Y. A. Udum, K. Pekmez, A. Yildiz, 'A new self-doped copolymer consisting of 3-methyl thiophene and aniline units,' *Journal of Solid State Electrochemistry* **2006**, *10*, 110.
[37] Y. Ikenoue, H. Tomozawa, Y. Saida, M. Kira, H. Yashima, 'Evaluation of electrochromic fast-switching behavior of self-doped conducting polymer,' *Synthetic Metals* **1991**, *40*, 333.
[38] M. I. Arroyovillan, G. A. Diazquijada, M. S. A. Abdou, S. Holdcroft, 'Poly(n-(3-thienyl)alkanesulfonates) – synthesis, regioregularity, morphology and photochemistry,' *Macromolecules* **1995**, *28*, 975.
[39] W. J. L. Albery, F. Li, A. R. Mount, 'Electrochemical polymerization of poly(thiophene-3-acetic acid), poly(thiophene-co-thiophene-3-acetic acid) and determination of their molar mass,' *Journal of Electroanalytical Chemistry* **1991**, *310*, 239.
[40] M. Chayer, K. Faid, M. Leclerc, 'Highly conducting water-soluble polythiophene derivatives,' *Chemistry of Materials* **1997**, *9*, 2902.
[41] K. Faid, M. Leclerc, 'Functionalized regioregular polythiophenes: towards the development of biochromic sensors,' *Chemical Communications* **1996**, 2761.
[42] K. Faid, M. Leclerc, 'Responsive supramolecular polythiophene assemblies,' *Journal of the American Chemical Society* **1998**, *120*, 5274.
[43] J. Lukkari, A. Viinikanoja, J. Paukkunen, M. Salomaki, M. Janhonen, T. Aaritalo, J. Kankare, 'Oxidation induced variation in polyelectrolyte multilayers prepared from sulfonated self-dopable poly(alkoxythiophene),' *Chemical Communications* **2000**, 571.
[44] J. Lukkari, M. Salomaki, A. Viinikanoja, T. Aaritalo, J. Paukkunen, N. Kocharova, J. Kankare, 'Polyelectrolyte multilayers prepared from water-soluble poly(alkoxythiophene) derivatives,' *Journal of the American Chemical Society* **2001**, *123*, 6083.
[45] C. A. Cutler, M. Bouguettaya, J. R. Reynolds, 'PEDOT polyelectrolyte based electrochromic films via electrostatic adsorption,' *Advanced Materials* **2002**, *14*, 684.
[46] C. A. Cutler, M. Bouguettaya, T. S. Kang, J. R. Reynolds, 'Alkoxysulfonate-functionalized PEDOT polyelectrolyte multilayer films: electrochromic and hole transport materials,' *Macromolecules* **2005**, *38*, 3068.
[47] B. S. Lee, V. Seshadri, H. Palko, G. A. Sotzing, 'Ring-sulfonated poly(thienothiophene),' *Advanced Materials* **2005**, *17*, 1792.
[48] P. G. Baeuerle, K. U. Gaudl, F. Wuerthner, N. S. Sariciftci, H. Neugebauer, M. Mehring, C. Zhong, K. Doblhofer, 'Thiophenes. 3. Synthesis and properties of carboxy-functionalized poly(3-alkylthienylenes)' *Advanced Materials* **1990**, *2*, 490.
[49] F. B. A. Li, W. J. Albery, 'Electrochemical deposition of a conducting polymer, poly(thiophene-3-acetic acid): the first observation of individual events of

polymer nucleation and two-dimensional layer-by-layer growth,' *Langmuir* **1992**, *8*, 1645.

[50] P. N. D. Bartlett, D. H. Dawson, 'Electrochemistry of poly(3-thiopheneacetic acid) in aqueous solution: evidence for an intramolecular chemical reaction,' *Journal of Material Chemistry* **1994**, *4*, 1805.

[51] S. C. P. Rasmussen, J. Pickens, J. E. Hutchison, 'Highly conjugated, water-soluble polymers *via* the direct oxidative polymerization of monosubstituted bithiophenes,' *Macromolecules* **1998**, *31*, 933.

[52] R. D. McCullough, P. C. Ewbank, R. S. Loewe, 'Self-assembly and disassembly of regioregular, water soluble polythiophenes: chemoselective ionchromatic sensing in water,' *Journal of the American Chemical Society* **1997**, *119*, 633.

[53] B. S. Kim, L. Chen, J. P. Gong, Y. Osada, 'Titration behavior and spectral transitions of water-soluble polythiophene carboxylic acids,' *Macromolecules* **1999**, *32*, 3964.

[54] D. J. Liaw, B. Y. Liaw, J. P. Gong, Y. Osada, 'Synthesis and properties of poly(3-thiopheneacetic acid) and its networks via electropolymerization,' *Synthetic Metals* **1999**, *99*, 53.

[55] A. Viinikanoja, J. Lukkari, T. Aaritalo, T. Laiho, J. Kankare, 'Phosphonic acid derivatized polythiophene: a building block for metal phosphonate and polyelectrolyte multilayers,' *Langmuir* **2003**, *19*, 2768.

5
Miscellaneous Self-Doped Polymers

Similarly to polyaniline and polythiophene, several other self-doped conducting polymers have been explored including polypyrrole, poly(3,6-(carbaz-9-yl) propanesulfonate), poly(p-phenylene), poly(indolecarboxylic acid), polyphenylenevinylene, and poly(2-ethynyl-N-(4-sulfobutyl)pyridinium betaine). The details of synthesis and properties are described in the following sections.

5.1 SELF-DOPED POLYPYRROLE

Polypyrrole was first prepared electrochemically by Dall'Olio *et al.* in 1968 [1]. The formation of conductive and highly stable polypyrrole films on a platinum electrode by the electrochemical oxidation of pyrrole in acetonitrile was first reported in 1979 by Diaz *et al.* [2, 3]. Subsequently, polypyrrole was synthesized chemically using various oxidizing agents such as ferric chloride, $SbCl_5$, H_2O_2, $Cu(ClO_4)_2$ [4–10]. The polymers obtained using these methods are neither thermally processible nor soluble in organic solvents due to chain stiffness, interchain interactions and crosslinking. To overcome the issue of solubility, substituents have been introduced on the nitrogen atom [11–20], or on the 3,4-positions [21–34] of the polymeric chains in order to inhibit the interchain interactions and degree of crosslinking, producing short chain polymers that can solubilize more easily in organic solvents. In addition, it has been possible to increase the solubility by increasing favorable

---

Self-Doped Conducting Polymers M.S. Freund and B.A. Deore
© 2007 John Wiley & Sons, Ltd

substituent–solvent interactions. The substitution of pyrrole at the N-position results in decreased conductivity (10^{-3}–10^{-5} S/cm) [11, 12, 35] relative to unsubstituted polypyrrole (≤ 100 S/cm) [36–40]. In addition, substitution on the nitrogen atom hinders their use in aqueous environments since they require more positive potentials in order to attain their oxidized conducting state. The polypyrroles substituted in the 3,4-positions have been found to be much more conductive (2–10 S/cm) than the corresponding N-substituted polymers.

A new class of water soluble polypyrrole has been prepared by self-doping of the polymer. These self-doped polypyrroles can be prepared electrochemically or chemically, using various dopant anions covalently bound to the polymer backbone. The self-doped sulfonated polypyrrole is most commonly synthesized electrochemically in nonaqueous media. Electrochemical synthesis in aqueous media and chemical synthesis are not typically used, presumably due to issues with overoxidation. The postpolymerization modification of polypyrrole, in a manner similar to that used to form sulfonated polyaniline is rare [41]. The various synthetic approaches and properties of the polymer are discussed in the following sections.

5.1.1 Electrochemical Polymerization

5.1.1.1 Nonaqueous Media

Self-doped polypyrrole was first prepared by Reynolds *et al.* [42] and Havinga *et al.* [43] in 1987. Their approach to self-doping of polymers was based on monomers that were easier to polymerize electrochemically. Reynolds *et al.* prepared the N-substituted pyrrole copolymer, poly(pyrrole-co-(3-(pyrrol-1-yl)propanesulfonate)) (Figure 5.1) in acetonitrile containing tetrabutyl ammonium tetrafluoroborate as a supporting electrolyte on a platinum electrode. The monomer, potassium

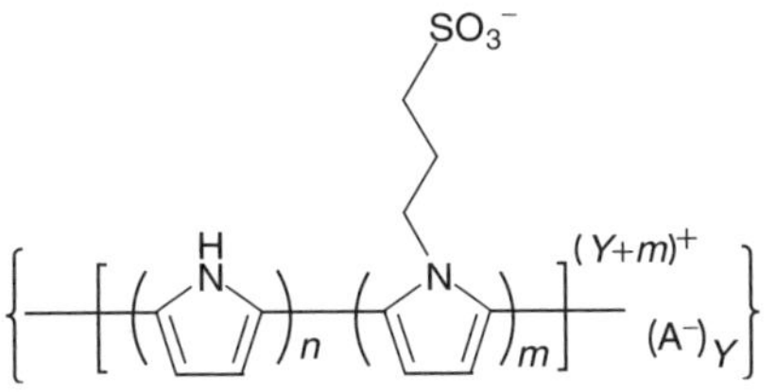

Figure 5.1 Chemical structure of self-doped poly(pyrrole-co-(3-(pyrrol-1-yl)propanesulfonate)). (*Chemical Communications*, 1987, 621, N. S. Sundaresan, S. Basak, M. Pomerantz, J. R. Reynolds. Reproduced by permission of The Royal Society of Chemistry.)

3-(pyrrol-1-yl)propanesulfonate, was prepared by the reaction of pyrrole with potassium and then with 1,3-propanesultone. The oxidation of the monomer in acetonitrile/tetrabutyl ammonium tetrafluoroborate solution was observed at 1.1 V vs Ag/Ag$^+$ as shown in Figure 5.2, a. After oxidation, in subsequent scans, the absence of anodic current suggested the formation of a nonconducting electrode surface or the passivation of the electrode surface. In an equimolar mixture of comonomer solution of potassium 3-(pyrrol-1-yl)propanesulfonate and pyrrole in the same medium, the cooxidation of both the monomers is observed

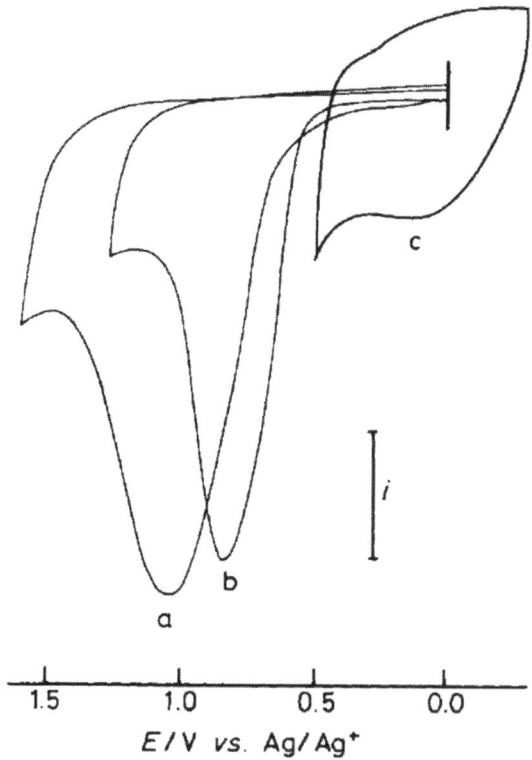

Figure 5.2 Cyclic voltammograms for the redox processes, (a) oxidation of 3-(pyrrol-1-yl)propanesulfonate (ca 7 mM) in MeCN-Bu$_4^n$NBF$_4$ electrolyte at a Pt electrode ($A = 1.8$ mm^2), scan rate 100 mV/s, $i = 10$ μA; (b) oxidation of a 1:1 mixture of pyrrole and 3-(pyrrol-1-yl)propanesulfonate under the same conditions, $i = 20$ μA; (c) redox reaction of poly(pyrrole-co-(3-(pyrrol-1-yl)propanesulfonate)) in MeCN–0.1 M LiClO$_4$ electrolyte, $i = 5$ μA. (*Chemical Communications*, 1987, 621, N. S. Sundaresan, S. Basak, M. Pomerantz, J. R. Reynolds. Reproduced by permission of The Royal Society of Chemistry.)

below 1.0 V (Figure 5.2, b). With subsequent scans, the magnitude of the peak current increased and film was deposited on the substrate. The cyclic voltammogram of copolymer film in acetonitrile/lithium perchlorate solution revealed an anodic peak around 0.2 V and a broad reduction peak in the cathodic region (Figure 5.2, c). The elemental analysis of the copolymer showed a pyrrole:pyrrolesulfonate ratio of 3:1 and a BF^-: pyrrole ratio of 1:16. Based on the low content of BF^- in the copolymer in comparison with unsubstituted polypyrrole doped with tetrafluoroborate (BF^-: pyrrole ratio 1:3), it was concluded that the copolymer is self-doped under these conditions. The charge and mass transport properties of the copolymer, poly(pyrrole-co-(3-(pyrrol-1-yl)propanesulfonate)), in acetonitrile with different supporting electrolytes, were studied in the potential range of −0.2 to 0.5 V in order to demonstrate the effect of ion size on the charge transport rate [42, 44]. The transport rate D/L^2 (D = apparent diffusion coefficient and L = thickness of the solvent swollen film) values of the copolymer film for lithium perchlorate, lithium tetrafluoroborate and tetrabutyl ammonium tetrafluoroborate were found to be 0.05 (±0.01), 0.28 (±0.07) and 0.08 (±0.02) s^{-1}, respectively. The ion mobility of the self-doped copolymer was restricted to cation species as opposed to unsubstituted polypyrrole, which showed more complex behavior. Similarly, N-alkylsulfonate polypyrroles were prepared by Bidan et al. [45, 46].

Havinga et al. electrochemically synthesized 3-alkylsulfonate (propyl, butyl and hexyl) substituted polypyrroles in acetonitrile without adding supporting electrolyte (Figure 5.3) [43, 47]. The monomer itself acted as a supporting electrolyte. The synthetic route for the monomers, the sodium salts of 3-alkylsulfonate pyrrole with various lengths of the

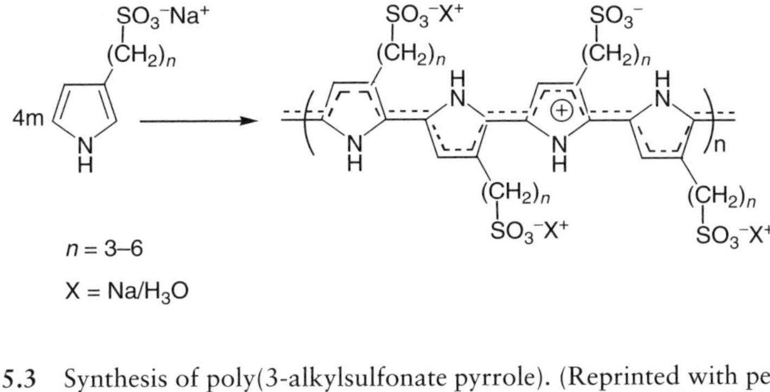

Figure 5.3 Synthesis of poly(3-alkylsulfonate pyrrole). (Reprinted with permission from *Chemistry of Materials*, **1**, 650. Copyright (1989) American Chemical Society.)

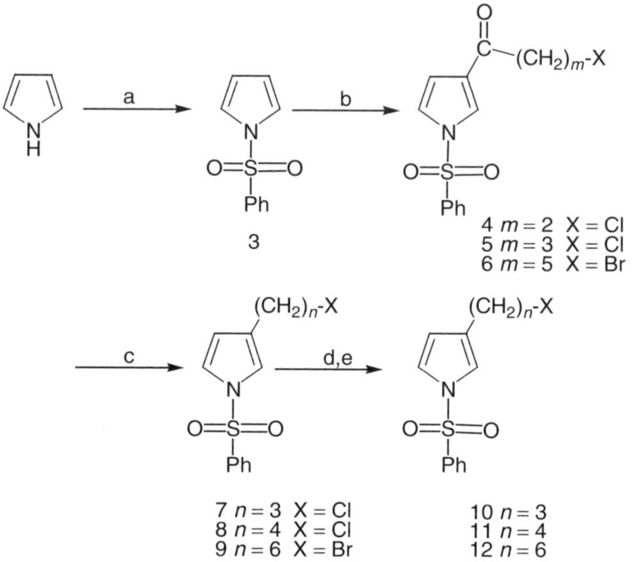

(a) PhSO$_2$Cl, PTC, H$_2$O, CH$_2$Cl$_2$; (b) X(CH$_2$)$_m$COCl, AlCl$_3$, CH$_2$Cl$_2$;
(c) (Hg)Zn, HCl, toluene; (d) H$_2$O, EtOH, Na$_2$SO$_3$, PTC, NaI; (e) NaOH

Figure 5.4 Synthesis of monomer 3-alkylsulfonate pyrrole. (Reprinted with permission from *Chemistry of Materials*, 1, 650. Copyright (1989) American Chemical Society.)

alkyl side chain, is shown in Figure 5.4. The monomers were prepared from N-protected pyrrole by acylation with ω-haloacid chloride under Freidel–Crafts conditions, followed by Clemmensen reduction, and treatment with Na$_2$SO$_3$, followed by N-deprotection with base. The saturated solutions of monomers (0.5 g/L) were polymerized at potentials below 1.7 V vs SCE. During polymerization, the monomer dissolved gradually and the formation of a insoluble bluish black self-doped polymer on the anode was reported. In the self-doped state, the delocalized positive charge on the polymer is balanced by the covalently bound sulfonate ions (Figure 5.3). The polymer prepared above a potential of 1.7 V was reportedly water insoluble due to slight crosslinking. Similarly to unsubstituted polypyrrole, X-ray diffraction analysis indicated the amorphous nature of poly(3-alkylsulfonate pyrrole) films.

The UV-Vis near infrared spectra shown in Figure 5.5 suggest that poly(3-alkylsulfonate pyrrole) films are highly doped. Differences in the low energy transitions associated with the bipolaron band are observed for propyl, butyl and hexyl sulfonate polypyrrole. These spectra are

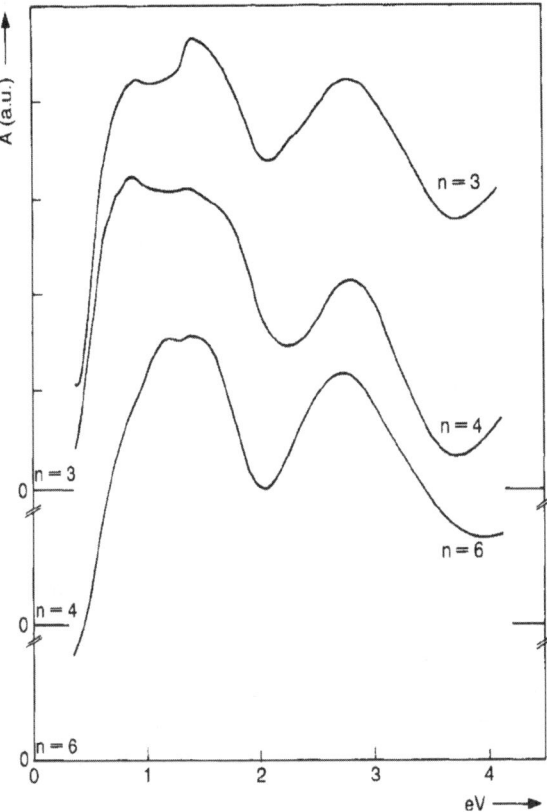

Figure 5.5 Characteristic UV-Vis NIR absorption spectra of self-doped poly(3-alkylsulfonate pyrrole) thin films with propyl ($n = 3$), butyl ($n = 4$), and hexyl ($n = 6$) derivatives, respectively, on ITO glass electrodes. (Reprinted with permission from *Chemistry of Materials*, 1, 650. Copyright (1989) American Chemical Society.)

reportedly different from unsubstituted polypyrrole due to the different nature of the charge carriers. In acetonitrile/tetrabutyl ammonium tetrafluoroborate solution, the self-doped poly(3-alkylsulfonate pyrrole) films were unable to be reduced over a short timescale. However, the films were dedoped when held at a potential of -0.6 V vs SCE for a long period of time (more than 1 h). These results indicated slow diffusion of bulky tetrabutyl cation into the film. In electrolyte with small cations such as $NaBF_4$ or in the presence of small amount of acid, the faster redox reaction was achieved due to rapid diffusion of small cations and/or protons. The poly(3-alkylsulfonate pyrroles) were found to be slightly soluble in water (0.1–0.5 g/L) in the self-doped state. The higher solubility was observed for hexylsulfonate polypyrrole. The conductivities of the polymers were in the range 0.002 to 0.5 S/cm. These polymers can also be prepared electrochemically in water and chemically using ferric chloride as an oxidizing agent. The chemically synthesized polymer formed a precipitate in an aqueous reaction mixture similarly to unsubstituted polypyrrole.

Zotti *et al.* [48] electrochemically synthesized cationic and anionic conducting polymers using various 2,2'-bipyrroles N-substituted with alkyl ammonium or alkylsulfonate moieties in acetonitrile (Figure 5.6). The oxidation peak potentials of monomers, redox potentials and conductivities of polymers in acetonitrile with tetrabutyl ammonium perchlorate are given in Table 5.1. The higher conductivities of poly(4) (Figure 5.6) and poly(8) (Figure 5.6) were attributed to the increased coplanarity of the pyrrole rings. These conductivity values are higher than those of N-alkyl-substituted polypyrroles, which have conductivities in the range 10^{-3}–10^{-5} S/cm [11, 12, 35]. FT-IR results confirmed that the coupling of dipyrroles instead of pyrroles minimizes the introduction of overoxidative defects into the polymer chain. The polymers were found to be insoluble in the common organic solvents, both in the dedoped and the oxidized forms and the protonated and deprotonated states for the sulfonated polymers except poly(2). During an electrochemical, quartz-crystal microbalance analysis of the oxidative doping process in acetonitrile, the alkyl ammonium substituted polymers increased in mass with doping, however, the alkylsulfonate-substituted

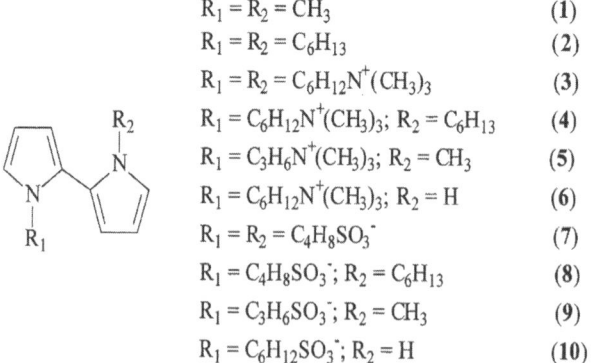

Figure 5.6 Structure of alkyl ammonium and alkylsulfonate substituted pyrrole monomers. (Reprinted with permission from *Chemistry of Materials*, **14**, 3607. Copyright (2002) American Chemical Society.)

Table 5.1 Oxidation peak potentials E_p for monomers, redox potentials $E°$ and conductivities of σ for polymers. (Reprinted with permission from *Chemistry of Materials*, 14, 3607. Copyright (2002) American Chemical Society.)

monomer	E_p (V)	$E°$ (V)	σ (S/cm)
1	0.42	0.13	0.005
2	0.39	0.15	0.007
3	0.39	0.15	0.03
4	0.19	−0.15	0.2
5	0.40	0.15	0.001
6	0.40	0.15	0.005
7	0.40	0.15	− a
8	0.10	−0.39	0.05

[a] Not measurable

polymers decreased in mass due to ejection of cations. These results indicate self-doping of alkylsulfonate substituted polymers.

Teixidor *et al.* [49] synthesized self-doped polypyrrole by using the nonconventional covalently bound low charge density anion [3,3'-Co(1,2-$C_2B_9H_{10}$)$_2$]$^-$ to a pyrrole unit via a spacer diether aliphatic chain. The electropolymerization of the potassium salt of the monomer, denoted as [1]$^-$, (Figure 5.7) was obtained in dry acetonitrile with tetrabutyl ammonium chloride in the potential range of −0.5 to 1.7 V vs Ag/AgCl. Similarly a copolymer of [1]$^-$ was prepared with pyrrole monomer (ratio 1:1) under identical conditions. The conductivity of poly[1]$^-$ and copolymer films measured using a four-point probe

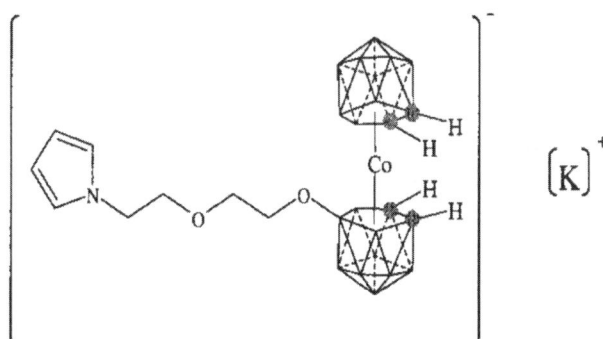

Figure 5.7 Pyrrole incorporating doping agent: [3,3'-Co(8-C_4H_4N-(CH_2)-O-(CH_2)-O-1,2-$C_2B_9H_{10}$)(1',2'-$C_2B_9H_{11}$)]$^-$ [1]$^-$. (Reproduced from *Advanced Materials*, 2002, 14, 826, C. Masalles, J. Llop, C. Vinas, F. Teixidor, with permission from Wiley-VCH.)

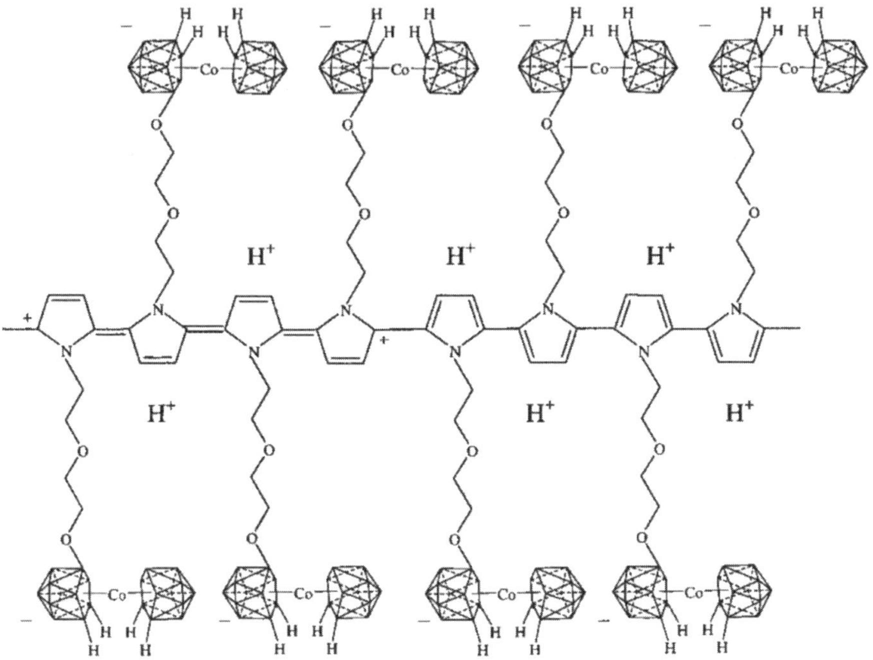

Figure 5.8 Proposed structure of poly[1]. (Reproduced from *Advanced Materials*, 2002, **14**, 826, C. Masalles, J. Llop, C. Vinas, F. Teixidor, with permission from Wiley-VCH.)

were reported to be approximately 1 and 9 S/cm, respectively. The proposed structure of self-doped poly[1]$^-$ based on X-ray photoelectron spectroscopy results is shown in Figure 5.8. The N$^+$/N ratio of 0.25 obtained from nitrogen-peak deconvolution suggested the presence of one charge carrier per roughly every four pyrrole units. The covalently bound [3,3'-Co(1,2-C$_2$B$_9$H$_{10}$)$_2$]$^-$ anion to poly[1]$^-$ is highly hydrophobic and did not allow electrochemical measurements in aqueous media. The overoxidation of unsubstituted polypyrrole films doped with common anions due to nucleophilic attack by OH$^-$ is well known [50, 51]. Teixidor *et al.* reported that the conventional polypyrrole doped with [Co(C$_2$B$_9$H$_{11}$)$_2$] shows over 300 mV overoxidation resistance compared with polypyrrole doped with common anions [52, 53]. The authors expected the increased overoxidation resistance of self-doped poly[1]$^-$ due to a covalently bound anionic hydrophobic cluster. As expected, the poly[1]$^-$ films are highly resistant to overoxidation. Figure 5.9 shows the dynamic overoxidation for polypyrrole/ClO$_4^-$ and poly[1]$^-$/polypyrrole

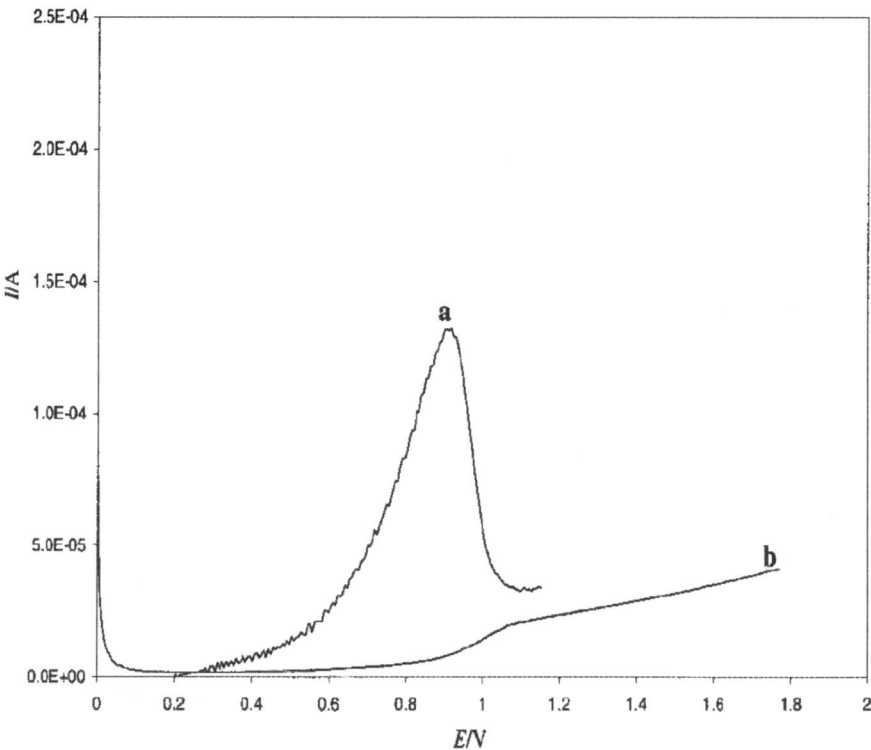

Figure 5.9 Dynamic overoxidation for: (a) polypyrrole[ClO$_4$] and (b) [1]$^-$/polypyrrole copolymer. (Reproduced from *Advanced Materials*, 2002, **14**, 826, C. Masalles, J. Llop, C. Vinas, F. Teixidor, with permission from Wiley-VCH.)

copolymer. The overoxidation of polypyrrole/ClO$_4^-$ starts around 0.6 V and a peak is observed around 0.9 V. However, in the case of poly[1]$^-$/polypyrrole copolymer, no defined peak was observed except for a slight increase in current around 0.95 V. It was suggested that this slight increase in current was probably due to two parallel processes, polypyrrole overoxidation and water discharge.

Reynolds *et al.* [54] reported the electrochemical synthesis of self-doped, water-soluble, N-propanesulfonated poly(3,4-propylenedioxypyrrole). The sulfonated monomer, N-propanesulfonate-substituted 3,4-propylenedioxypyrrole (N-PrS PProDOP) was prepared by treating 3,4-propylenedioxypyrrole with NaH in dry tetrahydrofuran and 1-propanesulfonate as shown in Figure 5.10. The poly(N-PrS PProDOP) films were synthesized in a mixture of propylene carbonate and water (94:6) with supporting electrolyte LiClO$_4$ in the potential range −0.4 to

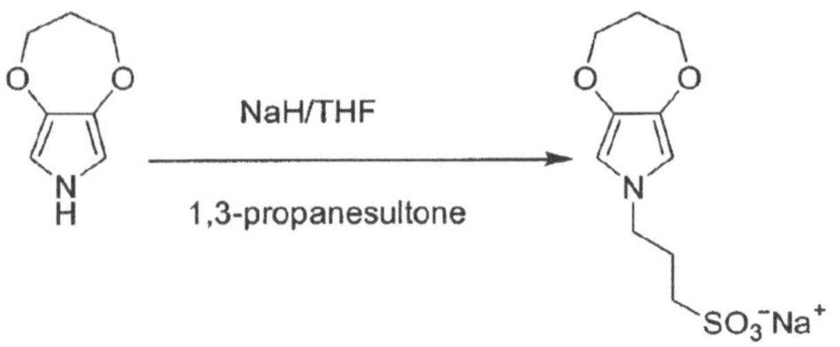

Figure 5.10 Synthesis of the monomer, *N*-propanesulfonate-substituted 3,4-propylenedioxypyrrole (*N*-PrS PProDOP). (Reprinted with permission from *Macromolecules*, **36**, 639. Copyright (2003) American Chemical Society.)

0.2 V vs Fc/Fc$^+$. Also, the polymer films were deposited galvanostatically at a current density of 0.04 mA/cm^2 in propylene carbonate:water (94:6) without supporting electrolyte. The monomer was reported to be insoluble in organic solvents such as acetonitrile, *N,N*-dimethylformamide or propylene carbonate; however, the addition of a small amount of water made the monomer soluble in these solvents. The cyclic voltammogram of a poly(*N*-PrS PProDOP) film prepared without supporting electrolyte as a function of scan rate is shown in Figure 5.11. The polymer exhibits

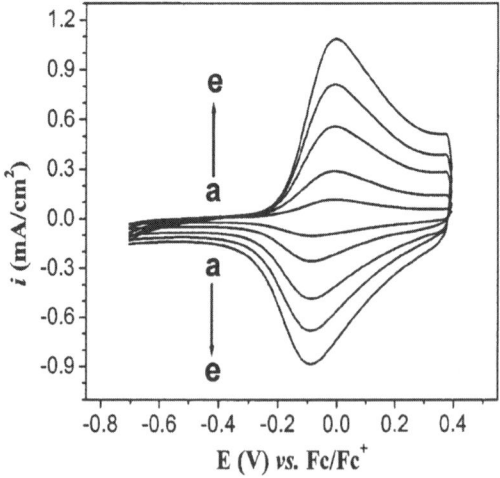

Figure 5.11 Cyclic voltammogram of a thin *N*-PrS PProDOP film in a monomer free solution of 0.1 M LiClO$_4$/propylene carbonate. Scan rates are (a) 20, (b) 50, (c) 100, (d) 150, and (e) 200 mV/s. (Reprinted by permission of ref. (Reprinted with permission from *Macromolecules*, **36**, 639. Copyright (2003) American Chemical Society.)

well-defined reversible redox behavior as a function of scan rate. The anodic and cathodic peak current ratio (i_{pa}/i_{pc}) of 0.98 and the peak separation (ΔE_p) of 6 mV indicate the reversibility of the redox process. The current densities observed for poly(N-PrS PProDOP) film prepared with and without supporting electrolyte are reported to be similar, suggesting that the polymer is self-doped. The scan rate dependence of anodic and cathodic peak currents of poly(N-PrS PProDOP) film are shown in Figure 5.12, A. The linear dependence suggests that the electrochemical

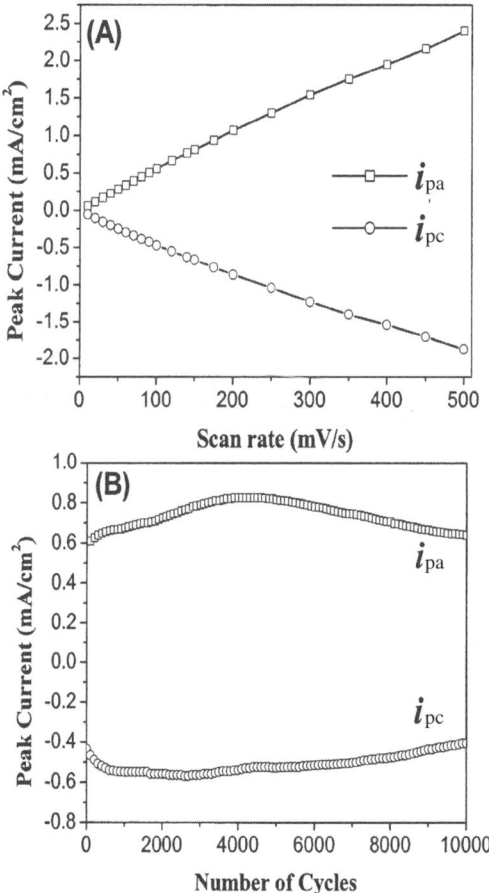

Figure 5.12 (A) Variation of anodic and cathodic peak currents as a function of the scan rate for a poly(N-PrS PProDOP) film in 0.1 M LiClO$_4$/propylene carbonate solution. (B) Variation of anodic (□) and cathodic (O) peak currents as a function of the number of cycles for 0.12 μm thick poly(N-PrS PProDOP) film, cycled 10 000 times at a scan rate of 100 mV/s. (Reprinted with permission from *Macromolecules*, **36**, 639. Copyright (2003) American Chemical Society.)

process is not diffusion controlled and is reversible up to 0.5 V/s. It was suggested that the unusually high reversibility at higher scan rates of poly(N-PrS PProDOP) is due to the thickness of the film (30 nm). In addition, the polymer shows redox stability over the course of 10 000 cycles with a scan rate of 100 mV/s in 0.1 M LiClO$_4$/propylenecarbonate (Figure 5.12, B). The conductivity of free standing films of poly(N-PrS PProDOP) prepared without supporting electrolyte were found to be in the range 10^{-4}–10^{-3} S/cm. The polymer films were electrochromic and underwent reversible transition from colorless in the neutral state to colored upon doping.

In situ electrophilic substitution of sulfonic acid on the polypyrrole backbone was carried out electrochemically by Sahin *et al.* [55]. They prepared self-doped sulfonated polyrrole electrochemically in acetonitrile-containing anhydrous fluorosulfonic acid (FSO$_3$H). Fluorosulfonic acid was used as both the sulfonation reagent for pyrrole and also as the supporting electrolyte. The degree of sulfonation of the polymer was controlled by varying FSO$_3$H and pyrrole concentrations in the solutions used for polymerization. Electrodeposition was performed by potential cycling in the range 0 to 1.5 V (vs Ag/AgCl) at a sweep rate of 100 mV/s. It was suggested that the electrophilic substitution (sulfonation) reaction can take place by two different routes, either before or after formation of polypyrrole, as shown in Figure 5.13. According to the first proposed route, pyrrole can polymerize first at the 2- and 5-positions and then the sulfonation reaction can take place at the 3- or 4-positions on the pyrrole rings (Figure 5.13, A). In another proposed mechanism, pyrrole can undergo electrophilic substitution, i.e., sulfonation reaction at the 2- and 5-positions before polymerization and then the sulfonated pyrrole monomer can be polymerized as shown in Figure 5.13, B. The sulfonation ratios and conductivities of polypyrrole films as a function of FSO$_3$H concentration during polymerization are given in Table 5.2. The comparable S/N ratios of oxidized and reduced polypyrrole films suggest that the sulfonate group is covalently bound to the polypyrrole backbone and polymer is self-doped.

5.1.1.2 Aqueous Media

The reports on electrochemical synthesis of self-doped sulfonated polypyrroles in aqueous media are limited in comparison with nonaqueous media. The first report on the synthesis of copolymer poly(pyrroleco-(3-(pyrrol-1-yl)propanesulfonate)) films in aqueous media for ion exchange was by Rajeshwar *et al.* [56]. The copolymer films were prepared without

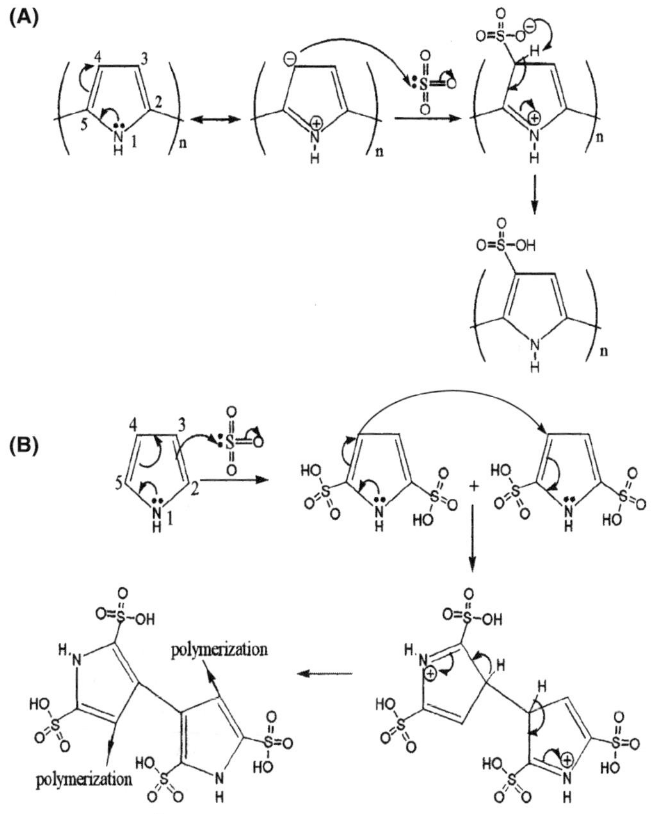

Figure 5.13 Proposed sulfonation mechanism. (Reproduced from *Journal of Applied Polymer Science*, 2004, **93**, 526. Reprinted with permission of John Wiley & Sons, Inc.)

supporting electrolyte by varying pyrrolesulfonate and pyrrole concentration and cycling the potential in the range −0.2 to 1.12 V vs Ag/AgCl. The self-doping properties of copolymer films are not reported; however, the films were used for binding of positively charged metal complexes such as $Ru(NH_3)_6^{3+}$ and $Ru(2,2'-bipyridyl)_3^{2+}$.

Collard *et al.* [57] reported the self-assembly of monomers, potassium 3-(3-alkylpyrrol-1-yl)propanesulfonates (Figure 5.14), and their subsequent electrochemical synthesis on gold electrodes to form lamellar structures of substituted polypyrrole. The fact that 3-(3-alkylpyrrol-1-yl)propanesulfonates possess amphiphilic character by virtue of the hydrophobic alkyl chain and hydrophilic ionic sulfonate head group, was used to form micellar solutions above critical concentrations. A

Table 5.2 Results of elemental analysis and dry conductivity of polypyrrole films[a]. (Reproduced from *Journal of Applied Polymer Science*, 2004, 93, 526. Reprinted with permission of John Wiley & Sons, Inc.)

FSO_3H Concentration (M)	$LiClO_4$ Concentration (M)	S/N ratio (Freshly prepared films)	S/N ratio (Reduced films)	Conductivity σ (S/cm)
0.010	–	0.29	0.39	65
0.025	–	0.39	0.42	52
0.050	–	0.51	0.46	27
0.075	–	0.54	0.54	22
0.100	–	0.53	0.56	6.5
–	0.100	–	–	26

[a] Prepared from acetonitrile solutions of 0.1 M pyrrole with different concentrations of FSO_3H and 0.1 M $LiClO_4$ at a constant potential of 1.0 V vs Ag/AgCl.

series of monomers, shown in Figure 5.14 with varying alkyl chain lengths, were synthesized by various steps such as preparing N-tosyl-3-alkanoylpyrroles, acylation of N-tosylpyrrole using Friedel–Crafts conditions, treatment with sodium hydroxide to hydrolyze the sulfonamide, reduction of 3-ketopyrrole, and finally treatment with potassium hexamethyldisilazide and 1,3-propanesultone. The majority of the monomers were soluble in water and formed micelles above a critical micelle concentration of $10^{-2}-10^{-5}$ M. The polymer was prepared on a gold electrode potentiostatically at 0.84 V vs Ag/AgCl in 0.1 M KNO_3 solution. Polymer deposition was unsuccessful when the concentration of monomer fell below a critical micelle concentration; however, above the critical micelle concentration, the rate of polymerization was insensitive to the monomer concentration. X-ray photoelectron spectroscopy studies indicated the absence of nitrate from the polymerization solution in the polymer. It was not mentioned whether the polymer was self-doped; however, it was assumed that the covalently bound sulfonate groups acted as charge balancing counterions in the oxidized polymer

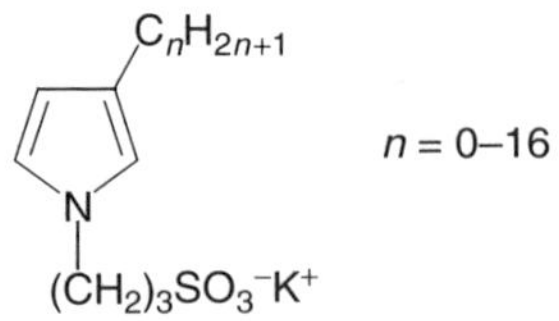

Figure 5.14 Structure of potassium 3-(3-alkylpyrrol-1-yl)propanesulfonates. (Reprinted with permission from *Chemistry of Materials*, 6, 850. Copyright (1994) American Chemical Society.)

and redox switching was a result of cation (K^+) migration. The films were electrochromic and underwent a reversible transition upon oxidation from pale green to red. X-ray diffraction studies showed several diffraction peaks in the range of $2\theta = 1-15°$ suggesting the presence of an ordered lamellar structure having an interlayer species with a length of the order of that of the alkyl chains (Figure 5.15). The straight line fit through the values for the higher homologs indicates that each additional methylene in the alkyl chain contributes approximately 2.2 Å to the spacing. Based on these results, it was suggested that the lamellar structure corresponds to head-to-head bilayer packing as shown in Figure 5.16. The conductivities of as-grown films were less than 10^{-6} S/cm; upon treatment with perchloric acid, conductivities increased to 10^{-3} S/cm. Wallace *et al.* [58] prepared water soluble homopolymer, poly(4-(3-pyrrolyl)butane sulfonate), from an aqueous solution using an electrohydrodynamic processing technique. The molecular weight and conductivity of optimized polymer were reported to be 10 500 g/mol and 0.01 S/cm.

5.1.2 Chemical Polymerization

5.1.2.1 Homopolymerization

Similarly to the electrochemical synthesis, chemical polymerization of the ring-substituted polypyrrole, sodium(3'-pyrrole)alkanesulfonate, has

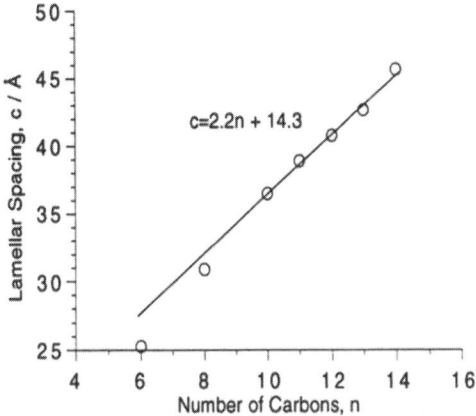

Figure 5.15 Plot of interlamellar unit cell spacing of poly(potassium 3-(3-alkylpyrrol-l-yl)propanesulfonate)s, vs the number of carbons in the 3-alkyl substituent. (Reprinted with permission from *Chemistry of Materials*, 6, 850. Copyright (1994) American Chemical Society.)

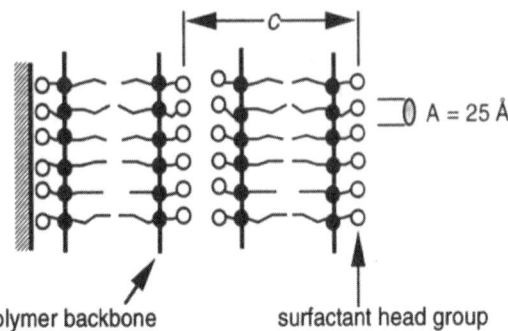

Figure 5.16 Schematic for lamellar phase of poly l deposited from micellar phases. (Reprinted with permission from *Chemistry of Materials*, 6, 850. Copyright (1994) American Chemical Society.)

been carried out by Havinga *et al.* [47]. The polymer was synthesized in aqueous media using ferric chloride as an oxidizing agent. The product was reportedly self-doped. Similarly, there is a report on the chemical synthesis of water-soluble, self-acid-doped, poly(4-(3-pyrrole)butanesulfonic acid) [59]. In this report, the chemically synthesized sodium salt of poly(4-(3-pyrrolyl)butanesulfonate) was ion exchanged with H^+ using an H^+-type ion exchanger (Figure 5.17). The conductivity of a free-standing film prepared from an aqueous solution of poly(4-(3-pyrrole)butanesulfonic acid) was reported to be approximately 10^{-4} S/cm. This conductivity value is two orders of magnitude higher than the sodium salt of poly(4-(3-pyrrolyl)butanesulfonate) reportedly due to the self-doping by the proton on the sulfonic acid side chain. Li [60] chemically synthesized water-soluble, self-doped, polypyrrole, starting from a butane sulfonyl substituted pyrrole monomer. This polymer was combined with collagen to obtain a hybrid biomaterial whose conductivity was 10^{-2} S/cm.

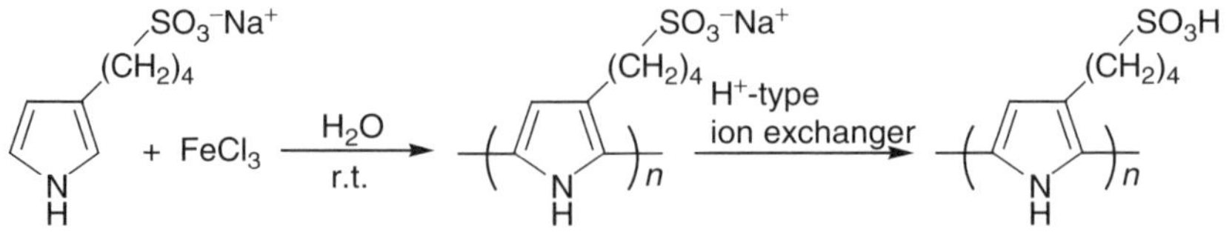

Figure 5.17 Synthesis of poly(4-(3-pyrrolyl)butanesulfonic acid). (Reproduced from *Journal of Polymer Research*, 5, 1998, 249, E. C. Chung, M. Y. Hua, S. A. Chen, with kind permission from Springer Science and Business Media.)

5.1.2.2 Copolymerization

In the above-mentioned electrochemical and chemical synthesis methods, the sulfonated pyrrole monomer used was first prepared by various synthetic approaches. The synthesis of the sulfonated pyrrole monomer is complex and therefore, the cost of the monomer becomes expensive. Hence, the alternative approach of 'graft copolymerization' has been investigated for the synthesis of self-doped conducting polypyrrole [61, 62]. Ruckenstein *et al.* synthesized a water-soluble, self-doped, graft polypyrrole copolymer [61]. The copolymer was prepared via grafting of pyrrole onto the *p*-aminodiphenylamine moieties of a highly water-soluble copolymer, poly(2-acrylamido-2-methyl-1-propanesulfonic acid-co-N-(4-anilinophenyl)-methylacrylate) (poly(AMP-co-APMA)). The synthesis of self-doped poly(pyrrole-co-AMP-co-APMA) is shown in Figure 5.18. The grafting of pyrrole onto (poly(AMP-co-AMPA)) was

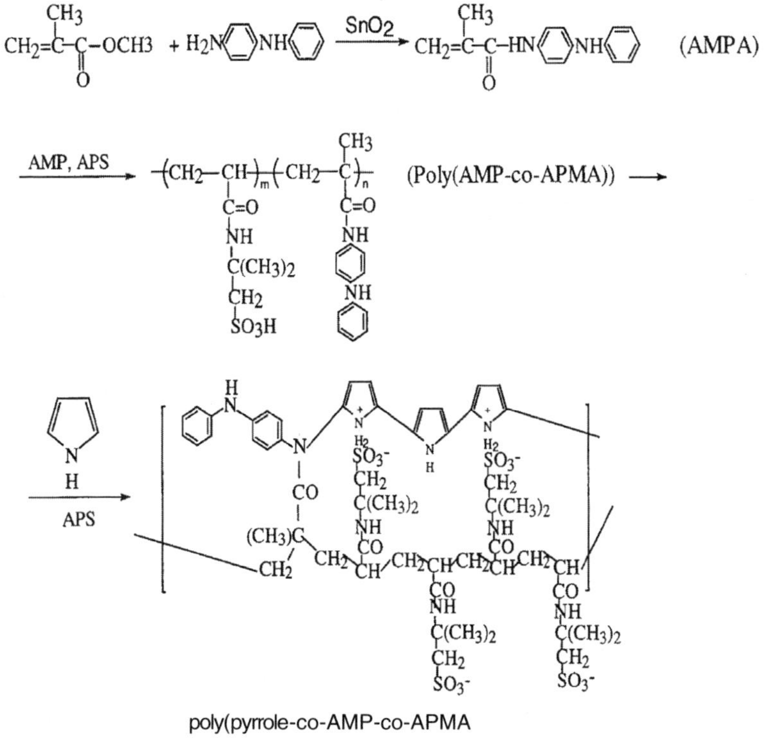

Figure 5.18 Synthesis of a water soluble self-doped polypyrrole copolymer, poly(pyrrole-co-AMP-co-APMA). (Reproduced from *Journal of Applied Polymer Science*, 2001, 79, 86. Reprinted with permission of John Wiley & Sons, Inc.)

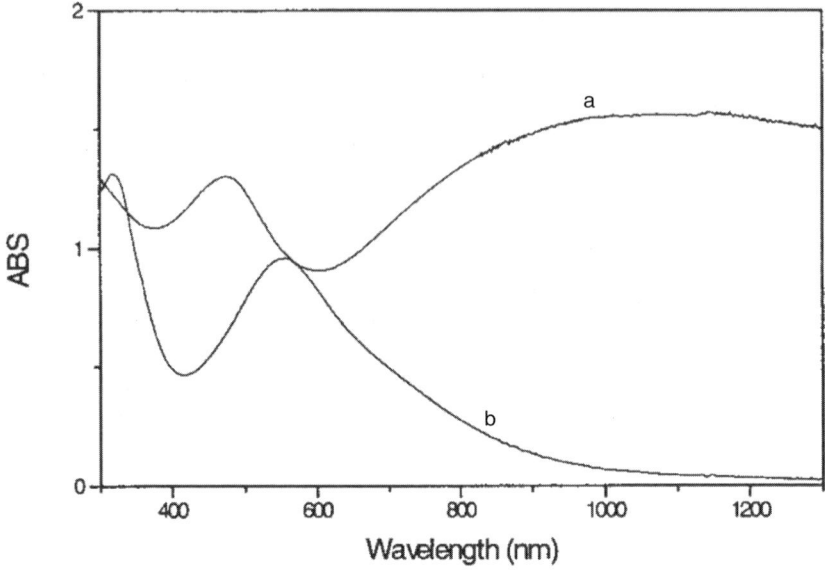

Figure 5.19 UV-Vis spectra of poly(PY-co-AMP-co-APMA): (a) aqueous solution and (b) 1 M NH$_4$OH solution. (Reproduced from *Journal of Applied Polymer Science*, 2001, 79, 86. Reprinted with permission from John Wiley & Sons, Inc.)

carried out in an aqueous solution. The polymer was synthesized using ammonium persulfate as an oxidizing agent and purified with an H$^+$ exchange resin column. A maximum conductivity of 3.4 S/cm for the graft copolymer was obtained at a polypyrrole/poly(AMP-co-APMA) weight ratio of 1:1. The UV-Vis spectrum of poly(pyrrole-co-AMP-co-APMA) in aqueous solution showed a bipolaron peak at 475 nm, and a free carrier tail in the near IR region with a peak at 1100 nm (Figure 5.19, A). These results suggest that copolymer is self-doped in aqueous solution. The copolymer was dedoped in alkaline solution (Figure 5.19, B).

Jo *et al.* [62] prepared a water soluble self-doped conducting polypyrrole graft copolymer using poly(sodium styrenesulfonate-co-pyrrolylmethylstyrene) (P(SSNa-co-PMS)) as the precursor. The synthetic method is described in Figure 5.20. The graft copolymerization was carried out using two oxidizing agents, ammonium persulfate and ferric chloride. It was suggested that the ammonium persulfate was more effective than ferric chloride for preparation of soluble graft copolymer. The molecular weight of the graft copolymer was 33 000 g/mol, as determined by size exclusion chromatography using water/dimethyl

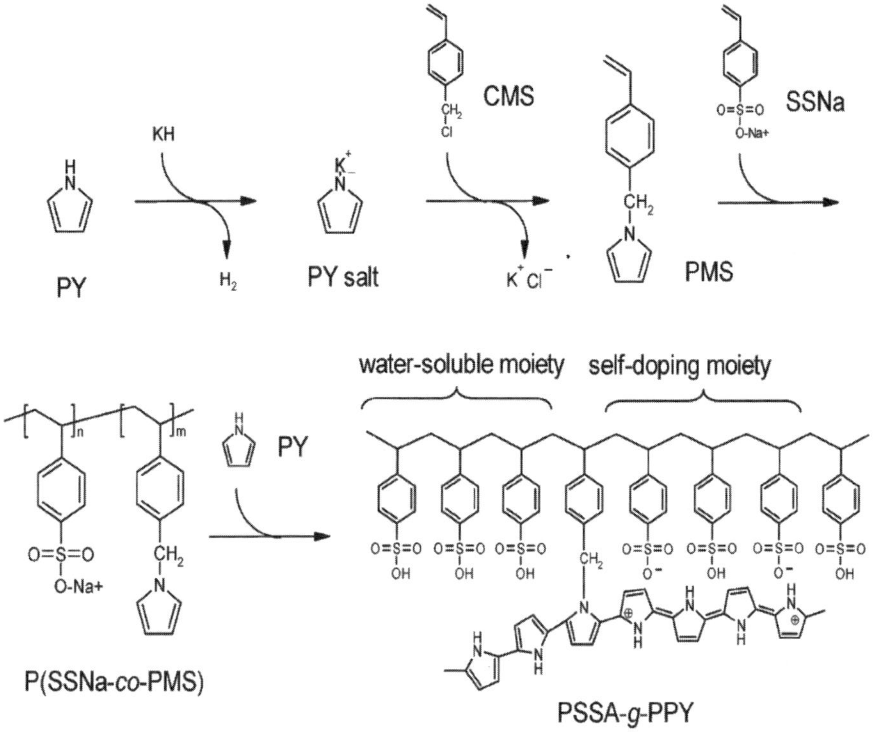

Figure 5.20 Synthetic route for PSSA-g-PPY. PY = pyrrole, CMS = chloromethylstyrene, PMS = pyrrolylmethylstyrene, SSNa = sodium styrenesulfonate, P(SSNa-co-PMS) = poly(sodium styrenesulfonate-co-pyrrolylmethylstyrene). (Reprinted with permission from *Macromolecules*, 38, 1044. Copyright (2005) American Chemical Society.)

formamide (v/v = 9.5/0.5) + 0.05 M NaOH as an eluant. The electrical conductivity of the polymer in the form of a pellet was approximately 0.5 S/cm. The graft copolymer PSSA-g-PPY was completely soluble in water and self-doped as suggested by the UV-Vis studies. The spectrum of PSSA-g-PPY in aqueous solution showed a bipolaron peak at 473 nm and free carrier tail in the near IR region with a peak at 1000 nm. In alkaline solution, the bipolaron absorption and the free carrier tail disappeared due to dedoping. The electrical conductivity of dedoped PSSA-g-PPY was 6.7×10^{-6} S/cm. The intrinsic viscosities of doped PSSA-g-PPY water solution and dedoped PSSA-g-PPY alkaline solution (1 M NH_4OH) were 0.23 and 0.12 dL/g, respectively. Based on these results, it was suggested that the chain conformation of PSSA-g-PPY in solution depends on the doping state, and that the polypyrrole chains are

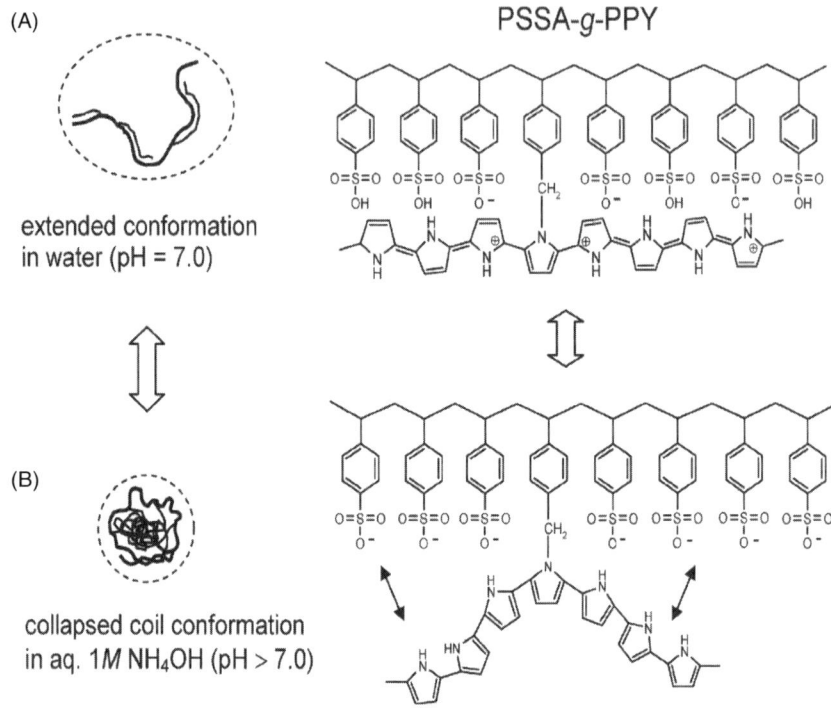

Figure 5.21 Illustration of the hydrodynamic volume and structural change of PSSA-g-PPY (A) in water and (B) in aqueous NH₄OH (1 M) solution. (Reprinted with permission from *Macromolecules*, 38, 1044. Copyright (2005) American Chemical Society.)

self-doped with polymeric dopant PSSA. The schematic representation of hydrodynamic volume and structural changes of PSSA-g-PPY in water and alkaline solution is shown in Figure 5.21. Jo *et al.* suggested that in the doped state, the extended conformation of PSSA-g-PPY is due to the intertwined or closely associated incorporating strands of the extended chain of polypyrrole and PSSA. At alkaline pH, the dedoped grafted polypyrrole, chains presumably have more freedom of movement and thus are easily collapsed in hydrophilic aqueous medium, forming a core–shell structure.

5.1.3 Polycondensation

Using a polycondensation approach, Ajayaghoh *et al.* [63] prepared water-soluble squaraine oligomers by condensation of squaric acid with

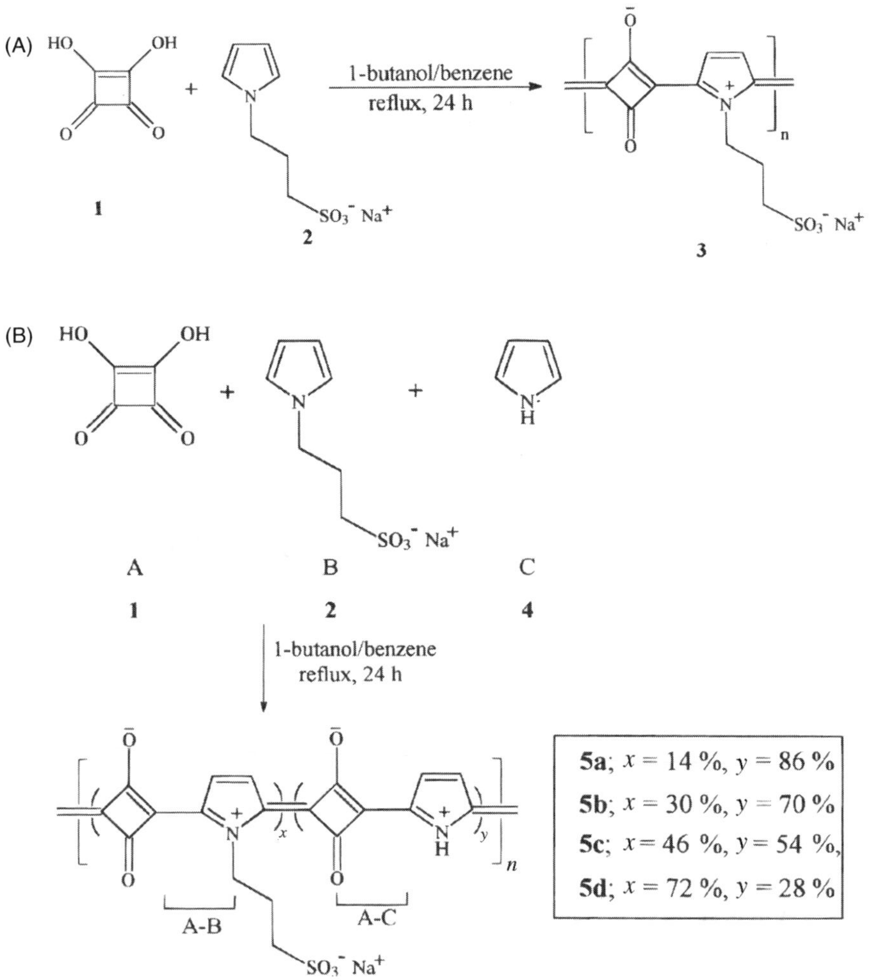

SELF-DOPED POLYPYRROLE

Figure 5.22 Synthesis of water soluble oligomer prepared by polycondensation of squaric acid and sodium 3-(pyrrole-1-yl)propane sulfonate (A) and in the presence of pyrrole (B). (Reprinted with permission from *Chemistry of Materials*, 10, 1657. Copyright (1998) American Chemical Society.)

N-propanesulfonate pyrrole both alone (Figure 5.22, A) and in the presence of pyrrole (Figure 5.22, B). A dark green precipitate was obtained after condensation in a 1:1 mixture of 1-butanol and benzene. The absorption spectra of the oligomers in water and dimethyl sulfoxide showed sharp peaks at around 569 and 584 nm, respectively. The number average molecular weights of oligomers 3 (Figure 5.22, A) and 5a–d (Figure 5.22, B) were reported to be in the range of 2800–3000.

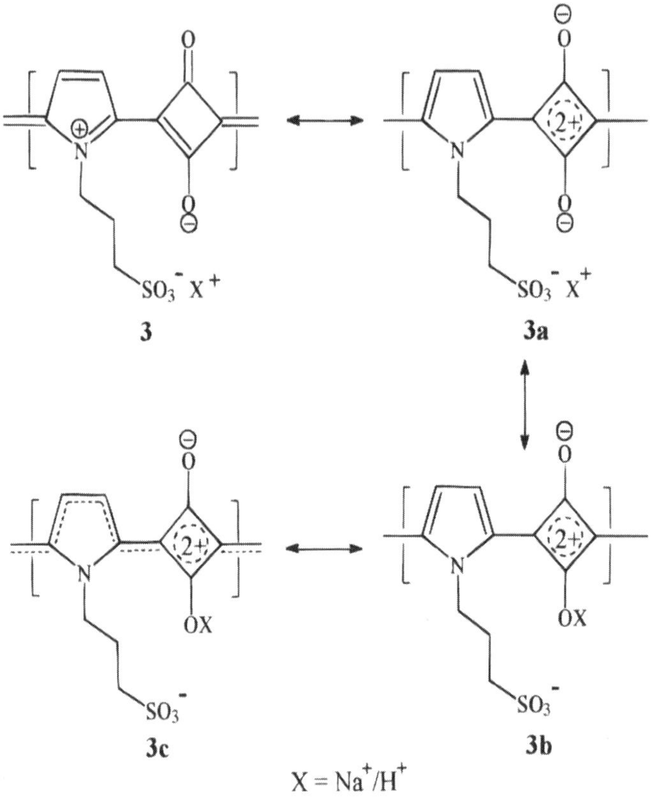

Figure 5.23 Proposed mechanism of self-doping of water soluble squaraine oligomers. (Reprinted with permission from *Chemistry of Materials*, **10**, 1657. Copyright (1998) American Chemical Society.)

Four-point probe conductivity of oligomers **5a–d** in the form of a pellet increased from 3.6×10^{-6} S/cm to 3.2×10^{-5} S/cm. The increased conductivity with increasing mole % of N-propanesulfonate pyrrole monomer was attributed to an intramolecular doping or self-doping of oligomers by pendant propanesulfonate groups. A proposed mechanism for improved conductivity or self-doping in squaraine oligomers is shown in Figure 5.23.

5.2 CARBOXYLIC ACID DERIVATIVES

Several groups have studied the synthesis of carboxylic acid derivatives of polypyrrole; however, the electrochemistry of very few carboxylic acid substituted polypyrroles has been reported. Smith *et al.* [64] reported that electrochemical synthesis of carboxylic-acid substituted pyrrole i.e., (3-methylpyrrole-4-carboxylic acid), in acetonitrile yields a soluble yellow polymer. In a subsequent study, Cross *et al.* [65] reported the formation of a golden film on the electrode surface by electrochemical polymerization of pyrrole-2-carboxylic acid. Pickup [66] was probably the first to study aqueous electrochemical behavior of a polymer with a carboxylic acid substituent in detail, with the objective of increasing the ion-exchange capacity of polypyrrole. The golden brown films of poly(3-methylpyrrole-4-carboxylic acid) were prepared electrochemically in acetonitrile using tetrabutyl ammonium perchlorate at a potential above 1.1 V vs SCE. The conductivity of the films was found to be in the range 10^{-5} to 10^{-7} S/cm. In aqueous solution, cyclic voltammograms of poly(3-methylpyrrole-4-carboxylic acid) showed a single redox couple. The potentials of the oxidation and reduction peaks (E_{pa} and E_{pc}) were found to depend upon the pH of the aqueous electrolyte (Figure 5.24).

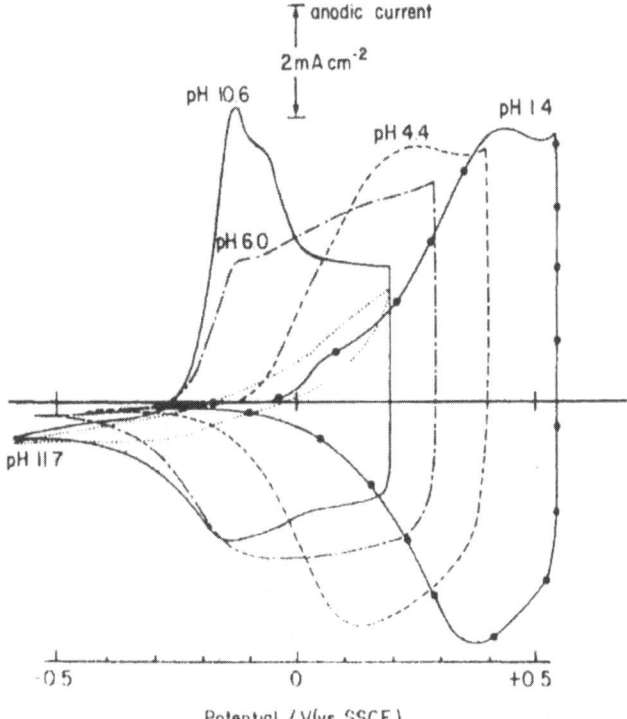

Figure 5.24 Cyclic voltammetry of poly(3-methylpyrrole-4-carboxylic acid) films as a function of pH. (Reprinted from *Journal of Electroanalytical Chemistry*, 225, P. G. Pickup, 273. Copyright (1987), with permission from Elsevier.)

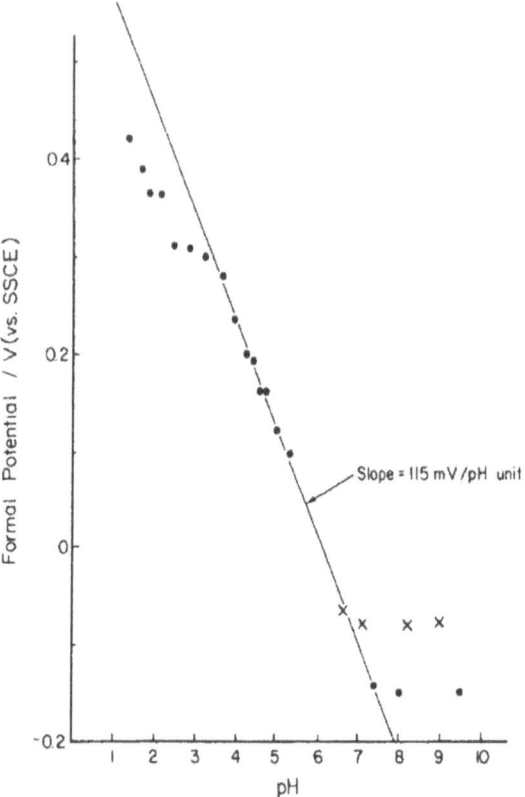

Figure 5.25 Plot of formal potential ($E^{0'} = (E_{pa} - E_{pc})/2$) vs pH for poly(3-methyl-pyrrole-4-carboxylic acid) in 1 M KNO_3 containing 0.2 M acetate buffer (●). Experiments in the pH range 6.5–9.5 were repeated with 0.1 M NH_4NO_3 added to the cell solution (X). (Reprinted from *Journal of Electroanalytical Chemistry*, 225, P. G. Pickup, 273. Copyright (1987), with permission from Elsevier.)

At pH values below 10, the polymer redox chemistry is reversible and rapid. The formal potential of the polymer [$E^{0'} = (E_{pa} + E_{pc})/2$] shifted by 115 mV/pH unit for pH values less than 2.5 and for pH values between 3.5 and 5 (Figure 5.25). This shift indicates that two protons per electron are involved in the oxidation and reduction of the polymer. Between pH 2.5 and 3.5, the formal potential was practically independent of pH. Above pH 6.5, the formal potential for the film was pH independent. This behavior was rationalized by assuming that, in acidic solution (pH < 2.5), the reduced polymer is fully protonated and that the removal of each electron during the oxidation of the polymer chain is accompanied by the deprotonation of two carboxylic acid groups. The break in the pH behavior between pH 2.5 and 3.5 was attributed to the deprotonation of carboxylic acid groups in the reduced polymer at around pH 3. Between pH 3.5 and 5, the oxidation of the polymer is again associated with the loss of two protons for each electron removed from the film to give, in the fully oxidized state, the fully deprotonated polymer. The redox behavior that occurs at different pH values is consistent with the following scheme [66, 67]:

pH < 2.5

$$[(MPCO_2H)_2(MPCO_2)_2]_n^{n-} + ne^- + 2nH^+ \rightleftharpoons \quad (5.1)$$
$$[(MPCO_2H)_4]_n$$

2.5 < pH < 3.5

$$[(MPCO_2H)_2(MPCO_2)_2]_n^{n-} + ne^- \rightleftharpoons \quad (5.2)$$
$$[(MPCO_2H)_2(MPCO_2)_2]_n^{2n-}$$

3.5 < pH < 5

$$[(MPCO_2)_4]_n^{3n-} + ne + 2nH^+ \rightleftharpoons \quad (5.3)$$
$$[(MPCO_2H)_2(MPCO_2)_2]_n^{2n-}$$

6.5 < pH

$$[(MPCO_2)_4]_n^{3n-} + ne^- \rightleftharpoons [(MPCO_2)_4]_n^{4-} \quad (5.4)$$

The behavior of poly(3-methylpyrrole-4-carboxylic acid) was not clearly identified as self-doping since the focus was on the electrostatic binding of cations such as $[Co(bpy)_3]^{2+}$ and methyl viologen, MV^{2+}.

Delabouglise and Garnier [68] prepared poly(3-carboxymethylpyrrole) in aqueous media (0.2 M KCl/H_2O). These films were found to be highly conducting ($\sigma = 37$ S/cm) and to exhibit pH dependent electrochemistry in aqueous solution (Figure 5.26). The anodic peak potential was found to shift on average by around 60 mV/pH unit in the region between pH 0 and pH 6 indicating that one proton was lost for each electron removed in the oxidation of the film. It was suggested that the linear behavior as a function of pH of solution is indicative of a self-doped conducting state. However, the plots of E_{pa} vs pH show some curvature, suggesting that the process may be more complex. The dissociation equilibria of the pendent alkylcarboxylic acid substituents contributes to the redox behavior of poly(3-carboxymethylpyrrole). Based on the interaction between the redox property of the polymer and the dissociation of the carboxylic groups, it was proposed that polymer could be used for electrochemically controlled release of protons [69].

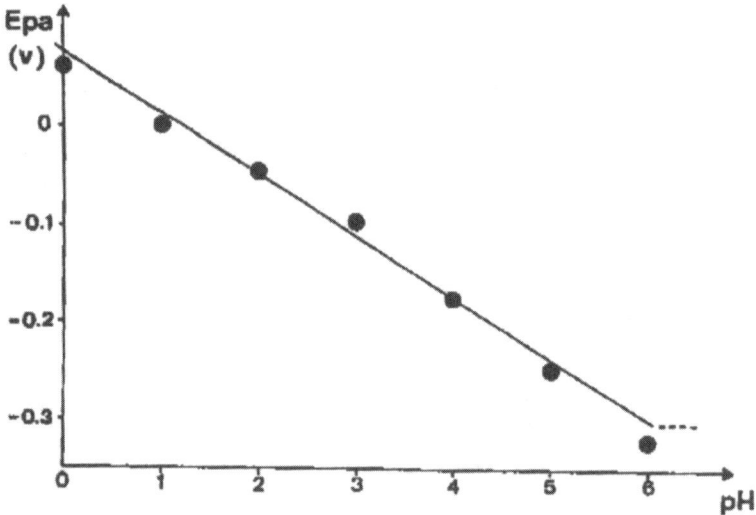

Figure 5.26 Variation of anodic peak potential (E_{pa} vs SCE) of poly(3-carboxymethyl pyrrole) as a function of pH determined by cyclic voltammetry at 25 mV/s. (Reprinted from *New Journal of Chemistry*, 15, D. Delabouglise, F. Garnier, 233. Copyright (1991), CNRS.)

There are also reports on the synthesis of N-substituted and ring-substituted pyrroles bearing an alkyl chain with terminal carboxylic acid groups for biological interfacial redox reactions involving biomolecular recognition. Cooper et al. [70] used carboxylic acid functionalized polypyrroles such as poly(3-methyl-4-pyrrolylcarboxylic acid), poly(3-pyrrolyacetic acid) and poly(3-pyrrolylpentanoic acid) for redox interactions with a small redox protein such as cytochrome c. The electrodes coated with these polymers showed well defined electron transfer reaction with cytochrome c; however, no response was observed with unsubstituted polypyrrole and blank electrodes. It was suggested that the acidic carboxylic groups are required to orient the protein prior to electron transfer.

Poly(pyrrole-N-propionic acid) can be prepared in a single step by the hydrolysis of 3-(pyrrol-1-yl)propionitrile (Figure 5.27) [71]. In addition, it could be polymerized electrochemically in propylene carbonate using sodium perchlorate. Bartlett et al. [72] electrochemically synthesized poly(3-(pyrrolyl)-carboxylic acid) (**1**), poly(3-(pyrrolyl)-butanoic acid) (**2**) and poly(3-(pyrrolyl)-pentanoic acid) (**3**) in acetonitrile/LiClO$_4$ (Figure 5.28). The redox behavior of poly(3-(pyrrolyl)-carboxylic acid) (**1**) and poly(3-(pyrrolyl)-butanoic acid) (**2**) was reported to be pH dependent similar to the results of Pickup [66], and Delabouglise and

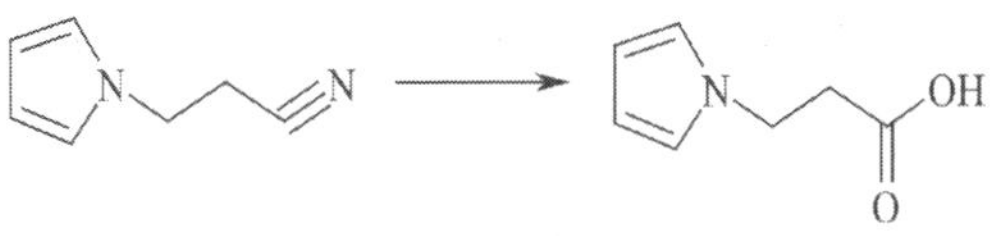

CARBOXYLIC ACID DERIVATIVES

Figure 5.27 Synthesis of the N-substituted pyrrole, poly(pyrrole-N-propionic acid). (Reprinted from *Solid State Ionics*, **169**, M. D. Ingram, H. Staesche, K. S. Ryder, 51. Copyright (2004), with permission from Elsevier.)

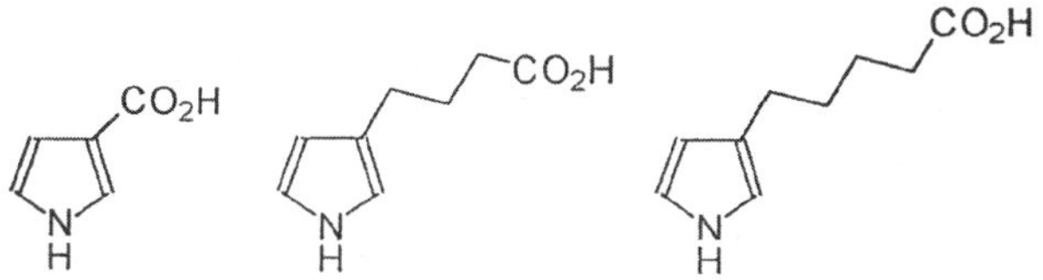

Figure 5.28 Structures of 3-(pyrrolyl)-carboxylic acid (**1**), 3-(pyrrolyl)-butanoic acid (**2**) and 3-(pyrrolyl)-pentanoic acid (**3**). (Reprinted from *Journal of Electroanalytical Chemistry*, **487**, P. N. Bartlett, M. C. Grossel, E. M. Barrios, 142. Copyright (2000), with permission from Elsevier.)

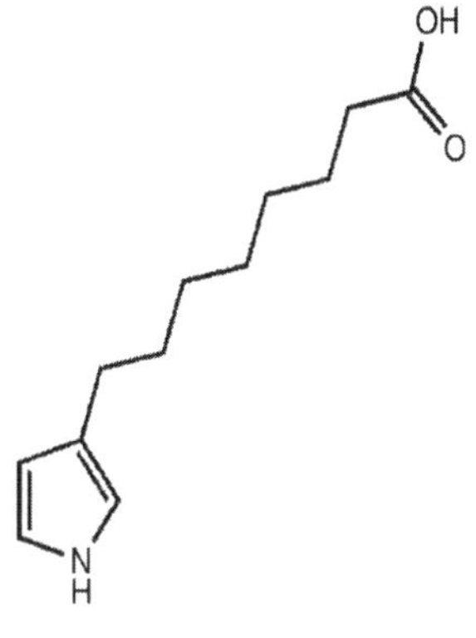

Figure 5.29 Structure of (β-pyrrolyl)-octanoic acid. (Reprinted from *Synthetic Metals*, **155**, J. M. Freitas, M. L. Duarte, T. Darbre, L. M. Abrantes, 549. Copyright (2005), with permission from Elsevier.)

Garnier [68]. Recently, poly(β-pyrrolyl)octanoic acid was synthesized electrochemically in propylene carbonate (Figure 5.29) [73]. Current transient analysis suggested a three-dimensional nucleation and growth mechanism for the film formation on platinum. The polymer films showed high redox stability under repetitive potential cycling.

5.3 SELF-DOPED POLY(3,6-(CARBAZ-9-YL) PROPANESULFONATE)

Polycarbazoles such as poly(N-vinylcarbazole) and numerous side chain polymers with pendent carbazolyl groups have been studied extensively because of their photoconductive properties and ability to form charge transfer complexes, given the electron donating character of the carbazole moiety [74–78]. In addition, they have received considerable attention due to specific applications in electroluminescent devices [79–82], and as coatings on optically transparent electrodes in electrochromic displays [83–86]. Qui and Reynolds were the first to extend the concept of self-doping to polycarbazoles. They prepared water-soluble, self-doped, poly(3,6-(carbaz-9-yl)propanesulfonate) with cation and anion exchange properties [87]. The monomer, tetrabutyl ammonium (3,6-(carbaz-9-yl)propanesulfonate) (Figure 5.30, **1**), was prepared by reacting carbazole with potassium in refluxing tetrahydrofuran, followed by reaction with propane sultone at room temperature. The product was washed with tetrahydrofuran and dried, and further reacted with an equimolar amount of tetrabutyl ammonium hydroxide to yield tetrabutyl ammonium poly(3,6-(carbaz-9-yl)propanesulfonate).

The polymer tetrabutyl ammonium poly(3,6-(carbaz-9-yl)propanesulfonate) (Figure 5.30, **2**) was synthesized electrochemically by galvanostatic deposition using 2 mA/cm^2 in methanol with 0.1 M tetrabutyl ammonium perchlorate. The brown coloured polymer obtained was

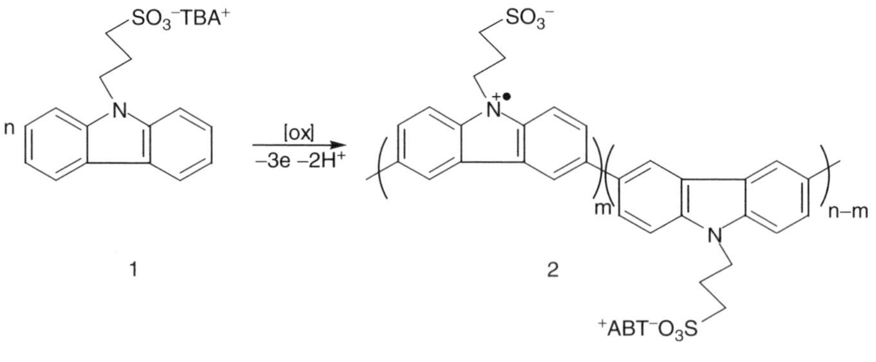

Figure 5.30 Synthesis of poly(3,6-(carbaz-9-yl)propanesulfonate). (Y. J. Qiu, J. R. Reynolds, *Journal of the Electrochemical Society*, 1990, **137**, 900. Reproduced by permission of The Electrochemical Society, Inc.)

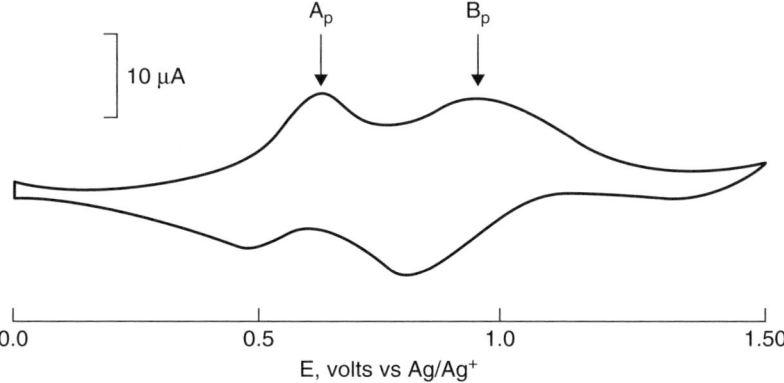

Figure 5.31 Cyclic voltammogram of poly(3,6-(carbaz-9-yl)propanesulfonate) on Pt in 0.1 M tetrabutyl ammonium perchlorate/acetonitrile. Scan rate = 100 mV/s. (Y. J. Qiu, J. R. Reynolds, *Journal of the Electrochemical Society*, 1990, **137**, 900. Reproduced by permission of The Electrochemical Society, Inc.)

soluble in water. It was suggested that the polymer acts as a polyelectrolyte due to the presence of non-doping sulfonate groups along the chain. The electrical conductivity of free-standing, brittle films obtained from the aqueous solution of the polymer was reported to be approximately 10^{-7} S/cm. The conductivity did not increase upon further doping with iodine or bromine vapor. The polymer tetrabutyl ammonium poly(3,6-(carbaz-9-yl)propanesulfonate) is electroactive and the cyclic voltammogram reveals two redox waves at 0.55 and 0.83 V, as shown in Figure 5.31. Electrochemical quartz crystal microbalance microgravimetry suggests that a mass decrease observed during the first oxidation wave (0.55 V) is due to the expulsion of counterions and solvent from the film, while a mass increase during the second oxidation wave (0.83 V) occurs as a result of ion/solvent flux into the film.

Based on the cyclic voltammetric and microgravimetric results, the proposed mechanism of oxidative doping and ion exchange is shown in Figure 5.32. It was postulated that during the first oxidation process (first wave, 0.55 V), mass decreases due to loss of tetrabutyl ammonium cations and results in a self-doped polymer. Upon further oxidation (second wave), the formation of a highly charged structure allows perchlorate anions and associated solvent to penetrate into the film. The polymer films show reversible electrochromic switching between transparent and dark green states.

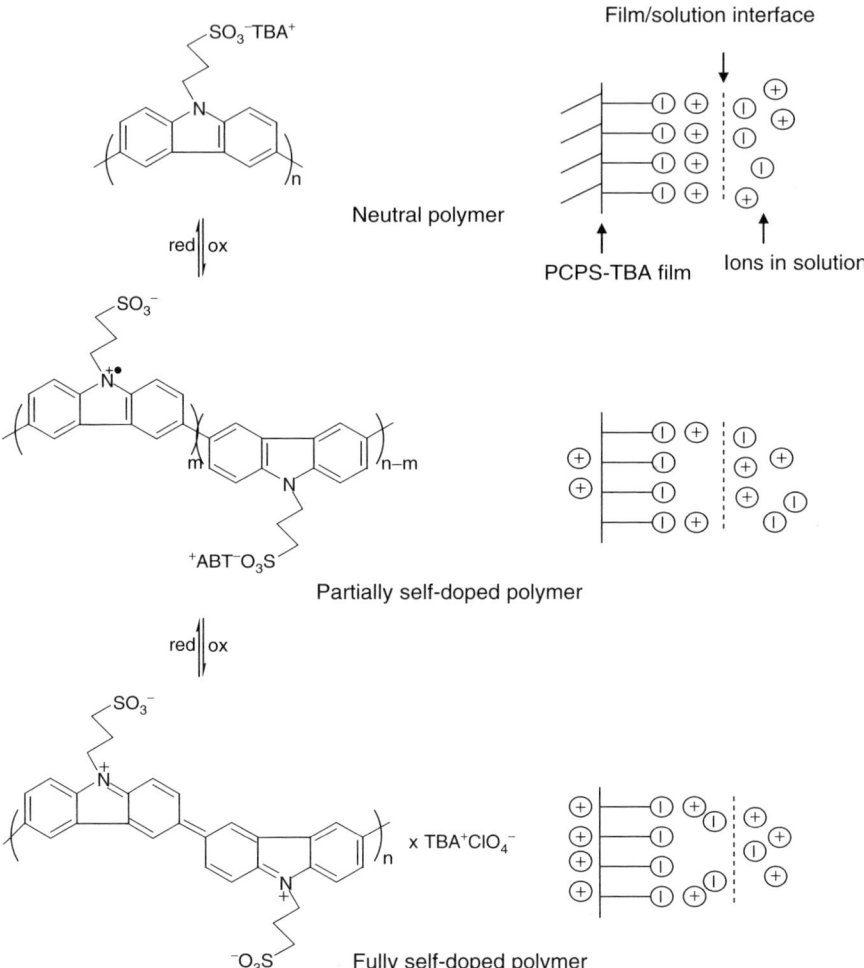

Figure 5.32 Self-doping and ion exchange mechanism of poly(3,6-(carbaz-9-yl) propanesulfonate). (Y. J. Qiu, J. R. Reynolds, *Journal of the Electrochemical Society*, 1990, **137**, 900. Reproduced by permission of The Electrochemical Society, Inc.)

5.4 SELF-DOPED POLY(*p*-PHENYLENE)S

Poly(*p*-phenylene) has attracted considerable attention since it can act as an excellent organic conductor upon doping and possesses a unique combination of physical properties, such as low density, high mechanical strength, excellent thermal stability and remarkable chemical resistance [88, 89]. It can convert from an electrical insulator in its pristine form ($\sigma = 10^{-12}$ S/cm) to a highly conductive charge transfer complex ($\sigma = 500$ S/cm) when treated with dopants such as AsF_5 [90, 91]. Poly(p-phenylene) also exhibits important optical properties such as electroluminescence [92] and other potentially useful material properties such as excellent radiation resistance and intrinsic paramagnetism [90]. However, its development for commercial applications has been hampered due to its insolubility and intractability in common solvents.

In order to increase the processability, poly(p-phenylene) derivatives have been synthesized with functional groups attached to, or incorporated within, the backbone. Schlüter and Wegner used Yamamoto and Suzuki coupling conditions to polymerize alkyl-derivatized benzenes [93–95]. By careful choice of catalyst and polymerization conditions, they were able to prepare soluble polymers with repeat units of approximately 50. Kim and Webster [96, 97] synthesized a trifunctional benzene-based monomer, (3,5-dibromophenyl)boronic acid, which could be self-condensed to a hyperbranched macromolecule that is water soluble. Subsequently, the synthesis of a water soluble, rigid rod poly(p-phenylene) derivative using a water soluble Pd(0) catalyst was reported by Novak *et al.* [98, 99]. This poly(p-phenylene) derivative contains two carboxylic acid groups directly attached to each quaterphenylene repeat unit and all *para* linkages along its backbone. Using zero valent nickel, Kaeriyama *et al.* [100, 101] synthesized poly(2-carboxyphenylene-l,4-diyl) containing one carboxylic acid group per ring. This polymer is soluble in aqueous base and can be converted to poly(p-phenylene) by treatment with CuO. Rehahn *et al.* [102–104] prepared a series of poly(p-phenylene)-based polyelectrolytes containing both carboxylate and tetraalkyl ammonium functionality. Wegner *et al.* [105] prepared directly sulfonated poly(p-phenylene)s via an organic soluble precursor. These were later used in the formation of blue light-emitting devices [106]. In these reports, the covalently attached sulfonate and carboxyl groups induced water solubility; however, it is not mentioned whether they serve as a charge compensating dopants or the polymer is self-doped.

The self-doped conducting alkoxy sulfonated poly(p-phenylene), poly [2,5-bis(3-sulfonatopropoxy)-1,4-phenylene-alt-l,4-phenylene] (polyphenylene-ORSO$_3$) was first reported by Reynolds *et al.* [107]. In this polymer, covalently attached sulfonate functionality reportedly induces water solubility to the rigid rod backbone and serves as the charge compensating dopant ion during redox switching. Polyphenylene-ORSO$_3$ (2) was synthesized using the homogeneous Suzuki coupling method by the reaction of 1,4-benzenediboronic acid and the disodium salt

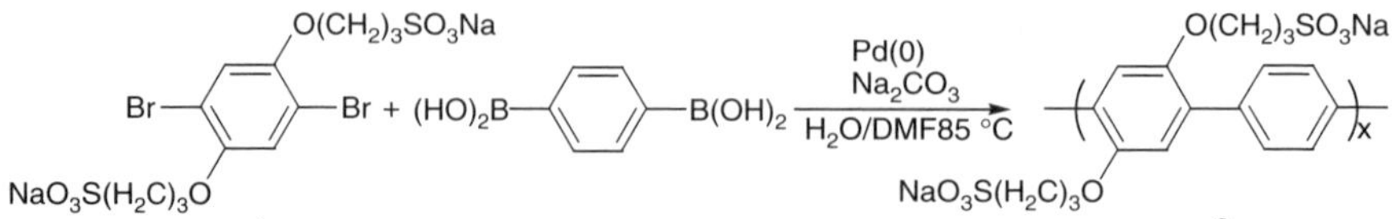

Figure 5.33 Synthesis of poly[2,5-bis(3-sulfonatopropoxy)-1,4-phenylene-alt-l,4-phenylene] (polyphenylene-ORSO$_3$). (Reprinted with permission from *Macromolecules*, 27, 1975. Copyright (1994) American Chemical Society.)

of 1,4-dibromo-2,5-bis(3-sulfonatopropoxy)benzene (**1**), as shown in Figure 5.33. The polymer was purified by dialysis of an aqueous solution using a 3500 g/mol cutoff membrane and the polymer structure (**2**) was confirmed by ^{1}H- and ^{13}C-NMR spectroscopy along with elemental analysis. The thermogravimetric analysis of polyphenylene-ORSO$_3$ under N$_2$ showed an onset of decomposition at around 250 °C. In the temperature range 300 to 350 °C, the polymer lost 30 % of its mass, attributed to initial side chain degradation from a highly thermally stable polymer backbone, and retained 60 % of its mass at 800 °C. The UV-Vis spectra of polyphenylene-ORSO$_3$ in aqueous solution shows an absorption maximum of the $\pi-\pi^*$ transition at 3.6 eV (344 nm) close to that observed for unsubstituted poly(*p*-phenylene) (3.5 eV).

Poly(*p*-phenylene)s are known for their ability to be reversibly electrochemically oxidized (p-type doping) and reduced (n-type doping). Polyphenylene-ORSO$_3$ thin films prepared from aqueous solution on indium tin oxide coated glass are found to be capable of p- and n-type doping. The cyclic voltammogram of polyphenylene-ORSO$_3$ in 0.1 M NaClO$_4$/acetonitrile showed an unresolved oxidation peak; however, upon cathodic dedoping, a peak was resolved at 0.62 V vs Ag wire. This potential value is reportedly close to that observed for poly(*p*-phenylene) in the same electrolyte [108]. The changes in the electronic band structure of the polyphenylene-ORSO$_3$ during electrochemical oxidation and reduction were studied by *in situ* spectroelectrochemistry in 0.1 M NaClO$_4$/acetonitrile. Spectra obtained in the range 2100 to 300 nm as a function of the applied potential are shown in Figures 5.34 and 5.35. For neutral polyphenylene-ORSO$_3$, a single strong absorption band is observed at 3.5 eV with an onset (typically denoted as the electronic band gap) at 3.0 eV (Figure 5.34). In subsequent stepwise electrochemical oxidation, two additional potential dependent absorptions are observed at 0.55 and 2.50 eV attributed to bipolaronic charge carriers, as shown by the band structure inset in Figure 5.34. This band structure is reported to be similar to the one calculated for p-doped poly(*p*-phenylene) by Bredas

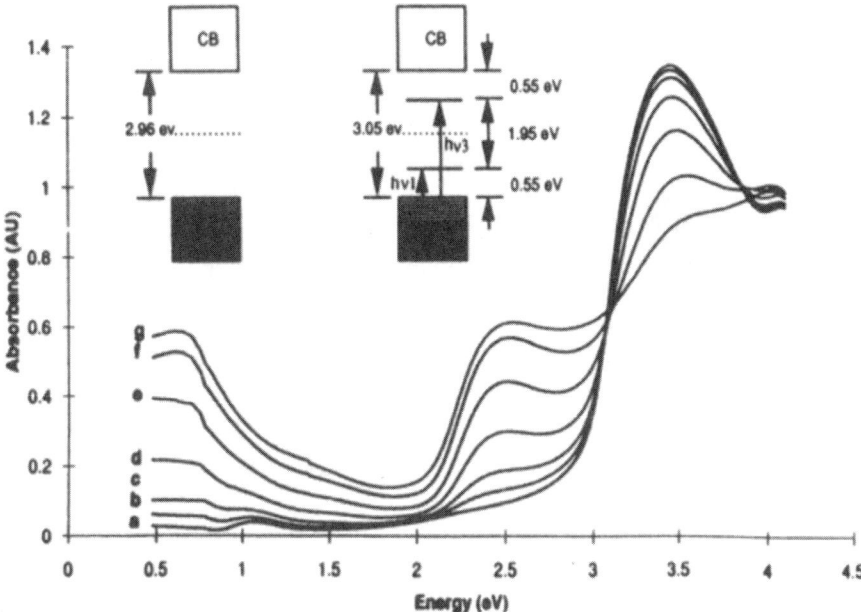

Figure 5.34 In situ spectroelectrochemistry spectra and band structure of polyphenylene-ORSO$_3$ during electrochemical oxidation: (a) 0.0, (b) 0.90, (c) 0.95, (d) 1.00, (e) 1.05, (f) 0 1.10, (g) 1.15 V vs Ag wire. (Reprinted with permission from *Macromolecules*, **27**, 1975. Copyright (1994) American Chemical Society.)

et al. [109, 110] where the energy gap between the valence band and the first bipolaron band (hν1) is 0.56 eV. However, the experimentally calculated hν1 for p-doped poly(p-phenylene) is slightly larger (0.8–1.3 eV), indicating a higher degree of distortion that may be due to Coulombic interactions between charge carriers [111, 112]. These interactions are suggested to be largely screened, in the case of polyphenylene-ORSO$_3$, by the pendant anionic groups.

The band structures observed based on *in situ* spectroelectrochemistry of n-doped polyphenylene-ORSO$_3$ are consistent with negative charge carrier formation as shown in Figure 5.35. The intragap states shown in the figure inset are slightly farther from the band edges in the n-doped polymer than in the p-doped polymer, and suggest a larger degree of distortion. Based on this difference in band structure, and the low stability of the n-doped polymer, it is suggested that the electron donating sulfonatoalkoxy substituents impart a greater electronic stabilizing effect on positive charge carriers compared with negative charge carriers.

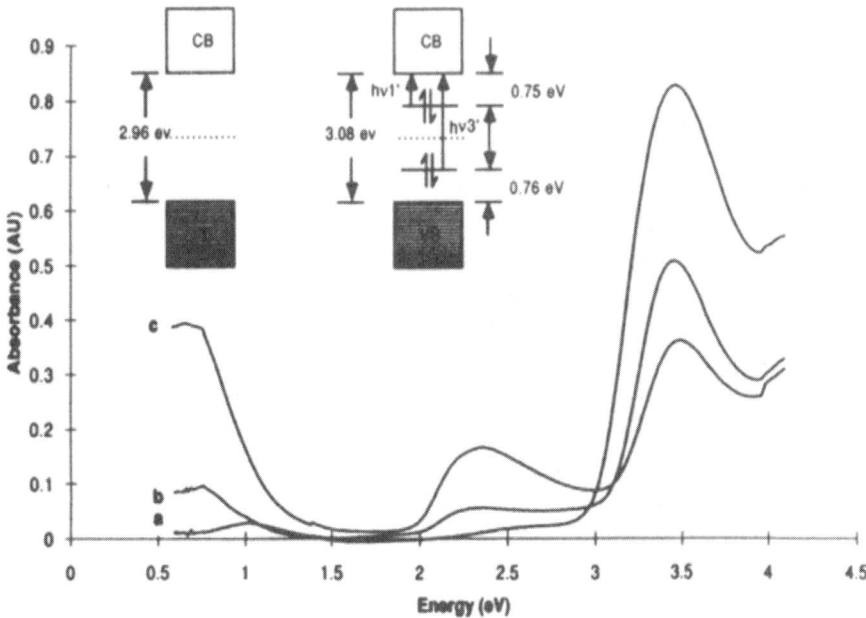

Figure 5.35 *In situ* spectroelectrochemistry spectra and band structure of polyphenylene-ORSO$_3$ during electrochemical reduction: (a) −2.20, (b) −2.30, (c) −2.40 V vs Ag wire. (Reprinted with permission from *Macromolecules*, 27, 1975. Copyright (1994) American Chemical Society.)

5.5 SELF-DOPED POLY(*p*-PHENYLENEVINYLENE)S

Poly(*p*-phenylenevinylene)s have attracted considerable interest due to their electrical [113–119] and nonlinear optical properties [120–123]. Substitution on the phenylene ring of poly(*p*-phenylenevinylene) with different types of electron donating or electron withdrawing groups has shown a significant influence on the electronic structure and electrochemical properties of the polymer [115, 116, 124, 125]. The first water soluble, self-doped and low band gap sulfonated poly(*p*-phenylenevinylene) was prepared by Shi and Wudl [126]. The polymer poly(5-methoxy-2-(3-sulfopropoxy)-1,4-phenylenevinylene) was prepared by a conventional precursor polymer approach as shown in Figure 5.36. In another method, the precursor polymer (**3**) solution in dimethyl formamide/H$_2$O and polymer (**4**) solution in water were heated to reflux under nitrogen in the presence of a small amount of hydrochloric acid. After water dialysis, a red aqueous solution of poly(5-methoxy-2-(3-sulfopropoxy)-1,4-phenylenevinylene) was obtained. The proposed mechanism for the acid catalyzed reaction is shown in Figure 5.37. The polymers (**5**) and (**6**)

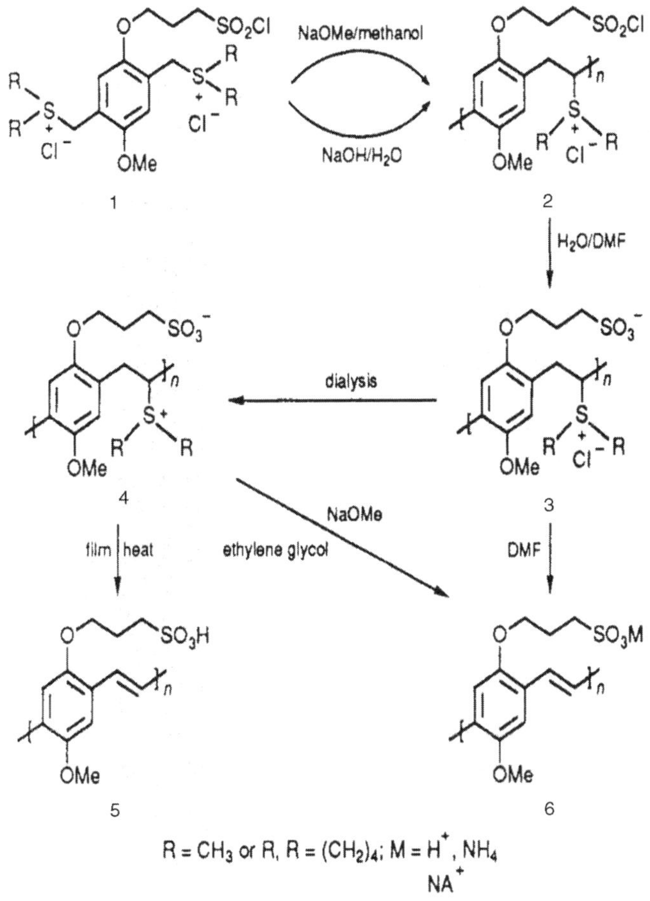

Figure 5.36 Synthesis of alkanesulfonate substituted poly(p-phenylenevinylene). (Reprinted with permission from *Macromolecules*, **23**, 2119. Copyright (1990) American Chemical Society.)

in Figure 5.36, prepared by both methods, exhibit similar FT-IR and UV-Vis characteristics. The characteristic peaks of doped conducting polymer were observed in IR spectra of polymers (**5**) and (**6**), suggesting that the films are partially doped. The UV-Vis spectra made from these polymer films show a major broad peak at 500 nm and small broad peak at 720 nm, as shown in Figure 5.38. The small peak at 720 nm indicates the partial doping of the polymer. This peak disappears after reaction of films with ammonia vapor and in an aqueous solution of the polymer. The conductivities of polymers (**5**) and (**6**) were reported to be around 10^{-6} and 10^{-4}–10^{-2} S/cm, respectively. The molecular weight

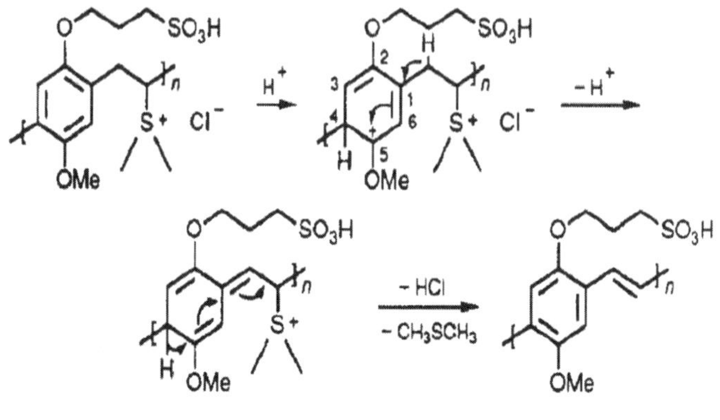

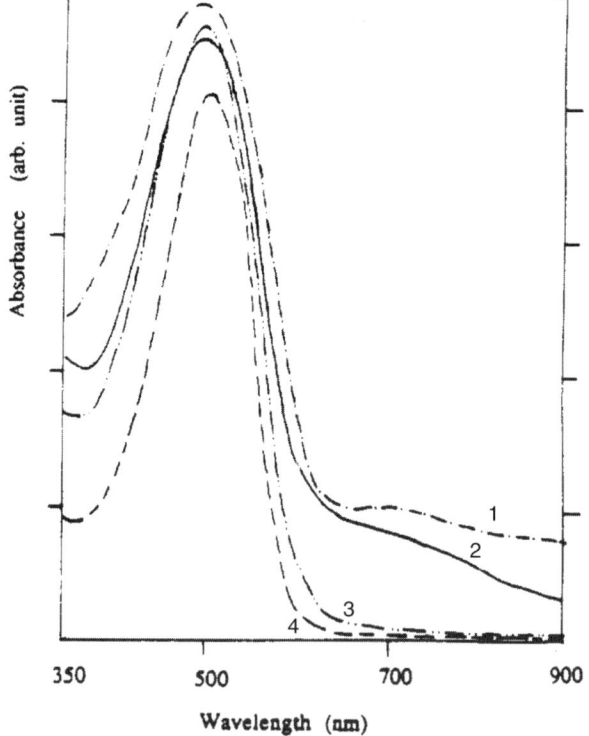

Figure 5.37 Proposed mechanism of synthesis of fully conjugated alkanesulfonate substituted poly(*p*-phenylenevinylene) in the presence of acid. (Reprinted with permission from *Macromolecules*, **23**, 2119. Copyright (1990) American Chemical Society.)

Figure 5.38 UV-Vis spectra of polymer (6) film (1), polymer (6) film (2), polymer (5) film after compensation with NH_3 (3), and polymer (5) aqueous solution (4) (see Figure 5.36). (Reprinted with permission from *Macromolecules*, **23**, 2119. Copyright (1990) American Chemical Society.)

of polymer (6) determined based on gel permeation chromatography using pullulan standard(s) was around 10^6 g/mol.

5.6 SELF-DOPED POLY(INDOLE-5-CARBOXYLIC ACID)

The chemical and electrochemical synthesis of polyindole and its derivatives has been widely studied in the literature [127–137]. However, there is uncertainty regarding the linkage sites of polymerization [127, 129, 138]. The different possibilities of monomer linkages involving the pyrrole ring in 1,3; 1,1-3,3; 2,2-3,3 and 2,3 positions have been suggested in the literature are shown in Figure 5.39 [138, 139]. The carboxylic acid functionalized polyindole, poly(indole-5-carboxylic acid) can be self-doped depending on the pH of solution [67, 134, 140]. Bartlett et al. prepared a micro pH sensor based on this polymer [134]. The polymer showed good stability and rapid response to changes in solution pH. This polymer can also be used for the direct oxidation and reduction of cyctochrome c [141]. The polymerization of poly(indole-5-carboxylic acid) is typically carried out electrochemically in an organic solvent such as acetonitrile. Attempts to grow films from aqueous solutions have been unsuccessful, resulting only in the formation of thin insulating layers on the electrode surface [67]. The electropolymerization of 5-substituted indole involves the formation and polymerization of an electroactive asymmetric trimer species (Figure 5.40) [135]. This trimer species probably acts as a site for linkage to form a polymer containing n connected trimers.

The cyclic voltammograms of the poly(indole-carboxylic acid) films as a function of solution pH are shown in Figure 5.41. The films are reported to be stable up to pH 5; however, the film dissolves above pH 7. Between pH 5 and 7, the voltammetric response decays slowly with repeated cycling. Two sets of redox waves are observed at all pH values, where the first and second redox processes are pH independent and dependent, respectively. The pH sensitivity of poly(indole-carboxylic acid) film was found to be different from that of poly(indole), due to the presence of carboxylate groups within the film [67]. Similarly, the cyclic voltammogram of the poly(indole-5-carboxylic acid) film in acetonitrile shows two redox processes [67, 140]. Shivkumar et al. suggested the self-doping mechanism of poly(indole-5-carboxylic acid), where the

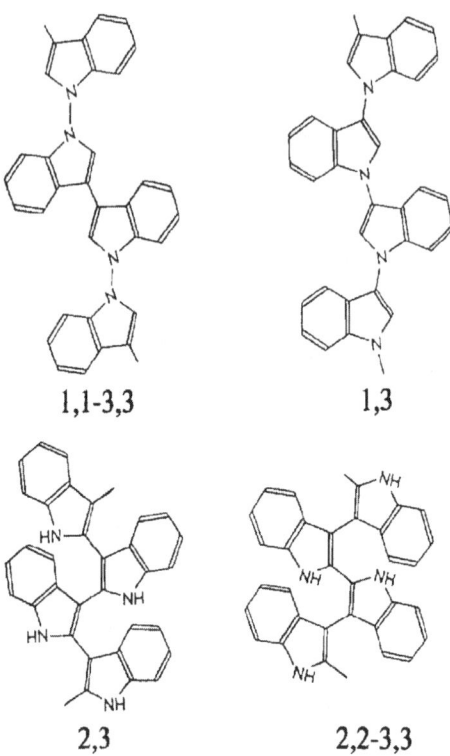

Figure 5.39 The different possible structures of polyindole. (Reprinted from *Journal of Physics and Chemistry of Solids*, 57, H. Talbi, E. B. Maarouf, B. Humbert, M. Alnot, J. Ehrhardt, J. Ghanbaja, D. Billaud, 1145. Copyright (1996), with permission from Elsevier.)

second redox process exchanges the cation between the oxidized and reduced forms of the polymer as shown in Figure 5.42. The poly(indole-5-carboxylic acid) can be used as a cathode material together with a Zn anode in rechargeable cells containing 1 M $ZnSO_4$ at pH 5 [140]. The cell exhibited an open-circuit voltage of 1.36 V and a specific capacity of 67 Ah kg^{-1}. An impedance study suggested that the diffusion of the cation is the rate limiting process as shown in Figure 5.43. The 45° linear behavior in the potential range 0.9 to 1.3 V is attributed to the diffusion of cations away from the polymer electrode. The absence of a semicircle in the high frequency region reportedly suggests that the diffusion of

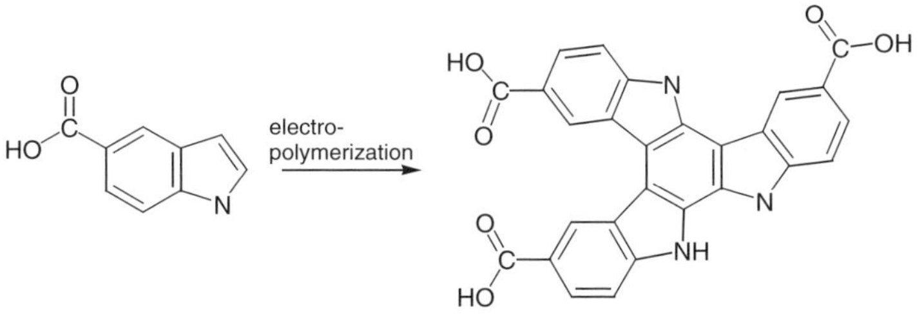

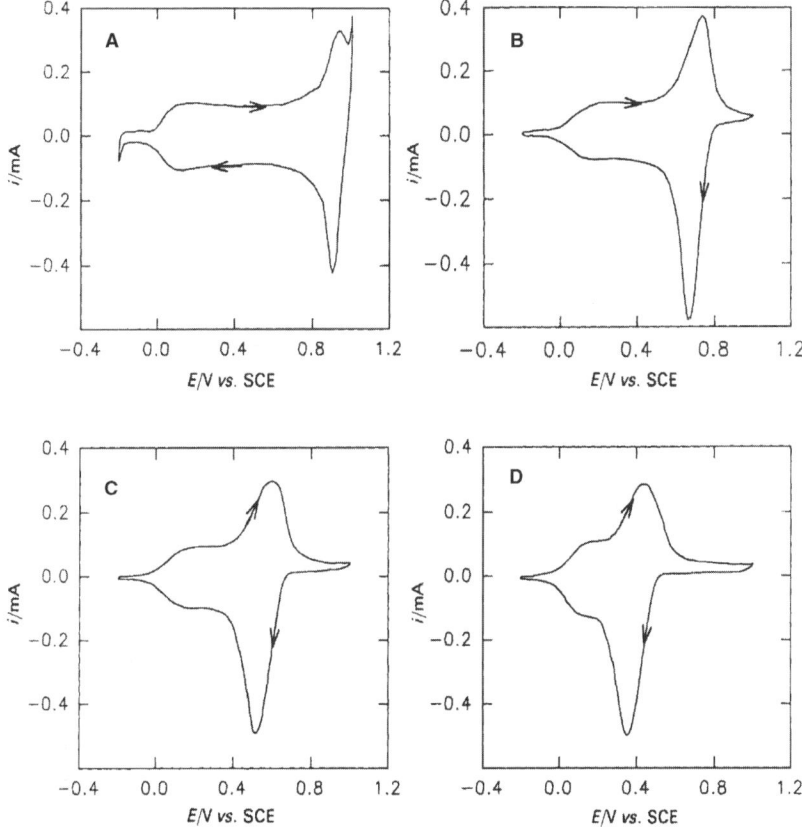

Figure 5.40 Molecular structure of the trimer formed during electropolymerization of (indole-carboxylic acid). (*Journal of the Chemical Society, Faraday Transactions*, 1994, 90, 1121, J. G. Mackintosh, A. R. Mount. Reproduced by permission of The Royal Society of Chemistry.)

Figure 5.41 Cyclic voltammograms of a poly(indole-5-carboxylic acid)-coated platinum electrode at different pH. The data for pH 1.2, 3.0 and 5.0 were recorded in a McIlvaine buffer containing 0.1 M NaCl. In all cases, the sweep rate was 10 mV/s. (A) 2.5 M HCl, (B) pH 1.2, (C) pH 3.0, (D) pH 5. (*Journal of the Chemical Society, Faraday Transactions*, 1992, 88, 2685, P. N. Bartlett, D. H. Dawson, J. Farrington. Reproduced by permission of The Royal Society of Chemistry.)

$$PSCO_2X \rightleftharpoons P^+5CO_2^- + X^+ + e$$

$$X = H^+ \text{ or } Li^+$$

Figure 5.42 Redox reaction in poly(indole-5-carboxylic acid) (P5CO2H). (Reproduced from *Journal of Applied Polymer Science*, 2005, 98, 917. Reprinted with permission of John Wiley & Sons, Inc.)

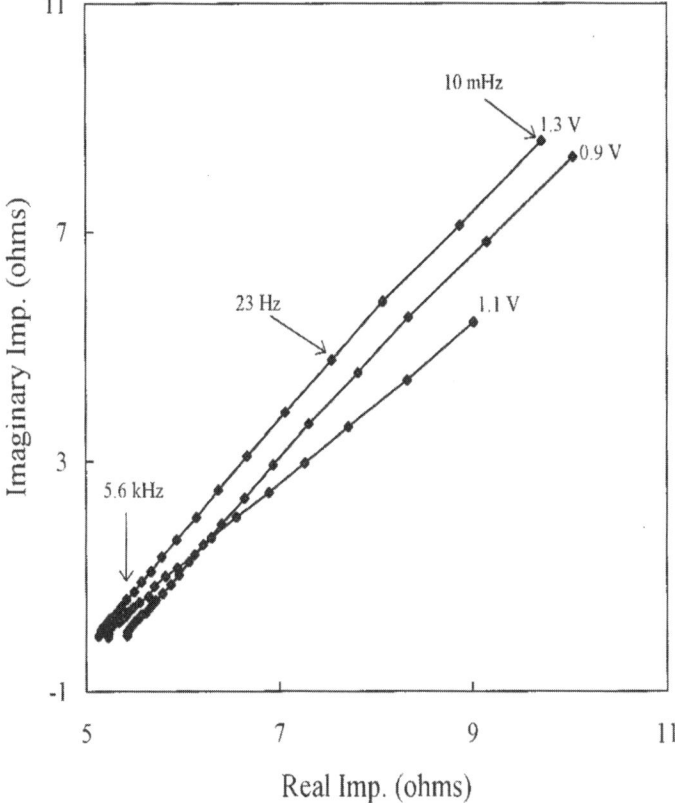

Figure 5.43 Impedance spectra of the cell Zn/1M ZnSO$_4$ (pH 5)/poly(indolecarboxylic acid) at various applied dc potentials in the frequency range of 10 mHz to 100 kHz, (Reproduced from *Journal of Applied Polymer Science*, 2005, 98, 917. Reprinted with permission of John Wiley & Sons, Inc.)

cations is playing a predominant role due to fast charge transfer kinetics. The self-doped conductivity of poly(indole-5-carboxylic acid) film was found to be approximately 10^{-3} S/cm [140, 142].

5.7 SELF-DOPED IONICALLY CONDUCTING POLYMERS

Synthesis of substituted polyacetylenes is highly desirable due to their greater oxidative stability compared with the unsubstituted polyacetylene, and improved processability [143]. However, substituted polyacetylenes have low conductivity and nonlinear optical susceptibility mainly due to loss of conjugation that arises from polymer backbone twisting and steric interactions [144]. A new class of mono and disubstituted ionic polyacetylenes, such as poly(ethynylpyridines) and derivatives, with extensive backbone conjugations, is unique [145–148]. These polyenes contain pyridinium ring substituents associated with halide, methanesulfonate or trifluromethanesulfonate counterions. The conjugation in these systems, despite the presence of substituents, is attributed to strong electrostatic interactions between the pyridinium ring substituents and the counterions, which dominate steric factors that are responsible for twisted backbones in the uncharged system.

Recently, Gal and Jin reported the synthesis of the self-doped ionic conjugated polymers poly(2-ethynylpyridinium-N-benzoylsulfonate) [149] and poly(2-ethynyl-N-(4-sulfobutyl)pyridinium betaine) [150]. Poly(2-ethynylpyridinium-N-benzoylsulfonate) was synthesized by the activated polymerization of 2-ethynylpyridine with the ring opening of 2-sulfobenzoic acid cyclic anhydride without any additional initiator or catalyst, as shown in Figure 5.44. The 1:1 mixture of 2-ethynylpyridine and 2-sulfobenzoic acid cyclic anhydride in dimethyl formamide solvent was stirred for 24 h at 85 °C under a nitrogen atmosphere. The polymer is reported to be very hygroscopic, and soluble in water and organic solvents. The intrinsic viscosity and conductivity of the polymer is around 0.21 dL/g at 25 °C in dimethyl formamide and 5.7×10^{-9} S/cm, respectively. Similarly, Gal and Jin synthesized poly(2-ethynyl-N-(4-sulfobutyl)pyridinium betaine) by activated polymerization of 2-ethynylpyridine with 1,4-butanesultone (Figure 5.45).

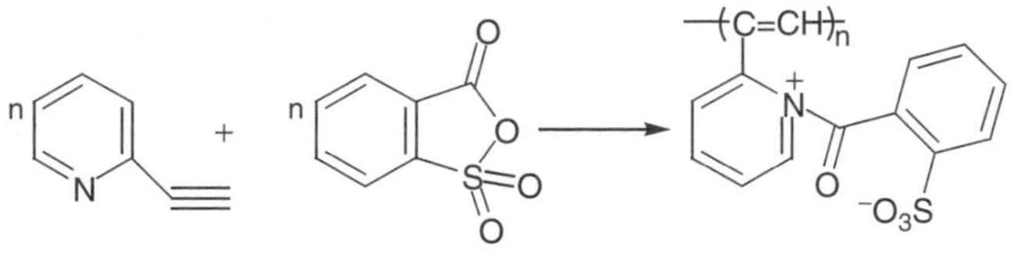

Figure 5.44 Synthesis of poly(2-ethynylpyridinium-N-benzoylsulfonate). (Reproduced from Bulletin of the Korean Chemical Society, 2004, 25, 777.)

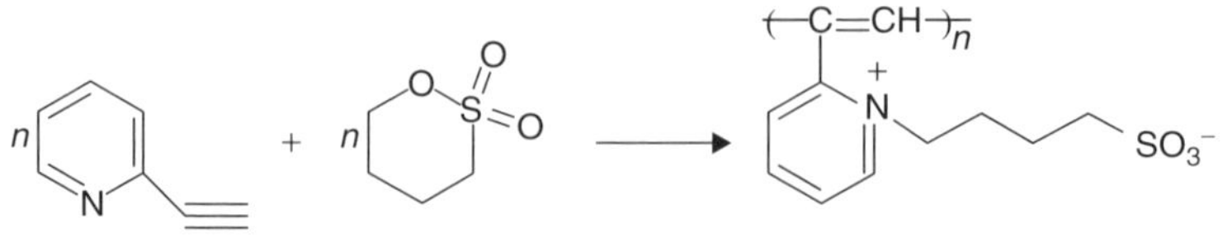

Figure 5.45 Synthesis of poly(2-ethynyl-N-(4-sulfobutyl)pyridinium betaine). (Reprinted from *Current Applied Physics*, 5, Y. S. Gal, S. H. Jin, K. T. Lim, S. H. Kim, K. Koh, 38. Copyright (2005), with permission from Elsevier.)

The conductivity of polymer is observed to be around 3.6×10^{-9} S/cm. These conductivity values of both the polymers (Figure 5.44 and 5.45) are reported to be higher than pyridine based conjugated polymers due to self-doping properties.

REFERENCES

[1] A. D. Dall'Olio, G. Dascola, V. Varacca, V. Bocchi, 'Resonance paramagnetique electronique et conductivité d'un noir d'oxypyrrol electrolytique,' *Comptes. Rendus de l'Académie des Sciences Série.* 1968, 267, 433.
[2] A. F. Diaz, K. K. Kanazawa, G. P. Gardini, 'Electrochemical polymerization of pyrrole,' *Chemical Communications* 1979, 14, 635.
[3] K. K. Kanazawa, A. F. Diaz, R. H. Geiss, W. D. Gill, J. F. Kwak, J. A. Logan, J. F. Rabolt and G. B. Street, 'Organic metals: polypyrrole, a stable synthetic metallic polymer,' *Chemical Communications* 1979, 854.
[4] M. Salmon, K. K. Kanazawa, A. F. Diaz, M. Krounbi, 'A Chemical route to pyrrole polymer films,' *Journal of Polymer Science Part C – Polymer Letters* 1982, 20, 187.
[5] A. Pron, Z. Kucharski, C. Budrowski, M. Zagorska, S. Krichene, J. Suwalski, G. Dehe, S. Lefrant, 'Mossbauer spectroscopy studies of selected conducting polypyrroles,' *Journal of Chemical Physics* 1985, 83, 5923.
[6] H. S. Nalwa, L. R. Dalton, W. F. Schmidt, J. G. Rabe, 'Electrical and optical studies of chemically synthesized polypyrrole,' *Polymer Communications* 1985, 26, 240.
[7] R. E. Myers, 'Chemical oxidative polymerization as a synthetic route to electrically conducting polypyrroles,' *Journal of Electronic Materials* 1986, 15, 61.
[8] S. P. Armes, 'Optimum reaction conditions for the polymerization of pyrrole by iron(III) chloride in aqueous solution,' *Synthetic Metals* 1987, 20, 365.
[9] S. Rapi, V. Bocchi, G. P. Gardini, 'Conducting polypyrrole by chemical synthesis in water,' *Synthetic Metals* 1988, 24, 217.
[10] K. G. Neoh, E. T. Kang, T. C. Tan, 'Effects of acceptor level on chemically synthesized polypyrrole–halogen complexes,' *Journal of Applied Polymer Science* 1989, 37, 2169.

[11] A. F. Diaz, J. Castillo, K. K. Kanazawa, J. A. Logan, M. Salmon, O. Fajardo, 'Conducting poly-N-alkylpyrrole polymer films,' *Journal of Electroanalytical Chemistry* **1982**, *133*, 233.
[12] A. F. Diaz, J. I. Castillo, J. A. Logan, W. Y. Lee, 'Electrochemistry of conducting polypyrrole films,' *Journal of Electroanalytical Chemistry* **1981**, *129*, 115.
[13] J. Heinze, 'Electronically conducting polymers' *Topics in Current Chemistry* **1990**, *152*, 1.
[14] N. C. Billingham, P. D. Calvert, 'Electrically conducting polymers – a polymer science viewpoint,' *Advances in Polymer Science* **1989**, *90*, 1.
[15] A. F. Diaz, J. F. Rubinson, H. B. Mark, 'Electrochemistry and electrode applications of electroactive conductive polymers,' *Advances in Polymer Science* **1988**, *84*, 113.
[16] J. R. Reynolds, 'Electrically conductive polymers,' *Chemtech* **1988**, *18*, 440.
[17] J. R. Reynolds, 'Advances in the chemistry of conducting organic polymers: a review,' *Journal of Molecular Electronics* **1986**, *2*, 1.
[18] A. Deronzier, J. C. Moutet, 'Functionalized polypyrroles – new molecular materials for electrocatalysis and related applications,' *Accounts of Chemical Research* **1989**, *22*, 249.
[19] P. Audebert, G. Bidan, 'Electrochemical study of poly(PHAS) in acetonitrile and water + acetonitrile electrolytes,' *Journal of Electroanalytical Chemistry* **1987**, *238*, 183.
[20] S. Basak, N. Kasinath, D. S. Marynick, K. Rajeshwar, 'Synthesis, characterization, theoretical modeling and polymerization of new fluorophore containing derivatives of thiophene and pyrrole,' *Chemistry of Materials* **1989**, *1*, 611.
[21] M. Salmon, A. F. Diaz, A. J. Logan, M. Krounbi, J. Bargon, 'Chemical modification of conducting polypyrrole films,' *Molecular Crystals and Liquid Crystals* **1982**, *83*, 1297.
[22] G. B. Street, T. C. Clarke, M. Krounbi, K. Kanazawa, V. Lee, P. Pfluger, J. C. Scott, G. Weiser, 'Preparation and characterization of neutral and oxidized polypyrrole films,' *Molecular Crystals and Liquid Crystals* **1982**, *83*, 1285.
[23] G. B. C. Street, T. C. Clarke, R. H. Geiss, V. Y. Lee, A. Nazzal, P. Pfluger, J. C. Scott, 'Characterization of polypyrrole,' *Journal de Physique, Colloque* **1983**, *C3*, 599.
[24] J. Ruhe, T. A. Ezquerra, G. Wegner, 'New conducting polymers from 3-alkylpyrroles,' *Synthetic Metals* **1989**, *28*, C177.
[25] J. Ruhe, T. Ezquerra, G. Wegner, 'Conducting polymers from 3-alkylpyrroles,' *Makromolekulare Chemie – Rapid Communications* **1989**, *10*, 103.
[26] S. Pugh, D. Bloor, 'The use of short chain oligomers as a model for soluble polypyrrole,' *Synthetic Metals* **1989**, *28*, C187.
[27] H. Masuda, S. Tanaka, K. Kaeriyama, 'Soluble conducting polypyrrole – poly(3-octylpyrrole),' *Chemical Communications* **1989**, 725.
[28] K. Kaeriyama, M. A. Sato, K. Hamada, 'Electrochemical preparation of poly (3-methylpyrrole),' *Makromolekulare Chemie – Rapid Communications* **1989**, *10*, 171.
[29] D. Delabouglise, J. Roncali, M. Lemaire, F. Garnier, 'Control of the lipophilicity of polypyrrole by 3-alkyl substitution,' *Chemical Communications* **1989**, 475.

[30] G. Zotti, G. Schiavon, A. Berlin, G. Pagani, 'Electrochemical polymerization of 3-alkylthiopyrroles,' *Synthetic Metals* **1989**, *28*, C183.
[31] A. Merz, R. Schwarz, R. Schropp, '3,4-Dimethoxypyrrole – monomer synthesis and conducting polymer formation,' *Advanced Materials* **1992**, *4*, 409.
[32] W. A. Goedel, G. Holz, G. Wegner, J. Rosenmund, G. Zotti, 'Electrochemical investigations of a substituted oxidation stable polypyrrole,' *Polymer* **1993**, *34*, 4341.
[33] F. Gassner, S. Graf, A. Merz, 'On the physical properties of conducting poly(3,4-dimethoxypyrrole) films,' *Synthetic Metals* **1997**, *87*, 75.
[34] C. L. Gaupp, K. W. Zong, P. Schottland, B. C. Thompson, C. A. Thomas, J. R. Reynolds, 'Poly(3,4-ethylenedioxypyrrole): organic electrochemistry of a highly stable electrochromic polymer,' *Macromolecules* **2000**, *33*, 1132.
[35] Y. Lvov, G. Decher, H. Mohwald, 'Assembly, structural characterization and thermal behavior of layer-by-layer deposited ultrathin films of poly(vinyl sulfate) and poly(allylamine),' *Langmuir* **1993**, *9*, 481.
[36] K. J. Wynne, G. B. Street, 'Poly(pyrrol-2-ylium tosylate) – electrochemical synthesis and physical and mechanical properties,' *Macromolecules* **1985**, *18*, 2361.
[37] R. S. Kohlman, J. Joo, Y. Z. Wang, J. P. Pouget, H. Kaneko, T. Ishiguro, A. J. Epstein, 'Drude metallic response of polypyrrole,' *Physical Review Letters* **1995**, *74*, 773.
[38] P. A. Calvo, J. Rodriguez, H. Grande, D. Mecerreyes, J. A. Pomposo, 'Chemical oxidative polymerization of pyrrole in the presence of *m*-hydroxybenzoic acid and *m*-hydroxycinnamic acid related compounds,' *Synthetic Metals* **2002**, *126*, 111.
[39] A. J. Heeger, 'Semiconducting and metallic polymers: the fourth generation of polymeric materials (Nobel lecture),' *Angewandte Chemie – International Edition* **2001**, *40*, 2591.
[40] A. G. MacDiarmid, ' "Synthetic metals": a novel role for organic polymers (Nobel lecture),' *Angewandte Chemie – International Edition* **2001**, *40*, 2581.
[41] E. T. Kang, K. G. Neoh, Y. L. Woo, K. L. Tan, 'Self-doped polyaniline and polypyrrole. A comparative study by X-ray photoelectron spectroscopy,' *Polymer Communications* **1991**, *32*, 412.
[42] N. S. Sundaresan, S. Basak, M. Pomerantz, J. R. Reynolds, 'Electroactive copolymers of pyrrole containing covalently bound dopant ions – poly(pyrrol-co-3-(pyrrol-1-yl)propanesulphonate),' *Chemical Communications* **1987**, 621.
[43] E. E. Havinga, L. W. Vanhorssen, W. Tenhoeve, H. Wynberg, E. W. Meijer, 'Self-doped water soluble conducting polymers,' *Polymer Bulletin* **1987**, *18*, 277.
[44] J. R. Reynolds, N. S. Sundaresan, M. Pomerantz, S. Basak, C. K. Baker, 'Self-doped conducting copolymers – a charge and mass transport study of poly(pyrrol-co-3-(pyrrol-1-yl)propanesulfonate),' *Journal of Electroanalytical Chemistry* **1988**, *250*, 355.
[45] P. Audebert, G. Bidan, M. Lapkowaki, D. Limosin, *Electronic Properties of Conjugated Polymers*, Springer-Verlag, Berlin, **1987**.
[46] G. Bidan, B. Ehui, M. Lapkowski, 'Conductive polymers with immobilised dopants: ionomer composites and auto-doped polymers – a review and recent advances,' *Journal of Physics D, Applied Physics* **1988**, *21*, 1043.

[47] E. E. Havinga, W. Ten Hoeve, E. W. Meijer, H. Wynberg, 'Water soluble self-doped 3-substituted polypyrroles,' *Chemistry of Materials* **1989**, *1*, 650.
[48] G. Zotti, S. Zecchin, G. Schiavon, A. Berlin, 'Low defect neutral, cationic and anionic conducting polymers from electrochemical polymerization of N-substituted bipyrroles. Synthesis, characterization, and EQCM analysis,' *Chemistry of Materials* **2002**, *14*, 3607.
[49] C. Masalles, J. Llop, C. Vinas, F. Teixidor, 'Extraordinary overoxidation resistance increase in self-doped polypyrroles by using nonconventional low charge density anions,' *Advanced Materials* **2002**, *14*, 826.
[50] G. Wegner, W. Wernet, D. T. Glatzhofer, J. Ulanski, C. Krohnke, M. Mohammadi, 'Chemistry and conductivity of some salts of polypyrrole,' *Synthetic Metals* **1987**, *18*, 1.
[51] A. Witkowski, M. S. Freund, A. Brajter-Toth, 'Effect of electrode substrate on the morphology and selectivity of overoxidized polypyrrole films,' *Analytical Chemistry* **1991**, *63*, 622.
[52] C. Masalles, S. Borros, C. Vinas, F. Teixidor, 'Are low coordinating anions of interest as doping agents in organic conducting polymers?,' *Advanced Materials* **2000**, *12*, 1199.
[53] C. Masalles, S. Borros, C. Vinas, F. Teixidor, 'Surface layer formation on polypyrrole films,' *Advanced Materials* **2002**, *14*, 449.
[54] G. Sonmez, I. Schwendeman, P. Schottland, K. W. Zong, J. R. Reynolds, 'N-substituted poly(3,4-propylenedioxypyrrole)s: high gap and low redox potential switching electroactive and electrochromic polymers,' *Macromolecules* **2003**, *36*, 639.
[55] Y. Sahin, A. Aydin, Y. A. Udum, K. Pekmez, A. Yildiz, 'Electrochemical synthesis of sulfonated polypyrrole in FSO_3H/acetonitrile solution,' *Journal of Applied Polymer Science* **2004**, *93*, 526.
[56] S. Basak, K. Rajeshwar, M. Kaneko, 'Ion binding by poly(pyrrole-co-3-(pyrrol-1-yl)propanesulfonate) thin films,' *Analytical Chemistry* **1990**, *62*, 1407.
[57] D. M. Collard, M. S. Stoakes, 'Lamellar conjugated polymers by electrochemical polymerization of heteroarene containing surfactants – potassium 3-(3-alkylpyrrol-1-yl)propanesulfonates,' *Chemistry of Materials* **1994**, *6*, 850.
[58] P. C. Innis, Y. C. Chen, S. Ashraf, G. G. Wallace, 'Electrohydrodynamic polymerisation of water soluble poly((4-(3-pyrrolyl))butane sulfonate),' *Polymer* **2000**, *41*, 4065.
[59] E. C. Chang, W. Y. Hua, S. A. Chen, 'Synthesis and properties of the water soluble self-acid-doped polypyrrole: poly(4-(3-pyrrolyl)butanesulfonic acid),' *Journal of Polymer Research – Taiwan* **1998**, *5*, 249.
[60] H. C. Li, E. Khor, 'A collagen–polypyrrole hybrid – influence of 3-butanesulfonate substitution,' *Macromolecular Chemistry and Physics* **1995**, *196*, 1801.
[61] W. S. Yin, E. Ruckenstein, 'A water soluble self-doped conducting polypyrrole based copolymer,' *Journal of Applied Polymer Science* **2001**, *79*, 86.
[62] W. J. Bae, K. H. Kim, W. H. Jo, Y. H. Park, 'A water soluble and self-doped conducting polypyrrole graft copolymer,' *Macromolecules* **2005**, *38*, 1044.

[63] C. R. Chenthamarakshan, A. Ajayaghosh, 'Synthesis and properties of water soluble squaraine oligomers containing pendant propanesulfonate moieties,' *Chemistry of Materials* **1998**, *10*, 1657.

[64] H. D. Tabba, K. M. Smith, 'Anodic oxidation potentials of substituted pyrroles derivation and analysis of substituent partial potentials,' *Journal of Organic Chemistry* **1984**, *49*, 1870.

[65] M. G. Cross, D. Walton, N. J. Morse, R. J. Mortimer, D. R. Rosseinsky, D. J. Simmonds, 'A voltammetric survey of steric and beta-linkage effects in the electropolymerization of some substituted pyrroles,' *Journal of Electroanalytical Chemistry* **1985**, *189*, 389.

[66] P. G. Pickup, 'Poly-(3-methylpyrrole-4-carboxylic acid) – an electronically conducting ion exchange polymer,' *Journal of Electroanalytical Chemistry* **1987**, *225*, 273.

[67] P. N. Bartlett, D. H. Dawson, J. Farrington, 'Electrochemically polymerized films of 5-carboxyindole – preparation and properties,' *Journal of the Chemical Society – Faraday Transactions* **1992**, *88*, 2685.

[68] D. Delabouglise, F. Garnier, 'Poly (3-carboxymethyl pyrrole), a pH sensitive, self-doped conducting polymer,' *New Journal of Chemistry* **1991**, *15*, 233.

[69] H. K. Youssoufi, F. Garnier, A. Yassar, S. Baiteche, P. Srivastava, 'A protonpump electrode based on poly(3-carboxymethylpyrrole),' *Advanced Materials* **1994**, *6*, 755.

[70] K. S. Ryder, D. G. Morris, J. M. Cooper, 'Tailored polymers to probe the nature of the bioelectrochemical interface,' *Langmuir* **1996**, *12*, 5681.

[71] M. D. Ingram, H. Staesche, K. S. Ryder, ' "Activated" polypyrrole electrodes for high-power supercapacitor applications,' *Solid State Ionics* **2004**, *169*, 51.

[72] P. N. Bartlett, M. C. Grossel, E. M. Barrios, 'Electrochemistry and contact angle measurements on polymer films of omega-(3-pyrrolyl)-alkanoic acids in aqueous solution,' *Journal of Electroanalytical Chemistry* **2000**, *487*, 142.

[73] J. M. Freitas, M. L. Duarte, T. Darbre, L. M. Abrantes, 'Electrochemical synthesis of a novel conducting polymer: the poly (3-pyrrolyl)-octanoic acid,' *Synthetic Metals* **2005**, *155*, 549.

[74] J. F. Morin, M. Leclerc, D. Ades, A. Siove, 'Polycarbazoles: 25 years of progress,' *Macromolecular Rapid Communications* **2005**, *26*, 761.

[75] J. V. Grazulevicius, P. Strohriegl, J. Pielichowski, K. Pielichowski, 'Carbazole-containing polymers: synthesis, properties and applications,' *Progress in Polymer Science* **2003**, *28*, 1297.

[76] M. S. J. H. Pearson, *Polymer Monographs*, Vol. 6, Gordon and Breach, New York, **1981**.

[77] M. Stolka, *Encyclopedia of Polymer Science and Engineering*, Vol. 11, Wiley, New York, **1988**.

[78] P. M. Boresenberger, D. S. Weiss, *Organic Photoreceptors in Xerography*, Marcel
Dekker, Inc., New York, **1998**.

[79] R. H. Partridge, 'Electro-luminescence from polyvinylcarbazole films. 1. Carbazole cations,' *Polymer* **1983**, *24*, 733.

[80] R. H. Partridge, 'Electro-luminescence from polyvinylcarbazole films. 3. Electroluminescent devices,' *Polymer* **1983**, *24*, 748.

REFERENCES

[81] R. H. Partridge, 'Electro-luminescence from polyvinylcarbazole films. 2. Polyvinylcarbazole films containing antimony pentachloride,' *Polymer* **1983**, *24*, 739.

[82] R. H. Partridge, 'Electro-luminescence from polyvinylcarbazole films. 4. Electro-luminescence using higher work function cathodes,' *Polymer* **1983**, *24*, 755.

[83] A. Desbenemonvernay, P. C. Lacaze, J. E. Dubois, 'Polaromicrotribometric (PMT) and IR, ESCA, electron-paramagnetic-res spectroscopic study of colored radical films formed by the electrochemical oxidation of carbazoles. 1. Carbazole and N-ethyl, N-phenyl and N-carbazyl derivatives,' *Journal of Electroanalytical Chemistry* **1981**, *129*, 229.

[84] J. E. Dubois, A. Desbenemonvernay, P. C. Lacaze, 'Polaromicrotribometric (PMT) and IR, ESCA, electron-paramagnetic-res spectroscopic study of colored radical films formed by the electrochemical oxidation of carbazoles. 2. N-vinylcarbazole,' *Journal of Electroanalytical Chemistry* **1982**, *132*, 177.

[85] P. C. Lacaze, J. E. Dubois, A. Desbenemonvernay, P. L. Desbene, J. J. Basselier, D. Richard, 'Polymer-modified electrodes as electrochromic material. 3. Formation of poly-N-vinylcarbazole films on transparent semiconductor ITO surfaces by electropolymerization of NVK in acetonitrile,' *Journal of Electroanalytical Chemistry* **1983**, *147*, 107.

[86] A. Desbenemonvernay, P. C. Lacaze, J. E. Dubois, P. L. Desbene, 'Polymer-modified electrodes as electrochromic material. 4. Spectroelectrochemical properties of poly-N-vinylcarbazole films,' *Journal of Electroanalytical Chemistry* **1983**, *152*, 87.

[87] Y. J. Qiu, J. R. Reynolds, 'Poly 3,6-(carbaz-9-yl)propanesulfonate – a self-doped polymer with both cation and anion exchange properties,' *Journal of the Electrochemical Society* **1990**, *137*, 900.

[88] G. Tourillon, *Handbook of Conducting Polymers*, Dekker, New York, **1986**.

[89] R. H. Baughman, J. L. Bredas, R. R. Chance, R. L. Elsenbaumer, L. W. Shacklette, 'Structural basis for semiconducting and metallic polymer dopant systems,' *Chemical Reviews* **1982**, *82*, 209.

[90] P. Kovacic, M. B. Jones, 'Dehydro coupling of aromatic nuclei by catalyst oxidant systems – poly(*para*-phenylene),' *Chemical Reviews* **1987**, *87*, 357.

[91] J. G. K. Speight, P. Kovacie, F. W. Koch, 'Synthesis and properties of polyphenyls and polyphenylenes,' *Journal of Macromolecular Science, Reviews in Macromolecular Chemistry* **1971**, *5*, 295.

[92] G. Grem, G. Leditzky, B. Ullrich, G. Leising, 'Realization of a blue-light-emitting device using poly(*para*-phenylene),' *Advanced Materials* **1992**, *4*, 36.

[93] M. Rehahn, A. D. Schlueter, G. Wegner, W. J. Feast, 'Soluble poly(*para*-phenylenes). 1. Extension of the Yamamoto synthesis to dibromobenzenes substituted with flexible side chains,' *Polymer* **1989**, *30*, 1054.

[94] M. Rehahn, A. D. Schlueter, G. Wegner, W. J. Feast, 'Soluble poly(*para*-phenylenes). 2. Improved synthesis of poly(*para*-2,5-di-n-hexylphenylene) via palladium catalyzed coupling of 4-bromo-2,5-di-n-hexylbenzeneboronic acid,' *Polymer* **1989**, *30*, 1060.

[95] M. Rehahn, A. D. Schluter, G. Wegner, 'Soluble poly(*para*-phenylene)s. 3. Variation of the length and the density of the solubilizing side chains,'

Makromolekulare Chemie – Macromolecular Chemistry and Physics **1990**, *191*, 1991.
[96] Y. H. Kim, O. W. Webster, 'Water-soluble hyperbranched polyphenylene – a unimolecular micelle,' *Journal of the American Chemical Society* **1990**, *112*, 4592.
[97] Y. H. Kim, O. W. Webster, 'Hyperbranched polyphenylenes,' *Macromolecules* **1992**, *25*, 5561.
[98] T. I. Wallow, B. M. Novak, 'In aqua synthesis of water soluble poly(*para*-phenylene) derivatives,' *Journal of the American Chemical Society* **1991**, *113*, 7411.
[99] T. I. N. Wallow, M. Bruce, 'Aqueous synthesis of soluble rigid chain polymers. An ionic poly(*p*-phenylene) analog,' *Polymer Preprints* **1991**, *32*, 191.
[100] V. Chaturvedi, S. Tanaka, K. Kaeriyama, 'Preparation of poly(*p*-phenylene) via processable precursors,' *Journal of the Chemical Society-Chemical Communications* **1992**, 1658.
[101] V. Chaturvedi, S. Tanaka, K. Kaeriyama, 'Preparation of poly(*p*-phenylene) via a new precursor route,' *Macromolecules* **1993**, *26*, 2607.
[102] I. U. Rau, M. Rehahn, 'Rigid-rod polyelectrolytes: carboxylated poly(*para*-phenylenes) via a novel precursor route,' *Polymer* **1993**, *34*, 2889.
[103] I. U. Rau, M. Rehahn, 'Towards rigid-rod polyelectrolytes via well defined precursor poly(*para*-phenylene)s substituted by 6-iodohexyl side chains,' *Acta Polymerica* **1994**, *45*, 3.
[104] G. Brodowski, A. Horvath, M. Ballauff, M. Rehahn, 'Synthesis and intrinsic viscosity in salt free solution of a stiff chain cationic poly(*p*-phenylene) polyelectrolyte,' *Macromolecules* **1996**, *29*, 6962.
[105] R. Rulkens, M. Schulze, G. Wegner, 'Rigid-rod polyelectrolytes – synthesis of sulfonated poly(*p*-phenylene)s,' *Macromolecular Rapid Communications* **1994**, *15*, 669.
[106] V. Cimrova, W. Schmidt, R. Rulkens, M. Schulze, W. Meyer, D. Neher, 'Efficient blue light emitting devices based on rigid-rod polyelectrolytes,' *Advanced Materials* **1996**, *8*, 585.
[107] A. D. Child, J. R. Reynolds, 'Water-soluble rigid-rod polyelectrolytes – a new self-doped, electroactive sulfonatoalkoxy substituted poly(*p*-phenylene),' *Macromolecules* **1994**, *27*, 1975.
[108] J. F. Fauvarque, M. A. Petit, A. Digua, G. Froyer, 'Electrochemical synthesis of poly(1,4-phenylene) films,' *Makromolekulare Chemie – Macromolecular Chemistry and Physics* **1987**, *188*, 1833.
[109] J. L. Bredas, R. R. Chance, R. Silbey, 'Comparative theoretical study of the doping of conjugated polymers – polarons in polyacetylene and polyparaphenylene,' *Physical Review B* **1982**, *26*, 5843.
[110] J. L. Bredas, B. Themans, J. G. Fripiat, J. M. Andre, R. R. Chance, 'Highly conducting polyparaphenylene, polypyrrole and polythiophene chains – an *ab initio* study of the geometry and electronic structure modifications upon doping,' *Physical Review B* **1984**, *29*, 6761.
[111] I. Rubinstein, 'Electrochemistry of polyphenylene films deposited anodically on platinum or glassy carbon electrodes in HF–benzene system,' *Journal of the Electrochemical Society* **1983**, *130*, 1506.

REFERENCES

[112] G. Froyer, Y. Pelous, F. Maurice, M. A. Petit, A. Digua, J. F. Fauvarque, 'Optical studies on poly(para-phenylene) thin film prepared by electroreduction,' *Synthetic Metals* **1987**, *21*, 241.

[113] I. Murase, T. Ohnishi, T. Noguchi, M. Hirooka, S. Murakami, 'Highly conducting poly(para-phenylene vinylene) prepared from sulfonium salt,' *Molecular Crystals and Liquid Crystals* **1985**, *118*, 333.

[114] D. R. Gagnon, J. D. Capistran, F. E. Karasz, R. W. Lenz, S. Antoun, 'Synthesis, doping and electrical conductivity of high molecular weight poly(paraphenylene vinylene),' *Polymer* **1987**, *28*, 567.

[115] I. Murase, T. Ohnishi, T. Noguchi, M. Hirooka, 'Alkoxy-substituent effect of poly(para-phenylene vinylene) conductivity,' *Polymer Communications* **1985**, *26*, 362.

[116] C. C. Han, R. W. Lenz, F. E. Karasz, 'Highly conducting, iodine doped copoly (phenylene vinylene)s,' *Polymer Communications* **1987**, *28*, 261.

[117] J. L. Jin, C. K. Park, H. K. Shim, Y. W. Park, 'Highly conducting poly(2-N-butoxy-5-methoxy-1,4-phenylene vinylene),' *Chemical Communications* **1989**, 1205.

[118] W. B. Liang, R. W. Lenz, F. E. Karasz, 'Poly(2-methoxyphenylene vinylene) – synthesis, electrical conductivity and control of electronic properties,' *Journal of Polymer Science Part A – Polymer Chemistry* **1990**, *28*, 2867.

[119] J. I. Jin, C. K. Park, H. K. Shim, 'Synthesis and electroconductivities of poly(2-methoxy-5-methythio-1,4-phenylene vinylene) and copolymers,' *Journal of Polymer Science Part A – Polymer Chemistry* **1991**, *29*, 93.

[120] T. Kaino, K. L. Kubodera, S. Tomaru, T. Kurihara, S. Saito, T. Tsutsui, S. Tokito, 'Optical 3rd harmonic generation from poly(para-phenylenevinylene) thin films,' *Electronics Letters* **1987**, *23*, 1095.

[121] B. P. Singh, P. N. Prasad, F. E. Karasz, '3rd order nonlinear optical properties of oriented films of poly(para-phenylene vinylene) investigated by femtosecond degenerate four wave mixing,' *Polymer* **1988**, *29*, 1940.

[122] D. McBranch, M. Sinclair, A. J. Heeger, A. O. Patil, S. Shi, S. Askari, F. Wudl, 'Linear and nonlinear optical studies of poly(para-phenylene vinylene) derivatives and polydiacetylene-4BCMU,' *Synthetic Metals* **1989**, *29*, E85.

[123] J. Swiatkiewicz, P. N. Prasad, F. E. Karasz, M. A. Druy, P. Glatkowski, 'Anisotropy of the linear and 3rd order nonlinear optical properties of a stretch oriented polymer film of poly-2,5-dimethoxy paraphenylenevinylene,' *Applied Physics Letters* **1990**, *56*, 892.

[124] S. H. R. Askari, D. Soonil, F. Wudl, 'Substituted poly(phenylene vinylene) conducting polymers: rigid-rod polymers with flexible side chains,' *Polymeric Materials Science and Engineering* **1988**, *59*, 1068.

[125] J. I. Jin, Y. H. Lee, H. K. Shim, 'Synthesis and characterization of poly(2-methoxy-5-nitro-1,4-phenylenevinylene) and poly(1,4-phenylenevinylene-co-2-methoxy-5-nitro-1,4-phenylenevinylene)s,' *Macromolecules* **1993**, *26*, 1805.

[126] S. Q. Shi, F. Wudl, 'Synthesis and characterization of a water soluble poly (para-phenylenevinylene) derivative,' *Macromolecules* **1990**, *23*, 2119.

[127] G. Tourillon, F. Garnier, 'New electrochemically generated organic conducting polymers,' *Journal of Electroanalytical Chemistry* **1982**, *135*, 173.

[128] F. Garnier, G. Tourillon, M. Gazard, J. C. Dubois, 'Organic conducting polymers derived from substituted thiophenes as electrochromic material,' *Journal of Electroanalytical Chemistry* **1983**, *148*, 299.
[129] R. J. Waltman, A. F. Diaz, J. Bargon, 'Substituent effects in the electropolymerization of aromatic heterocyclic compounds,' *Journal of Physical Chemistry* **1984**, *88*, 4343.
[130] K. M. Choi, C. Y. Kim, K. H. Kim, 'Polymerization mechanism and physicochemical properties of electrochemically prepared polyindole tetrafluoroborate,' *Journal of Physical Chemistry* **1992**, *96*, 3782.
[131] K. Jackowska, A. Kudelski, J. Bukowska, 'Spectroelectrochemical and EPR determination of the number of electrons transferred in redox processes in electroactive polymers – polyindole films,' *Electrochimica Acta* **1994**, *39*, 1365.
[132] D. Billaud, E. B. Maarouf, E. Hannecart, 'An investigation of electrochemically and chemically polymerized indole,' *Materials Research Bulletin* **1994**, *29*, 1239.
[133] R. Holze, C. H. Hamann, 'Electrosynthetic aspects of anodic reactions of anilines and indoles,' *Tetrahedron* **1991**, *47*, 737.
[134] P. N. Bartlett, J. Farrington, 'Electrochemically polymerized films of 5-carboxyindole: possible application as a micro pH sensor,' *Bulletin of Electrochemistry* **1992**, *8*, 208.
[135] J. G. Mackintosh, A. R. Mount, 'Electropolymerization of indole-5-carboxylic acid,' *Journal of the Chemical Society – Faraday Transactions* **1994**, *90*, 1121.
[136] J. G. Mackintosh, C. R. Redpath, A. C. Jones, P. R. R. Langridgesmith, A. R. Mount, 'The electropolymerization and characterization of 5-cyanoindole,' *Journal of Electroanalytical Chemistry* **1995**, *388*, 179.
[137] H. Talbi, B. Humbert, D. Billaud, 'Polyindole and poly(5-cyanoindole): electrochemical and FT-IR spectroscopic comparative studies,' *Synthetic Metals* **1997**, *84*, 875.
[138] G. Zotti, S. Zecchin, G. Schiavon, R. Seraglia, A. Berlin, A. Canavesi, 'Structure of polyindoles from anodic coupling of indoles – an electrochemical approach,' *Chemistry of Materials* **1994**, *6*, 1742.
[139] H. Talbi, E. B. Maarouf, B. Humbert, M. Alnot, J. J. Ehrhardt, J. Ghanbaja, D. Billaud, 'Spectroscopic studies of electrochemically doped polyindole,' *Journal of Physics and Chemistry of Solids* **1996**, *57*, 1145.
[140] S. R. Sivakkumar, N. Angulakshmi, R. Saraswathi, 'Characterization of poly(indole-5-carboxylic acid) in aqueous rechargeable cells,' *Journal of Applied Polymer Science* **2005**, *98*, 917.
[141] P. N. Bartlett, J. Farington, 'The electrochemistry of cytochrome *c* at a conducting polymer electrode,' *Journal of Electroanalytical Chemistry* **1989**, *261*, 471.
[142] P. S. Abthagir, R. Saraswathi, 'Electronic properties of polyindole and polycarbazole schottky diodes,' *Organic Electronics* **2004**, *5*, 299.
[143] J. C. W. Chien, *Polyacetylene: Chemistry, Physics, and Material Science.*, Academic Press, Orlando, FL, **1984**.
[144] W. Deits, P. Cukor, M. Rubner, H. Jopson, 'Analogs of polyacetylene – preparation and properties,' *Industrial and Engineering Chemistry Product Research and Development* **1981**, *20*, 696.

REFERENCES

[145] S. Subramanyam, A. Blumstein, 'Conjugated ionic polyacetylenes. 2. A new polymerization method for substituted acetylenes,' *Makromolekulare Chemie – Rapid Communications* **1991**, *12*, 23.

[146] S. Subramanyam, A. Blumstein, 'Conjugated ionic polyacetylenes. 3. Polymerization of ethynylpyridinium salts,' *Macromolecules* **1991**, *24*, 2668.

[147] S. Subramanyam, A. Blumstein, 'Conjugated ionic polyacetylenes. 5. Spontaneous polymerization of 2-ethynylpyridine in a strong acid,' *Macromolecules* **1992**, *25*, 4058.

[148] S. B. Clough, X. F. Sun, S. Subramanyam, N. Beladakere, A. Blumstein, S. K. Tripathy, 'Molecular dynamics simulation of substituted conjugated ionic polyacetylenes,' *Macromolecules* **1993**, *26*, 597.

[149] Y. S. Gal, S. H. Jin, 'A self-doped ionic conjugated polymer: poly(2-ethynylpyridinium-N-benzoylsulfonate) by the activated polymerization of 2-ethynylpyridine with ring-opening of 2-sulfobenzoic acid cyclic anhydride,' *Bulletin of the Korean Chemical Society* **2004**, *25*, 777.

[150] Y. S. Gal, S. H. Jin, K. T. Lim, S. H. Kim, K. Koh, 'Synthesis and electro-optical properties of self-doped ionic conjugated polymers: poly 2-ethynyl-N-(4-sulfobutyl)pyridinium betaine,' *Current Applied Physics* **2005**, *5*, 38.

Index

Note: Figures and Tables are indicated by *italic numbers*

3-(3-alkylpyrrol-1-yl)propanesulfonates, potassium salt
  polymerization of 276–7
  structure *276*
amine sensors 196–9
aminonaphthylenes, substituted, homopolymerization of 87
3-aminophenylboronic acid
  electropolymerization of 158–65
  in presence of NADH and NAD$^+$, ^{11}B NMR spectra *192*
  see also poly(anilineboronic acid) (PABA)
ammonia gas sensor 197
'aniline black' 1
aniline-based derivatives see polyaniline...
anthranilic acid (2-aminobenzoic acid) 123
  see also poly(aniline-co-anthranilic acid)
aromatic π-conjugated polymers, ground state structure 14, *15*
auto doping 38–42, 219
avidin, detection of 246

band theory 11
batteries 55–6, 133–4, 219
biocatalyzed synthesis, ICPs 22
biochromic polymers 245–7
biosensors 50, 52–3, 187–99, 288
biotin–avidin interactions 245
bipolaron bands (in UV-Vis spectra)
  poly(anilineboronic acid) 174
    interaction with RNA 202, *203*
  polyphenylenes 294, *295*, *296*
  polypyrrole sulfonic acid derivatives 281
  polythiophene sulfonic acid derivatives 225, 240
bipolaron(s) 14
  in polypyrrole
    binding energy 16
    formation of 15–16, *15*
  in polythiophene derivatives 221, *222*
    formation of *34*
boronate ester formation 156, *157*
  infrared vibrations due to 192
  tetrahedral boronate ester 170–1, *171*
boronic acid substituted polyanilines 156–218
  see also poly(anilineboronic acid)
butylamine, detection threshold(s) for 197
butylamine sensor 197–9

carboxylic acid derivatives see polyaniline...; polypyrrole...; polythiophene, carboxylic acid derivatives
charge injection 19–20
chemical doping 10, *11*
chemical synthesis
  ICPs 20
  limitations 20
  polyaniline derivatives 27–8, 77–91, 123–8, 130
  polypyrrole derivatives 277–82
  polythiophene derivatives 234–48, 249–50, 253, 266–6

circular dichroism spectra,
    poly(2-methoxyaniline-5-sulfonic
    acid) films 93
[3,3′-Co(8-C$_4$H$_4$N-(CH$_2$)-O-(CH$_2$)-O-
    1,2-C$_2$B$_9$H$_{10}$)(1′,2′-C$_2$B$_9$H$_{11}$)]$^-$,
    as doping agent in polypyrroles
    269–71
colloidal dispersions, processable ICPs
    24
composite polymers, polyanilines
    96–7
composites (polymer + filler) 3
conducting polymer composites 3
    drawbacks 3
conducting polymers
    comparison with self-n-doped
        polymers 37
    concept of doping 9–10
    conduction mechanism 11–20
    expansion in research 2, 7
    history 1–9
    processability 22–5
    synthesis 20–2
conduction mechanism 10–20
conductivity
    conjugated polymers 1
    effect of substituents on properties
        44–5
    ionically conducting polymers 303,
        304
    polyaniline carboxylic acid
        derivatives 124, 128
    polyaniline copolymers 33, 88, 90,
        91, 95, 124
    polyaniline phosphonic acid
        derivatives 130
    polyaniline sulfonic acid derivatives
        102–6
        pH dependence 102, 104–5
        temperature dependence 105–6
    poly(anilineboronic acid),
        temperature dependence 185–7
    polycarbazoles 291
    polyindole derivatives 302
    poly(*p*-phenylenevinylene)s 297
    polypyrrole copolymers 280, 281
    polypyrrole derivatives 27, 263, 268,
        274, 285
    polythiophene derivatives 221,
        228–9, 230, 239
        variation with temperature 225–8
    squaraine oligomers 284
conjugated conducting polymers,
    history 1–9

conjugated metallopolymers 5
conjugated polymers
    chemical structure 8
    doped 3, 8–9
    electrical conductivity 1
    self-doping in 25–6
controlled release applications
    layer-by-layer fabrication technique
        199–206
        advantages over encapsulation
            strategies 204
        potential-induced release technique
            205–6
copolymerization
    aniline sulfonic acid derivatives
        87–91, 94–6, 97
    pyrrole sulfonic acid derivatives
        279–82
counterion-induced processability
    ICPs 23–4
    polyaniline derivatives 29–30
critical micelle concentration, potassium
    3-(3-alkylpyrrol-1-yl)propanesul-
    fonates 276
Curie's Law 222
cyclic voltammograms
    electropolymerization of
        3-aminophenylboronic acid
        159, *161*
    polyaniline *110*
    poly(anilineboronic acid)
        in presence of fructose and fluoride
            166, *167*, *169*
        in presence of NADH and NAD$^+$
            190
    polycarbazoles *291*
    polyindole derivatives 299, *301*
    polyphenylenes 294, *295*, *296*
    polypyrrole derivatives *264*, *272*, *285*
    polythiophene derivatives *222*, *223*
cytochrome *c*, mediation of
    electrochemistry 138, 140, 288,
    299

DC conductivity
    effect of substituents on properties
        44–5
    sulfonated polyanilines, pH
        dependence *105*
    see also conductivity
degenerate ground state 12
diazonium salts, coupling with,
    polyaniline 84–5

diblock copolymers, polyaniline derivatives 90, 122
dip-pen nanolithography 56–7
dopant ions, sources 21
doped conjugated polymers 3
doping 8, 9–10
 (conventional) semiconductors 9–10
 intrinsically conducting polymers 10
 see also oxidative (p-type) doping; reductive (n-type) doping
doping mechanism, self-doped conducting polymers 33–42

electrochemical copolymerization
 advantages 94
 aniline sulfonic acid derivatives 94–6
electrochemical doping 10
electrochemical synthesis
 advantages 22, 94
 ICPs 21–2
 polyaniline sulfonic acid derivatives 92–8
  in aqueous media 92–7
  in non-aqueous media 97–8
 polythoiphene derivatives 220–34
 polypyrrole and derivatives 263–77
  in aqueous media 274–7
  in non-aqueous media 263–74, 285
electrochromic behavior
 polycarbazoles 291
 polypyrrole derivatives 55, 274, 277
electrochromic devices 54–5, 138, 219, 290
electrochromism, meaning of term 54
electroluminescence, polyphenylene derivatives 293
electroluminescent devices 290
electron spin resonance spectroscopy
 sulfonated polyanilines 116–18
 sulfonated polythiophenes 222
electron-beam lithography 53–4
 disadvantages 53
electronic properties, effect of substituents 46–7
electrophilic substitution, polyaniline and derivatives 27–8, 77–9, 97–8
enzymatic synthesis
 ICPs 24
 polyaniline derivatives 98–100
  in presence of polyelectrolytes 30, 32

enzyme-based sensors 187

field effect transistors 50, 51, 219
fuel cells, polymer electrolyte membranes in 47

gel permeation chromatography, molecular weight measurements 118, 175, 249, 280–1
glucose sensors 187
graft copolymers
 polyaniline derivatives 33, 88–9, 90–1, 122
 polypyrrole derivatives 32, 33, 280–2

Heeger, Alan J. 1, 2
homopolymerization
 aniline sulfonic acid derivatives 86–7, 92–4, 97
 pyrrole sulfonate derivatives 277–8
hyperbranched sulfonated polydiphenylamine (H-PSDA) 91
 morphology 91, 122–3
 self-doping mechanism 91

indium-doped tin oxide (ITO) coated glass electrodes 21
intrinsically conducting polymers (ICPs) 6–9
 applications 7
 commercial suppliers 7
 doped, anions to balance polarons 19
 doping of 9–10
 processability 22–5
  colloidal dispersions 24
  counterion-induced processability 23–4
  by enzymatic synthesis 24
  by in situ polymerization 24
  by reduction to non-conducting state 23
  by self-doping 24–5
  by substitution of alkyl chains 23
 properties 2–3
 synthesis 20–2
  biocatalyzed synthesis 22
  chemical synthesis 20
  electrochemical synthesis 21–2
  photochemical synthesis 22
ion exchangers 55
ionically conducting polymers 6, 302–4

light-emitting diodes (LEDs)   49, 219
lithium secondary batteries   55, 133–4
lithography *see* dip-pen nanolithography; electron-beam lithography

MacDiarmid, Alan G.   1, 2
matrix assisted laser desorption ionization (MALDI) measurements   256
mechanical properties
 crosslinked poly(anilineboronic acid) 181–2
 effect of substituents 47
metal oxide semiconductor field effect transistors (MOSFETs)   50, 51
molecular-level processing   48–50
 for controlled release of RNA 199–206
 poly(anilineboronic acid) used 199–206
multilayer films
 fabrication of 49–50, 199, 201–2
 P3TOPP films 257–8
 PABA–PABA bilayers 204
 PABA/RNA films 203–4
 PAH–PEDOT-S films 248

n-doping, ICPs   10
n-type self-doped conducting polymers, doping mechanism   36–8
nanocomposites, polyaniline-containing 133–4
nanofabrication, dip-pen nanolithography used   57
nanostructures, self-doped polyanilines 132–40
β-nicotinamide adenine dinucleotide (NADH)
 complexes with boronate esters *193*
 detection of 189–96
 interconversion to oxidized form (NAD$^+$) *189*
NMR spectroscopy
 ^{11}B NMR
  3-aminophenylboronic acid 160–2, *162, 170*
  complexation of PABA with nucleotides 191–2
  poly(anilineboronic acid) 182–4
 ^{19}F NMR, 3-aminophenylboronic acid *170*

magic angle spinning (MAS) NMR 182–3
nucleophilic substitution, polyaniline 79–83, *84*, 96
nucleotide sensors   189–96

olfaction, amine sensor compared with human ability   197
oligo(2,3-dicarboxyaniline)   126
 redox behavior 126, *127*
open circuit potential measurement, complexation of PABA with nucleotides studied by   194–6
oxidative (p-type) doping
 polyacetylene 10, *11*, 12, *13*
 polyaniline 17–19
 polypyrrole 14–16

p-doping, ICPs   10
p-type self-doped conducting polymers, doping mechanism   33–6
PABA *see* poly(anilineboronic acid)
PAH–PEDOT-S multilayer films   248
 scan rate dependence *248*
paramagnetism, poly(*p*-phenylene) derivatives   293
Pauli susceptibility   116, 117
phosphonic acid derivatives *see* polyaniline...; polythiophene, phosphonic acid derivatives
photobleaching, poly(n-(3′-thienyl)-alkanesulfonate)s   240–1
photochemical synthesis, ICPs   22
photodoping   19
photolithography, water-based photoresist developed for   237–8
photovoltaic cells   219
π-conjugated polymers, ground states 12
polarization modulated infrared reflection absorption spectroscopy (PM-IRRAS)
 complexation of PABA with nucleotides studied by 192, 194
 RNA interactions with PABA studied by 203
polaron bands (in UV-Vis spectra)
 phosphonated polythiophenes 256, 257
 poly(anilineboronic acid) 173–4
 sulfonated polyanilines 110, 111
 sulfonated polythiophenes 225

INDEX

polaron(s) 14
 in polypyrrole
  binding energy 16
  formation of 14, *15*
 in polythiophene derivatives,
  formation of *34*
polyacetylene(s) 2
 chemical structure *8*
 cis–trans isomerization 12
 crystalline films 2, 7
 ground state structure 12
 iodine-doped 8–9
 as ionically conducting polymers 303
 n-doped 2
 oxidative (p-type) doping of 10, *11*, 12, *13*
 p-doped 2
 reductive (n-type) doping of 10, *11*, *13*
poly(2-acrylamido-2-methyl-1-propane-sulfonic acid), polyaniline grafted on 89
poly(2-acrylamido-2-methyl-1-propane-sulfonic acid-co-*N*-(4-anilino phenyl)-methylacrylate) (poly(AMP-co-APMA)), polypyrrole grafted on 279–80
poly(3-(4-alkanesulfonate)thiophene), self-p-doping of *33*, *34*
poly(3-alkylsulfonatepyrrole)s
 solubility 268
 synthesis 265–6
 UV-Vis-NIR spectra 266, *267*
poly(3-alkylthiophene) 23
poly(allylamine hydrochloride)
 in PAH–PEDOT-S multilayer film 248
 structure *247*
poly(aminobenzenesulfonic acid)
 molecular weights 118
 nanoparticles 139–40
 solubility 101
 synthesis 92
poly(*o*-aminobenzenesulfonic acid-co-aniline) 'microflowers' 136, *137*
poly(2-aminobenzoic acid) [poly(*o*-aminobenzoic acid)] 123, 126, 128
poly(4-aminobenzoic acid) [poly(*p*-aminobenzoic acid)] 99, 100, 126

poly(2-aminobenzoic acid)/polyaniline copolymer *see* poly(aniline-co-anthranilic acid)
poly(*o*-aminobenzylphosphonic acid)
 conductivity affected by neutralization 130
 sodium salts 130, *132*
 spectroscopic studies 131–2
 synthesis 130, *131*
poly(5-aminonaphthalene-2-sulfonic acid) 87
polyaniline
 applications 75
 biocatalyzed synthesis of 22
 chemical structure *8*
 commercial suppliers 7
 composites 96–7
 conductivity, variation with temperature *186*
 coupling with diazonium salts 84–5
 deprotonated base 17, *18*, 23
 doping of 17–19, *18*
 effect of sulfonic acid group 35–6
 electrophilic aromatic substitution, sulfonated derivatives 27–8, 77–9
 emeraldine base form 17, *18*, 23
  nucleophilic substitution reactions 79–80, 81, 83
  reaction with propanesultone 79, *81*
  sulfonation of 27–8, 77–9
 emeraldine base sulfonated
  conductivity 102, *106*
  electron spin resonance data *117*
  X-ray photoelectron spectra 114, *115*
 emeraldine salt 18
 HCl-doped
  thermal stability 48, 184
  UV-Vis spectra 110, *111*
  X-ray photoelectron spectra 114, *115*
 leucoemeraldine base form 17
  nucleophilic substitution reaction 80
  sulfonation of 78
 leucoemeraldine base sulfonated
  conductivity 102, *106*
  electron paramagnetic resonance behavior 117
  X-ray photoelectron spectra 116
 limitations as conducting polymer 75
 nanostructures 132–40

polyaniline (*continued*)
  nucleophilic substitution, sulfonated derivatives 79–83, *84*
  oxidation states 17, *18*, 109
  pernigraniline base form 17
    nucleophilic substitution reaction 82–3
    sulfonation of 78–9
  post-polymerization modification 27–8, 77–85
  processability 23
  redox activity 107
  ring-sulfonated
    chemical structure(s) *28*, *35*, *76*
    conductivity 44, *103*, *104*, 106
    first reported 27, 76
    formation of 27–9
    solubility 101
    UV-Vis spectra 110, 111, *111*
  self-doped
    in biosensors 52–3
    in dip-pen nanolithography 57
  thermal stability 47
polyaniline, alkyl-substituted, molecular weight 45
polyaniline, boronic acid substituted *see* poly(anilineboronic acid)
polyaniline, carboxylic acid derivatives 123–9
  chemical synthesis 123–8
  electrochemical synthesis 128–9
polyaniline derivatives, self-doped, examples 27–8, *30*
polyaniline diblock copolymers 90, 122
polyaniline graft copolymers 33, 88–9, 90–1, 122
polyaniline, mercaptopropanesulfonic acid substituted
  synthesis 83
  thermal decomposition mechanism 120–1
  thermal stability 119–20, 184
polyaniline, phosphonic acid derivatives 130–2
polyaniline, propylthiosulfonated, thermal stability 47
polyaniline, sulfonic acid derivatives
  acid–base equilibrium 35–6
  chemical synthesis 27–8, 77–91
    polymerization of monomers 86–91
  post-polymerization modification 27–8, 77–85
  electrochemical synthesis 92–8
    in aqueous media 92–7
    in non-aqueous media 97–8
  electron spin resonance spectroscopy 116–18
  enzymatic synthesis 98–100
  HCl-doped 79, *80*, 101
  interchain interactions 34–5, *35*
  intrachain interactions 34, *35*
  multilayer film 49–50
  oxidation states *108*
  properties 100–23
    conductivity 102–6
    electronic and spectroscopic properties 46–7, 109–18
    molecular weight 118
    morphology 121–3
    pH-dependent redox behavior 107–9
    solubility 44, 76, 101–2
    thermal stability 119–21, 184
  in rechargeable batteries 56
  self-n-doping of 37
  self-p-doping of 33–4
  synthesis by copolymerization 87–91, 94–6, 97
  synthesis by homopolymerization 86–7, 92–4, 97
  synthesis by post-polymerization modification 27–8, 77–85, 96, 97–8
  thermal decomposition mechanism 120–1
  thermal stability 47, 119–20
  UV-VIS spectroscopy 109–12
  X-ray photoelectron spectroscopy 112–16
poly(aniline-co-2-acrylamido-2-methyl-1-propanesulfonic acid) 89
poly(aniline-N-alkylsulfonates) 97
poly(aniline-co-*o*-aminobenzenesulfonic acid) nanostructures 136, 137–8, *137*, 140
poly(aniline-co-2-aminobenzoic acid) 123–5
poly(aniline-co-3-aminobenzoic acid) 125
poly(aniline-co-aminonaphthalene-sulfonic acid) nanostructures 134–5, *135*, 136
poly(aniline-co-anthranilic acid)
  morphology *129*

structure *124*
synthesis 123–4, 125, 128
poly(aniline-co-N-benzoylsulfonic acid-aniline) 82, *83*
poly(anilineboronic acid) (PABA) 156–218
　applications 187–206
　　amine sensors 196–9
　　molecular level processing 199–206
　　nucleotide sensors 189–96
　　saccharide sensors 187–9
　chemical synthesis 165
　complexation with nucleotides 189–90
　　^{11}B NMR and 191–2
　　cyclic voltammetry and 190–1
　　PM-IRRAS spectra and 192, 194
　crosslinked 177–87
　　^{11}B NMR 182–4
　　characterization of 179–80
　　hardness measurement 181–2
　　mechanical properties 181–2
　　synthesis 179
　　temperature dependent conductivity 185–7
　　thermal properties 184–5
　electrochemical synthesis 158–65
　emeraldine base form 166, 168, 199
　emeraldine salt form 167, *179*
　first synthesized 157
　interactions with RNA 200–1
　investigation of 202–3
　leucoemeraldine base form 166
　post-polymerization modification of 176
　properties 166–77
　　conductivity 165
　　molecular weight 45–6, 175–7
　　pH-dependent redox behavior 166–71
　　redox behavior 158, 166–71
　　spectroscopy 172–5
　　thermal stability 48, 184–5
　self-doping mechanism 157–8
　self-doping properties 177–8
　sensitivity to various saccharides *188*
　water-soluble form, synthesis of 165, 179, 201
poly(aniline-N-butanesulfonic acid) 97
poly(aniline-N-butyl sulfonate), in electrochromic devices 54–5
poly(aniline-2,5-disulfonic acid) 92

conductivity 102, *104*
　in polyaniline composites 96–7
poly(aniline-co-metanilic acid) 95, 109
poly(anilinepropanesulfonic acid) 80–1
poly(aniline-N-propanesulfonic acid) 97
poly(aniline-co-N-propanesulfonic acid-aniline) 81–2, *82*
poly(aniline-2-sulfonic acid), in polyaniline composites 96–7
poly(aniline-2-sulfonic acid-co-aniline) 90–1
poly(2,5-bis(3-sulfonatopropoxy)-1,4-phenylen-alt-1,4-phenylene) 293–6
　see also poly(p-phenylene), alkoxysulfonated
poly(3,6-(carbaz-9-yl)propanesulfonate) 290–2
　ion exchange properties 55, 291, *292*
　redox behavior 291
　self-doping mechanism 291, *292*
　synthesis 290–1
poly(3-(ω-carboxyalkyl)thiophene)s
　doping mechanism for *251*
　properties 250
　synthesis 249–50
poly(4-carboxy-2,2′-bithiophene), synthesis 253
poly(3-carboxymethylpyrrole)
　pH-dependent anodic peak potential characteristics 287, *288*
　synthesis 287
poly(2-carboxyphenylene-1,4-diyl) 293
polycondensation, squaraine oligomers prepared via 282–4
poly(4-(4H-cyclopenta[2,1-b:3,4-b′]-dithienyl)butanesulfonate)
　conductivity 228–9, *230*
　self-doping mechanism 229, *231*
　synthesis 228
　UV-Vis spectra 228, *229*
poly(2,5-diaminobenzenesulfonate) 99–100
poly(4-(2,3-dihydrothien[3,4-b]-[1,4]dioxin-2-yl-methoxy)-1-butanesulfonic acid, sodium salt)
　in PAH–PEDOT-S multilayer film 248
　structure *247*
　synthesis 247

polydiphenylamine
  hyperbranched sulfonated 91
  morphology 91, 122–3
  self-doping mechanism 91
poly(dipropargyl-N-hexyl-N-methyl-
    ammonium triflate) 36, 37
polyelectrolytes, as dopants 30
poly(ethylaniline-co-sulfoanisidine)
    nanoparticles 135–6
poly(3,4-ethylenedioxythiophene),
    commercial supplier 7
poly(3,4-ethylenedioxythiophene)-sulfo-
    nate (polyEDTS)
  self-doped, structure 233
  synthesis 229–33
poly(2-ethynylpyridinium-N-benzoyl-
    sulfonate) 303
poly(2-ethynyl-N-(4-sulfobutyl)pyri-
    dinium betaine) 303–4
poly(3-hexylthiophene) (P3HT),
    photobleaching of 242
poly(2-hydroxy-1,4-phenylene) 38
  tetrabutyl ammonium salt 38
polyindole, structures 299, 300
poly(indole-5-carboxylic acid)
    299–302
  cyclic voltammograms 299, 301
  self-doping mechanism 299, 302
  synthesis 299, 301
polymer electrolytes 6
polymeric acids, as dopants 30
poly(2-methoxyaniline-5-sulfonic acid)
    86, 92–4
  conductivity 86, 93
  optically active films 93–4
  UV-Vis-NIR spectra 111–12, 113
poly(3-methylpyrrole-4-carboxylic acid)
  in biosensors 288
  redox behavior 285–7
  synthesis 285
poly(2-(4-methyl-3-thienyloxy)ethane-
    sulfonate)
  oxidation potential 244
  synthesis 242–4
poly(2-(4-methyl-3-thienyloxy)ethane-
    sulfonic acid)
  conductivity 244
  reaction with biocytin hydazine 246
poly(3-methylthiophene)
  morphology 234, 235
  sulfonated
    morphology 234, 235
    solubility 234
    synthesis 233–4

poly(p-phenylene)
  chemical structure 8
  ground state structure 15
poly(p-phenylene), alkoxysulfonated
    293–6
  band structure
    during electrochemical oxidation
        295–6, 296
    during electrochemical reduction
        295, 296
  optoelectrochemical spectra
    during electrochemical oxidation
        294–5
    during electrochemical reduction
        295, 296
  synthesis 293–4
  UV-Vis spectra 294
poly(p-phenylene)s 292–6
  synthesis 293–4
poly(p-phenylenevinylene)s 296–9
  chemical structure 8
poly(p-phenylenevinylene)s,
    alkanesulfonate substituted
  conductivity 297
  IR spectra 297
  synthesis 296, 297, 298
  UV-Vis spectra 297, 298
poly(potassium 3-(3-alkylpyrrol-1-yl)
    propanesulfonate)s
  electrochromic behavior 277
  lamellar structure 277, 278
  synthesis of 276–7
poly(3,4-propylenedioxypyrrole),
    N-propanesulfonated
  conductivity 274
  electrochromic behavior 55, 274
  redox behavior 272–4
  synthesis 271–2
  synthesis of monomer 271, 272
poly(3′-propylsulfonate
    2,2′:5′,5″-terthienyl)
  self-doped, UV-Vis spectra 224, 225
  synthesis 26–7, 224
polypyrrole
  band structure upon doping 16–17,
      17
  2,2′-bipyrrole based 268–9
  chemical structure 8
  electrochemical synthesis 21, 262,
      263
  oxidative (p-type) doping 14–16
  photochemical synthesis 22
  ring-sulfonated, conductivity 44–5
  self-doped 27, 262–84

INDEX

water-soluble forms 263, 278
polypyrrole, alkylammonium
    substituted 268
polypyrrole, alkylsulfonate substituted
    268–9, 277
polypyrrole, carboxylic acid derivatives
    284–9
polypyrrole derivatives
    overoxidation of 270–1
    self-doped, examples *31*
polypyrrole graft copolymer, self-doped
    32, 33
polypyrrole, sulfonated derivatives
    chemical synthesis
        by copolymerization 279–82
        by homopolymerization 277–8
    electrochemical synthesis
        in aqueous media 274–7
        fluorosulfonic acid as sulfonation
            agent 274, 275, 276
        in non-aqueous media 263–74
    synthesis by polycondensation 282–4
poly(pyrrole-co-AMP-co-APMA)
    synthesis of 279–80
    UV-Vis spectra 280
poly(pyrrole-N-propionic acid),
    synthesis 288, *289*
poly(pyrrole-co(3-(pyrrol-1-yl)propane-
    sulfonate))
    cyclic voltammogram *264*, *265*
    synthesis 27, 263–4, 274–5
poly(3-pyrrolylacetic acid) 288
poly(4-(3-pyrrolyl)butanesulfonate)
    277, 278
poly(4-(3-pyrrolyl)butanesulfonic acid)
    278
poly(3-(pyrrolyl)butanoic acid) 288,
    289
poly(3-(pyrrolyl)carboxylic acid) 288,
    289
poly(β-(pyrrolyl)octanoic acid) 289
poly(3-(pyrrolyl)pentanoic acid) 288,
    289
poly(sodium
    (3-pyrrolyl)alkanesulfonate) 27,
    28, 277–8
poly(styrenesulfonic acid-g-aniline)
    89–90
poly(styrenesulfonic acid-g-pyrrole)
    chain conformation in solution
        281–2
    conductivity 281
    hydrodynamic volume *282*
    structure *32*, *281*

synthesis 33, 280, *281*
UV-Vis spectra 281
poly(4-sulfobenzeneazo-(N-methyl-
    aniline)) 85
poly(sulfonic diphenyl
    aniline)/poly(ethylene oxide)
    diblock copolymer 90
poly(sulphur nitride) 2
poly(thieno[3,4-b]thiophene)
    sulfonated 249
    synthesis 249
poly(ω-(3-thienyl)alkanesulfonates)
    219
poly(n-(3'-thienyl)-alkanesulfonate)s
    aggregation model 240, *241*
    conductivities 239
    configurational isomers 239, *240*
    photobleaching 240–1
    synthesis 237
    UV-Vis spectra 239–40
    X-ray diffraction spectra 239
poly(n-(3'-thienyl)alkanesulfonic acids)
    (P3TASHs) 38
    auto-doping in 39, *40*, 224
    basic units *42*
    effect of side chain length on
        properties 39–40, *41*, 45,
        224–5
    hygroscopicity 39
    protonation studies 40–2
poly(2-(3'-thienyl)decanesulfonic acid)
    (P3TDSH) 39
poly(2-(2-thienyl)3-(4-dimethyldodecyl-
    ammoniumphenyl)thiophene
    triflate) 36–7
poly(2-(3'-thienyl)ethanesulfonate),
    sodium salt, IR spectra *41*
poly(2-(3'-thienyl)ethanesulfonic acid)
    (P3TESH) 39–40
    conductivity changes with
        temperature 225–8
    IR spectra *41*
    protonation studies 40–1, *42*
    thermal dedoping behavior and
        mechanism 225, *226*–8
    UV-Vis-NIR spectral changes with
        temperature 225, *226*
poly(2-(3'-thienyl)hexanesulfonate),
    sodium salt
    IR spectra *41*
    photobleaching of *242*
poly(2-(3'-thienyl)hexanesulfonic acid)
    (P3THSH) 39–40
    IR spectra *41*

324 INDEX

poly(2-(3′-thienyl)hexanesulfonic acid) (P3THSH) (continued)
  photobleaching of 242
  protonation studies 40–1, 42
poly(8-(3-thienyl)octanoic acid) films, self-doping in 250–1
poly(2-(3-thienyloxy)ethanesulfonate)
  oxidation potential 244
  structure 243
  synthesis 242–4
poly(2-(3-thienyloxy)ethanesulfonic acid), conductivity 244
poly(3-(3′-thienyloxy)propanephosphonate)
  in multilayer films 257–8
  self-acid-doping mechanism 257
  synthesis 255–6
  UV-Vis spectra 256, 257
poly(3-(3′-thienyloxy)propanesulfonate) 55
poly(3-(3′-thienyl)propanesulfonate)
  synthesis 38–9, 234, 236
  UV-Vis spectra 236, 237
poly(3-(3′-thienyl)propanesulfonic acid) (P3TPSH) 38–9
  electrochemical behavior 236–7, 238
  in electrochromic devices 54
  time-dependent UV-Vis-NIR spectra 237, 239
  UV-Vis spectra 236, 237
polythioenylene, electrochemical synthesis of 21–2
polythiophene, chemical structure 8
poly(thiophene-3-acetic acid)
  crosslinked film 253, 255
  electrochemical passivation of 251–3
  synthesis 251, 253, 254
poly(3-thiophene alkanesulfonate)s, sodium salt and acid forms
  conductivity 221
  electron spin resonance measurements 222
  self-doping mechanism 222
  synthesis 220–1
  UV-Vis spectra 221–2
polythiophene, alkyl-substituted
  first synthesized 219
  molecular weight 45
poly(3-thiophene butanesulfonate), sodium salt, cyclic voltammogram 222
poly(3-thiophene butanesulfonic acid)
  cyclic voltammogram 223

UV-Vis absorption spectra 221
polythiophene, carboxylate derivatives
  synthesis 249–55
  thin conducting layers 49
polythiophene derivatives 219–61
  commercial supplier 7
  n-type doping of 36–7
  p-type doping of 33, 34
  processability 23
  self-doped, examples 31
  self-doping in 25–6, 33, 34, 36–7
  superconductivity in 20
  synthesis 219
polythiophene, phosphonate derivatives 255–8
2,5-poly(thiophene-3-propionic acid) 253, 254
polythiophene, sulfonate derivatives
  chemical synthesis 234–48
  electrochemical synthesis 220–34
  post-polymerization modification 249
  self-p-doping of 33, 34
  solubility 44
post-polymerization modification
  poly(anilineboronic acid) 176
  polyanilines 27–8, 45, 77–85
  polythiophenes 249
protonic acid doping 18–19
pyrrole-based derivatives see polypyrrole...
3-(pyrrole-1-yl)propane sulfonate, polycondensation with squaric acid 282–3

quartz crystal microbalance measurements 195, 229, 268

radiation resistance, poly(p-phenylene) derivatives 293
rechargeable batteries 55–6, 300
redox behavior
  effect of substituents 46
  oligo(2,3-dicarboxyaniline) 126, 127
  poly(anilineboronic acid) 166–71
  sulfonated polyanilines 107–9
redox polymers 3–5
reductive (n-type) doping, polyacetylene 10, 11, 13
ribonucleic acid(s) (RNA)
  bilayer interactions with PABA 201
  controlled release from thin films 199–206

INDEX

saccharide sensors 187–9
scanning electron micrographs (SEMs)
  polyaniline nanostructures 135, 137, 138, 139
  polyanilines 129, 135
self-acid-dosing 38–42
self-doped conducting polymers
  applications 48–57
    biosensors 52–3
    dip-pen nanolithography 56–7
    electrochromic devices 54–5
    electron-beam lithography 53–4
    ion exchangers 55
    molecular-level processing 48–50
    rechargeable batteries 55–6
    transistors 50, 51–2
  background history 25–9
  doping mechanism 33–42
    auto dosing 38–42
    n-type dosing 36–8
    p-type dosing 33–6
  driving force behind research 25
  effect of substituents on properties 42–8
    DC conductivity 44–5
    electronic and spectroscopic properties 46–7
    mechanical properties 47
    molecular weight 45–6
    redox properties 46
    solubility 43–4
    thermal properties 47–8
  first reported 25
  processability 24–5
  types 29–33
semiconductors
  doping of 8–10
  energy band gap in 11
Shirakawa, Hideki 1, 2
solitons 12, 13, 14
solubility
  effect of substituents 46
  polyaniline sulfonic acid derivatives 44, 76, 101–2
  polypyrrole derivatives 268
  polythiophene sulfonic acid derivatives 44, 234
solvatochromism,
  poly(n-(3'-thienyl)alkanesulfonic acids) 39
spectroscopic properties
  effect of substituents 46–7

poly(anilineboronic acid) 172–5
sulfonated polyanilines 46–7, 109–18
squaraine oligomers, synthesis 282–4
sulfonic acid derivatives see polyaniline...; polypyrrole...; polythiophene, sulfonic acid derivatives
superconductors
  poly(sulphur nitride) 2
  polythiophene derivative(s) 20
surface plasmon resonance spectroscopy, polyaniline copolymers studied by 128–9
'synthetic metals' 3
  see also intrinsically conducting polymers

thermal stability
  effect of substituents 47–8
  polyaniline and derivatives 47–8, 119–21, 184–5
thieno[3,4-b]thiophene
  structure 249
  see also poly(thieno[3,4-b]thiophene)
thiophene-based derivatives see polythienyl...; polythiophene...
transistors 50, 51–2, 219
transmission electron micrographs (TEMs), polyanilines 135
types, self-doped conducting polymers 29–33

UV-Vis spectra
  polyaniline sulfonic acid derivatives 109–12, 113
  poly(anilineboronic acid) 172–5
    interaction with RNA 202
  polyphenylenes 294
  poly(p-phenylenevinylene)s 297, 298
  polypyrrole sulfonic acid derivatives 280, 281
  polythiophene phosphonic acid derivatives 256, 257
  polythiophene sulfonic acid derivatives 229, 237, 246
UV-Vis-NIR spectra
  polyaniline sulfonic acid derivatives 111–12, 113
  polypyrrole sulfonic acid derivatives 266, 267

polythiophene sulfonic acid
 derivatives 225, 226, 236, 237,
 239–40

Vickers hardness, crosslinked
 poly(anilineboronic acid) *181*,
 182

X-ray photoelectron spectroscopy
 carboxylated polyanilines 124
 PABA/RNA multilayers 203
 sulfonated polyanilines 112–16